LANGUAGES
FOR AUTOMATION

MANAGEMENT AND INFORMATION SYSTEMS

MANAGEMENT AND OFFICE INFORMATION SYSTEMS
Edited by Shi-Kuo Chang

LANGUAGES FOR AUTOMATION
Edited by Shi-Kuo Chang

A Continuation Order Plan is available for this series. A continuation order will bring delivery of each new volume immediately upon publication. Volumes are billed only upon actual shipment. For further information please contact the publisher.

LANGUAGES FOR AUTOMATION

Edited by

SHI-KUO CHANG

Illinois Institute of Technology
Chicago, Illinois

SPRINGER SCIENCE+BUSINESS MEDIA, LLC

Library of Congress Cataloging in Publication Data

Main entry under title:

Languages for automation.

(Management and information systems)
"Originally presented at the 1983 IEEE Workshop on Languages for Automation
held at Chicago, and the 1984 IEEE Workshop on Languages for Automation held at
New Orleans"—P.
Includes bibliographies and index.
1. Automation—Congresses. 2. Office practice—Automation—Congresses. 3. Pro-
gramming languages (Electronic computers)—Congresses. I. Chang, S. K. (Shi Kuo),
1944-   . II. IEEE Workshop on Languages for Automation (1983: Chicago, Ill.)
III. IEEE Workshop on Languages for Automation (1984: New Orleans, La.) IV.
Series.
T59.5.L36   1985   005   85-17032
ISBN 978-1-4757-1390-9   ISBN 978-1-4757-1388-6 (eBook)
DOI 10.1007/978-1-4757-1388-6

© 1985 Springer Science+Business Media New York
Originally published by Plenum Press, New York in 1985
Softcover reprint of the hardcover 1st edition 1985

CONTRIBUTORS

MARTIN L. BARIFF • *Stuart School of Business Administration, Illinois Institute of Technology, Chicago, Illinois*

ALEJANDRO P. BUCHMANN • *IIMAS, National University of Mexico, Mexico, D.F., Mexico*

STEVEN D. BURD • *Anderson School of Management, University of New Mexico, Albuquerque, New Mexico*

C. R. CARLSON • *Computer Science Department, Illinois Institute of Technology, Chicago, Illinois*

G. CASTELLI • *Instituto di Cibernetica, Università di Milano, Milano, Italy*

P. CHAN • *Department of Electrical and Computer Engineering, Illinois Institute of Technology, Chicago, Illinois*

S. K. CHANG • *Department of Electrical and Computer Engineering, Illinois Institute of Technology, Chicago, Illinois*

MING-YANG CHERN • *Department of Electrical Engineering and Computer Science, Northwestern University, Evanston, Illinois*

F. DE CINDIO • *Instituto di Cibernetica, Università di Milano, Milano, Italy*

G. DE MICHELIS • *Instituto di Cibernetica, Università di Milano, Milano, Italy*

T. D. DONNADIEU • *Department of Electrical and Computer Engineering, Illinois Institute of Technology, Chicago, Illinois*

CLARENCE A. ELLIS • *Xerox Corporation, Palo Alto California and Stanford University, Stanford, California*

K. S. FU • *School of Electrical Engineering, Purdue University, West Lafayette, Indiana*

S. FUJITA • *Toshiba Research and Development Center, Kanagawa, Japan*

J. MIGUEL GERZSO • *IIMAS, National University of Mexico, Mexico, D.F., Mexico*

JOEL D. GOLDHAR • *Stuart School of Business Administration, Illinois Institute of Technology, Chicago, Illinois*

KLAUS D. GÜNTHER • *Gesellschaft für Mathematik und Datenverarbeitung, Institut für Systemtechnik, Darmstadt, Federal Republic of Germany*

PEI-CHING HWANG • *Phoenix Data Systems, Albany, New York*

THEA IBERALL • *Department of Computer Science, University of Massachusetts, Amherst, Massachusetts*

MATTHIAS JARKE • *Graduate School of Business Administration, New York University, New York, New York*

TOHRU KIKUNO • *Faculty of Engineering, Hiroshima University, Higashi-Hiroshima, Japan*

K. KISHIMOTO • *Department of Circuits and Systems, Hiroshima University, Hagashi-Hiroshima, Japan*

GARY D. KIMURA • *Department of Computer Science, University of Washington, Seattle, Washington*

HIROSHI KOTERA • *Visual Communication Development Division, Yokosuka Electronic Communications Laboratory, Kanagawa, Japan*

JAMES A. LARSON • *Honeywell Computer Sciences Center, Bloomington, Minnesota*

v

C. S. GEORGE LEE • *School of Electrical Engineering, Purdue University, West Lafayette, Indiana*

Y. C. LEE • *School of Electrical Engineering, Purdue University, West Lafayette, Indiana*

L. LEUNG • *Gould Research Center, Rolling Meadows, Illinois*

DAMIAN LYONS • *Department of Computer Science, University of Massachusetts, Amherst, Massachusetts*

MARK C. MALETZ • *Inference Corporation, Los Angeles, California*

LEO MARK • *Department of Computer Science, University of Maryland, College Park, Maryland*

JAVIER MARTINEZ • *Departamento de Automatica, Universidad de Zaragoza, Zaragoza, Spain*

F. MONTENOISE • *Department of Electrical and Computer Engineering, Illinois Institute of Technology, Chicago, Illinois*

KAZUNARI NAKANE • *Visual Communication Development Division, Yokosuka Electronic Communications Laboratory, Kanagawa, Japan*

K. ONAGA • *Department of Circuits and Systems, Hiroshima University, Higashi-Hiroshima, Japan*

S. OYANAGI • *Toshiba Research and Development Center, Kanagawa, Japan*

SHUH-SHEN PAN • *Bell Communications Research, Holmdel, New Jersey*

CHRISTOS A. PAPACHRISTOU • *Computer Engineering and Science Department, Case Western Reserve University, Cleveland, Ohio*

NICK ROUSSOPOULOS • *Department of Computer Science, University of Maryland, College Park, Maryland*

H. SAKAI • *Toshiba Research and Development Center, Kanagawa, Japan*

YOSHINORI SAKAI • *Visual Communication Development Division, Yokosuka Electronic Communications Laboratory, Kanagawa, Japan*

SOL M. SHATZ • *Department of Electrical Engineering and Computer Science, University of Illinois, Chicago, Illinois*

ALAN C. SHAW • *Department of Computer Science, University of Washington, Seattle, Washington*

MANUEL SILVA • *Departamento de Automatica, Universidad de Zaragoza, Zaragoza, Spain*

C. SIMONE • *Instituto di Cibernetica, Università di Milano, Milano, Italy*

EDWARD STOHR • *Graduate School of Business Administration, New York University, New York, New York*

KAZUO SUGIHARA • *Faculty of Engineering, Hiroshima University, Higashi-Hiroshima, Japan*

A. TANAKA • *Toshiba Research and Development Center, Kanagawa, Japan*

T. TANAKA • *Toshiba Research and Development Center, Kanagawa, Japan*

MASAO TOGAWA • *Visual Communication Development Division, Yokosuka Electronic Communications Laboratory, Kanagawa, Japan*

JON TURNER • *Graduate School of Business Administration, New York University, New York, New York*

H. UTSUNOMIYA • *Department of Circuits and Systems, Hiroshima University, Higashi-Hiroshima, Japan*

YANNIS VASSILIOU • *Graduate School of Business Administration, New York University, New York, New York*

JENNIFER B. WALLICK • *Honeywell Computer Sciences Center, Bloomington, Minnesota*

ANDREW B. WHINSTON • *Krannert Graduate School of Management, Purdue University, West Lafayette, Indiana*

NORMAN WHITE • *Graduate School of Business Administration, New York University, New York, New York*

NORIYOSHI YOSHIDA • *Faculty of Engineering, Hiroshima University, Higashi-Hiroshima, Japan*

STANLEY B. ZDONIK • *Department of Computer Science, Brown University, Providence, Rhode Island*

PREFACE

Two central ideas in the movement toward advanced automation systems are the office-of-the-future (or office automation system), and the factory-of-the-future (or factory automation system).

An office automation system is an integrated system with diversified office equipment, communication devices, intelligent terminals, intelligent copiers, etc., for providing information management and control in a distributed office environment. A factory automation system is also an integrated system with programmable machine tools, robots, and other process equipment such as new "peripherals," for providing manufacturing information management and control.

Such advanced automation systems can be regarded as the response to the demand for greater variety, greater flexibility, customized designs, rapid response, and "just-in-time" delivery of office services or manufactured goods. The economy of scope, which allows the production of a variety of similar products in random order, gradually replaces the economy of scale derived from overall volume of operations.

In other words, we are gradually switching from the production of large volumes of standard products to systems for the production of a wide variety of similar products in small batches. This is the phenomenon of "demassification" of the marketplace, as described by Alvin Toffler in *The Third Wave*.

A common theme for such integrated, advanced automation systems is the need for "languages"—languages for man–machine communication, languages for describing office procedures, languages for describing design processes, languages for specifying manufacturing processes, languages for modeling, and languages for controlling robots. The study of languages for automation, therefore, becomes a vital research topic in advanced automation.

This book is the second volume in the series of books on *Management and Information Systems*. It includes several invited tutorials, and selected papers that were originally presented at the 1983 IEEE Workshop on Languages for Automation held at Chicago, and the 1984 IEEE Workshop on Languages for Automation held at New Orleans.

Twenty-five papers written by international experts cover the various aspects of languages for automation: office automation languages, database query languages, natural language query processing, object management, database management, communication management for distributed systems, robotics languages, CAD/CAM languages, and implications of advanced automation for corporate management.

S. K. Chang

CONTENTS

ix

PART VI
MANAGEMENT AND AUTOMATION

OFFICE AUTOMATION

OFFICE INFORMATION SYSTEMS OVERVIEW

Clarence A. Ellis

1. Introduction

Today's office is a changing environment. The days of manual type-writers, adding machines, and hand sorting, copying, and mailing are being replaced by integrated electronic office systems containing mixed media document editors, electronic mail systems, electronic files, and numerous personal electronic aids such as electronic spreadsheets and tickler files. Studies have shown that well over half of today's office workers use some form of data processing or telecommunications workstation on the job. The need for these systems is apparent: increasing administrative overhead and costs, coupled with increasing need for information and information processing, are causing many organizations to search for methods to increase efficiency (amount of work getting done per dollar expended) and effectiveness (extent to which work done actually meets goals and needs of the organization). Organizations are finding that careful design and imple-mentation of office information systems, taking into account important human factors of organizational design and office sociology, meet these needs, and can make some aspects of office work more pleasant and convenient.

The initial emphasis of the office automation thrust was concerned with clerical work and "structured tasks." A prime example is the word processor, a computer-based office aid which embodies the typewriter-with-memory notion with features to make the composition and editing of documents easy. The primary target of this machine was the secretary/typ-ist. Only 6% of the office dollars are spent on secretary/typist salaries, and only 20% of these people's time is consumed by typing.[77] The word pro-cessing market is nevertheless a huge market. Two other categories are

CLARENCE A. ELLIS • Xerox Corporation, 3333 Coyote Hill Road, Palo Alto, California 94304; and Stanford University, Computer Science Department, Stanford, California 94305

identified within the office work force. These are the "professional and technical workers" and the "managers and administrators." The combination of these two occupational categories is designated as "knowledge workers." These people typically spend a lot of their time performing work classified as "unstructured tasks." Henry Mintzberg, who conducted one of the most thorough investigations of managerial work, found that managers spend 66%–80% of their time in oral communications.[59] Other studies have verified this. One study reported that knowledge workers, on the average, spend their time approximately as follows:

Formal meetings—>20%;
Writing—10%;
Phone conversations—5%–10%;
Reading—8%;
Analysis—8%
Travel—10%.

The remaining time, classified as miscellaneous and idle, includes activities such as informal meetings, expediting, scheduling, waiting in airports, seeking people or information, copying, filing, and transcribing. Given this mix of structured and unstructured work, it seems that the emphasis within office information systems for this latter category of workers should be on using the system as an augmenter and assistant rather than as a replacement for people.

Today's advanced workstations have gone a long way toward making the office environment more productive, and the computer interface more friendly. A misspelled word in a document can be automatically detected and easily replaced without retyping the entire page or document. A user interface consisting of pictoral (iconic) objects such as pictures of file cabinets and mail boxes, displayed on a high-resolution screen, and voice input/output is easy to learn, use, and remember.[74] All of this is augmentation for the individual. It appears that the next big productivity breakthrough may occur when we can achieve community augmentation which works as well as systems of today work to augment the individual. Perhaps this will emerge after we have successfully embraced the concept of integration of the components of the augmented office to allow them to function together smoothly and effectively. Proponents of this view also stress the need for links between the augmented office and external systems, including the paper-based operations, including telephones and databases, and including management information systems and decision support systems. I believe that artificial intelligence and knowledge-based systems are an important ingredient in this next big step.

As previously mentioned, office automation was primarily concerned with automating the clerical functions within the office. On the other hand, management information systems have traditionally been concerned with data processing aids for managers, where data access systems or data monitoring systems would periodically generate reports. For top executives and senior managers, decision support systems were developed as interactive real-time systems to provide the top decision makers with rapid access to data and to models that would summarize and extrapolate from the raw data. These tools range from simple cross tabulation to on-line models which estimate the cash flow of an organization under varying business strategies or varying economic climate assumptions. Recently there has been a realization that there is a large need for information and a lot of exception handling and problem solving at all levels of the office. This is true because the office handles those aspects of the business which were not amenable to easy automation within the data processing department. Thus, office information systems try to integrate and share information between all three of the segments (clerical worker, professional, and manager) in a manner which benefits problem solvers at all levels of the corporate structure.

There are two primary technological innovations which form the basis for this change in the office. One is advanced workstations, e.g., personal computers and word processors driven by the inexpensive microprocessor boom; and the other is advanced communications systems, e.g., new technological developments in the areas of PBX, local area networks, and satellites. This overview attempts to provide a description of these various components and tools within a modern office information system in the context of a coherent framework. This overview also addresses office modeling, office technologies, and the building blocks that help form an integrated office information system. It is assumed that readers have a basic understanding of computer hardware and software, but not necessarily of specialized subareas.

2. People in Offices

The office is not so much a place as a locus of common activity. Many people are following trends of making their home their office, utilizing the many available modes of electronic communication. Many office tasks are routinely performed in remote locations (by salespersons on airplanes, for example) not at all resembling the stereotypical room with office desks and furniture. For these modes of work to be feasible, the social structures and organizational structures of the work organization must be explicitly recognized and nurtured. Potential social conflict can arise in the division of time between family and work when working at home. Potential organizational conflict can arise when a manager cannot accurately calibrate and evaluate the effectiveness of his/her subordinates because the subordinates are too frequently traveling. The office can be viewed as a conglomerate of people, information sources, and information manipulation tools which are drawn together by common goals and structures. We must recognize the presence and the importance of social structures and organizational structures in the office, and be sure that introduction of automated aids does not put these factors into negative imbalance.

At the level of individual and social needs, work within the field of social psychology provides useful information for office design. Work by Maslow has become famous because it elucidates layers of need within people.[53] This theory holds that there are five layers of needs. At layer 1 are some very basic *physiological needs,* such as the need for food. One can inevitably observe the hunger drive which results if this need is not fulfilled. Above this layer, there are *safety needs.* The cave men present an example of this in their relentless striving for a shelter safe from the animals of the wild. A third layer is called the *belonging need.* Because man is innately a social creature, there is a drive to find a group to which the individual can, at least to some extent, belong. This may be a family, an ethnic group, or an organization. Above these layers, which are called hygienic layers, are two further layers called motivators. These layers pinpoint the need for *self-esteem* (layer 4) and the need for *self-actualization* (layer 5). In our modern western society, office workers tend to be above the threshold level for the hygienic layers, so Hertzberg found in his studies that an increase in these had no noticeable effect on the workers' productivity or effectiveness. On the other hand, his studies found that increase or decrease in motivators had significant effects.

At the organizational level, Professor Ouchi and others have shown that successful organizations have organizational control structures which fit the size, style, and needs of the organization.[64] One prevalent structure is the *organizational hierarchy,* also referred to as the bureaucracy. In this structure, there are levels of management, and rules concerning what employees should do and how employees at each level should interact with the levels above and below them. For a large organization this allows delegation of work via standard channels. However, this structure can have a large amount of overhead from middle level managers, and from the amount of time and energy expended to get information from the top level to the actual workers at the bottom. Another organizational structure, which is associated with Japanese business structure, is the *clan.* Within this structure, employees within an organization feel that they are part of a family, and the employer is looked upon as the benevolent parent. There is a lot of company loyalty, and a lot of behavior is determined by peer group pressure. This has the advantage of a lot of devoted work being done without the need for a large bureaucratic overhead of managers to oversee or to pressure workers to get the work done. It has disadvantages of stifling individuality and not encouraging creative inventiveness. A third organizational structure is called the *marketplace*. In the marketplace the workers interact according to resource needs and resource availability. Clients or buyers search in the marketplace for contractors or sellers to perform the tasks or to provide the goods that are needed. If I need paper supplies, I would get them from the supplier which can provide them to me at the lowest price within the timeframe I need. Similarly, if I must deal with one of the accountants in the organization, I will choose to deal with the one who will charge the lowest price for services, but who will give me good service in a short waiting time. The market structure tends to be fair in its evaluation and rewards provided that the notions of cost of information can be concretely established. Although clients have some prior knowledge of the various contractors and capabilities, they must generally communicate with a number of potential contractors in order to determine the specific capabilities and current availability of the contractors. This tends to add a great deal of overhead. Notice that a pure market structure for a corporation is not practical, but the availability of specially tailored electronic mail facilities makes some aspects, like the cost of bargaining and keeping informed, more feasible.[83] Thus this organizational structure and the two preceding ones illustrate the pure ideals, whereas any actual organization would utilize a mixture of the above.

Given these observations concerning social and organizational structures, we next look at technology within the office. Afterward, we can put these two ingredients together to suggest how technologies can and should fit into office structures.

3. Technology in Offices

In this section, we briefly describe some of the technological innovations and systems which are typically included within integrated office information systems. Workstations, document editors, spreadsheets, networks, electronic mail, conferencing systems, and database systems are a few of the components. Others not discussed here are electronic diaries, calendars, tickler files, smart cards, FAX, OCR, and TELETEX.

3.1. Workstations

Workstations are computer-based systems which today typically sit on top of a worker's desk. Others are portable, including the briefcase computer, and the hand-held computer. Many have an auxiliary disk which may be a floppy disk holding thousands of bytes, or a rigid disk holding millions of bytes. Many workstations have a screen to display documents (text and graphics), sometimes in color. It may be a touch screen so that items on the screen can be selected by pointing with your finger, or it may have a mouse or other pointing device. Movement of the pointing device controls the movement of a cursor visible on the screen, and controls the selection of objects on the screen. Other facilities may include input–output connections to devices such as a local printer, or to a network or PBX. The workstation may be integrated with the telephone, and provide enhanced telephony facilities such as phone number lookup and autodial and timed recall. The user interface to the workstation is an important aspect, and may include, for example, voice input and output. Window packages are also available which divide the screen up into rectangular regions called windows. These present a user interface that supports and explicitly makes visible the concept of multitasking and the execution of parallel processes. Each window can be associated with a process, and each can be independently created, sized (windows can usually overlap), started, interrupted, stopped, and destroyed by the user. Thus, while my computer program is compiling in one window, I can be reading my electronic mail in a second window, and writing a paper in a third.

3.2. Document Editors

Document editors have become more powerful and elaborate than simple word processors. They share with word processors the ability to create, edit, and store electronically letters, forms, and reports. But they also may have the ability to create and manipulate other types of objects within documents such as graphics, voice, scanned-in images, and facsimile machine output. Full-page screen editors have been replacing the older line editors. Some editors present a WYSIWYG image on the screen. WYSIWYG stands for *what you see is what you get*, and means that the image on the display screen is identical or very similar to the output that one would get at the printer, so no extra codes or editing marks are visible, and so that margins, tabs, and justifications are immediately visible. Some editors are more properly called document processors, because they include facilities for creating, structuring, spell checking, and formatting documents. Sometimes these editors manipulate documents as structured objects, so that a document is a combination of paragraphs, sections, frames, and boxes, which in turn may be composed of other objects such as sentences or curved lines, or tables. These editors provide facilities for manipulation of the document structure as well as the document content.

3.3. Spreadsheets

Spreadsheets are two-dimensional tables partially or fully displayed on screen-based workstations whose entries can be specified as functions of other entries in the same or different tables. After the rules of dependency for each table entry are specified, the alteration of a particular entry immediately results in the updating of all other entries which depend upon the altered entry. Manual spreadsheets were used for many functions long before display-based computerized spreadsheets were constructed. This is a good example of technological improvement (the low-cost display-based workstation) making feasible computer-based tools (the computerized spreadsheet). Spreadsheets are useful for numerous calculations such as tax calculations, depreciation, standard deviations, and net present value calculations.

3.4. Networks

Local area networks and PBXs (standing for private branch exchanges) have been instrumental in allowing workstations to communicate with each other and with various mainframes and services. Transmission media over which this information may be transmitted include wire cable, coaxial cable, microwave, communication satellites, and optical fiber. These allow all workstations in a building or a floor of a building, or in some cases a campus to be interconnected. These can in turn be connected via long haul networks such as phone lines or satellite, to other branches of the same organization or other organizations. These long-haul connections may connect distant parts of the country, or connect far away countries.

3.5. Electronic Mail

Message systems (also called electronic mail) and teleconferencing are asynchronous facilities allowing messages and documents to be sent over computer networks, usually delivered in a matter of seconds or minutes. The fact that these are asynchronous means that the recipient of the message does not need to be present at the time the message is sent. The next time that the recipient signs on to his or her workstation, s/he will be notified that messages are waiting to be read. Modern systems allow arbitrary documents which may contain graphics, voice, etc., to be mailed. Numerous features such as priority mail, distribution lists, and automatic forwarding are also available within modern systems. Computer-based teleconferencing systems differ from mail systems in that an ordered transcript is kept of all contributions to the conference, and the default is a many-to-many conversation rather than a one-to-one conversation. Modern mail systems and conferencing systems are feature-rich enough to be used either for one-to-one mailing or for conferencing. The term "teleconferencing" is also sometimes used to denote a synchronous distributed session where all parties must be simultaneously present at their workstations. By contrast, audio and video teleconferencing are synchronous conferencing systems in which dedicated transmission facilities connect special conference rooms for real-time meetings. All of these technologies are destined to partially replace some face-to-face meetings, although the initial reception has not been as overwhelming as predicted.

3.4. Databases

Many workstations have personal databases as part of their local environment. The presence of communication facilities means that access to larger shared databases is also feasible. The advent of reliable efficient distributed databases means that the personal database can communicate with or merge with the shared communal database. Thus conferencing sessions or mail which a community of users may be interested in saving can be stored and retrieved on many keys such as date, topic, author, etc. Similarly, large collections of documents can be stored, retrieved, and searched very naturally if they are stored within a database facility. Databases appear to be an excellent base upon which many of the higher-layer components of an office information system can be structured. This is particularly true if the database has efficient facilities for handling triggers, alerters, and varying record structures containing nonstandard data types such as image and voice.[18,29,33]

4. A View of the Office

Given the social and organizational considerations of Section 2 and the technological considerations of Section 3, we can formulate a view of the office which takes into account both of these considerations. This will also pinpoint the unique aspects of office information systems. Some prominent features of offices and office systems which distinguish them from other (e.g., data processing) systems are as follows.

4.1. People Systems

An office is a social environment to which any introduction of procedural changes, goal changes, or automated equipment causes perturbations. If people who need to exchange information are moved so that they are no longer close to the coffee machine, then needed information, which used to get transferred in this informal way, may no longer get exchanged. Explicit consideration should be given beforehand to analyzing likely effects of changes. Many technologically successful systems have failed due to ignorance of human and social factors. For example, the mass movement of secretaries away from individual managers to word processing pools violated social maxims. The managers tended to feel a loss of a most valued resource, and the secretary tended to feel less directed and less motivated. In many cases, this change broke important social ties which had helped to keep many office organizations healthy.

4.2. Dynamic Systems

Change is frequent and expected in most domains of the office. An employee's vacation days, for instance, force others to change their routines accordingly. Change also results from promotions, employee turnover, competition's changes, sickness, changing government regulations, etc.

4.3. Concurrent Systems

The office is a highly parallel, highly asynchronous system. In many cases, this structure has grown from years of experiential learning by doing. There is much to be learned by systematically observing the informational and social checks and balances of a smoothly working organization. But also one may observe habits and impediments which are unnecessary relics of a past way of doing business. Thus, some of our mathematical office analyses which automatically detect potential parallelism have produced extremely useful results. Inherent in the media of the electronic office is the much greater potential for parallel processing than the previous age of unautomated offices. As a word of caution, we have observed that it is not always easy to discern whether an activity falls into the category of unnecessary relic or necessary redundant checks and balances.

4.4. Ill-Structured Systems

In terms of monetary investment, we have mentioned that most of the people resource is within the "knowledge worker" category. These are the professionals, the managers, and the executives who are highly paid, and therefore are a prime set of candidates for office augmentation aids. Much of the work performed by this group is unstructured or ill structured.

These workers need augmenters and aids rather than the structured data processing systems which are prevalent in more structured parts of a business. For example, a sales manager in a typical company may need to search diverse data, read between the lines of a report, and have a confidential lunch meeting with a colleague in order to track down the information necessary to salvage the account of a big customer. A useful office information system should be able to assist this person in all of these aspects of work.

4.5. Open-Ended Systems

Another important group of people is the "clerical/secretarial" category. One might assume that the work of people in this category is all structured, but office studies have shown that even within this category, the amount of problem solving, exception handling, and customer interfacing (all three are unstructured activities) is high. Just the activity of interpreting a customer's handwriting may involve significant problem solving. Thus systems and models must be capable of handling a diverse spectrum of activities with high proportions of semistructured and unstructured activities. We must insist that systems be inherently open ended with escape hatches to handle unanticipated exceptions and emergencies. This is much more tenable than aiming toward notions of total automation of procedures, total removal of paper from the office, or totally peopleless offices.

5. Office Models

Office models can be used by many different persons in many different ways. We cannot avoid the use of models; each worker in the office carries an informal model of his/her activities. Furthermore, there is high utility in explicitly devising and recognizing models within the office domain. Formal models, which tend to be mathematical and explicit, are useful for analysis and design; informal models, which tend to be implicit and ad hoc, are useful for capturing social and behavioral aspects. Before studying various office modeling topics of interest and importance, we will attempt to answer some fundamental questions of what modeling is and why models need to be built.

5.1. What Is a Model?

Models are limited abstractions of reality. The limitation is expressed by focusing upon a subset of the attributes and structures. Informal models can help workers to decide upon appropriate actions when they encounter unforeseen or exceptional situations.[76] Explicit awareness of user models can help to improve upon the acceptance and error-free use of office systems. For example, whereas a computer person has a computer model of disk storage which implies that retrieval of a page of information does not destroy the original copy on the disk, an office worker (noncomputer person) has a file cabinet image which says that when a page of information is taken out of a file, it obviously is no longer in the file. A user-friendly system design may choose to somehow take this into account; for example, there may be a DESTRUCTIVE READOUT command, and a NONDESTRUCTIVE READOUT. Formal models may range from social interaction models of matrix management to queuing and simulation models of bottlenecks and throughput in the flow of transactions within standard office procedures at peak hours. Both formal and informal models can be extremely valuable aids in the process of office systems design.

5.2. Why Modeling?

Models, using the power of abstraction, frequently provide insight into important characteristics of complex systems. Simple "back of the envelope" models can be quite useful for first-order approximations. Certain telephone operations models ignore much switching complexity, and give good estimates of number of lines which one can expect a PBX to be able to service. Generally speaking, mathematical models are used to provide rigorous mathematically tractable characterizations of, or approximations to, office systems. These models frequently provide insights into important characteristics of complex systems. One reason for this is because models can frequently be built at a high enough level of abstraction that properties, generalizations, individual differences, and inconsistencies can often be discovered which would not otherwise be apparent. This argument in favor of modeling is based upon the thesis that insights obtained by modeling lead to understanding and theories, which lead to consistent and well-structured implementations. In addition to models having a potentially strong impact on systems design and implementation, practical data input and existing implementations provide good foundations, motivation, and direction for the theoretician who is concocting a model. Indeed, some models demand that collected data or observations from existing implementations (or simulations) be used to drive the model (e.g., trace driven models, stochastic and statistical models). The aforementioned benefits of modeling must be traded off against the amount of time and effort needed for the model development. Modeling done hastily, haphazardly, and imprecisely has been known to produce misleading and in some cases grossly incorrect results. In the design of static, well-understood systems, modeling may not be cost effective. In considering the development of office information systems there are compelling arguments in favor of modeling: (1) the technology of the systems is still in the formative stage; (2) these systems are quite dynamic (changes to office procedures, office personnel or office requirements are frequent); and (3) there is no comprehensive theory of office information systems. Indeed, there is strong reason to believe that the office of the future will need to lean heavily on modeling and analysis. Both qualitative and quantitative analysis can be performed using mathematical models. In this section we define a family of models which are useful formally and informally in a variety of office situations to aid understanding and to perform analysis. In this way it is hoped that the value of modeling becomes apparent to readers. Furthermore, we firmly believe that data gathering, simulation, and measurement/evaluation are pragmatic activities which go hand-in-hand with mathematical modeling. Thus these activities should be tightly coupled with the theoretical modeling effort. In summary, the design of a system as complex and dynamic as an office information system can benefit from insights and theories developed through studies of office models.

5.3. Types of Models

In a previous section we defined an office model as a limited abstraction. This reflects our feelings that no one model spans all of our modeling needs, and that it is not possible to completely capture all aspects and nuances of a complex office situation. Factors of psychological make-up of office workers, acoustics, and even desk size may be very important to a particular office's success, and may not all be taken into account by a single model. Just capturing within a model all of the exceptional conditions which may occur may be an endless task. Thus it is useful to classify models and explore the utility of different types of models. One classification[63] of office models is based upon different possible views of the office adopted by the modeler.

1. *Information Flow Models.* These models seek to represent office work in terms of units of information (forms, memos, etc.) that flow between offices; they allow us to define, generally by some type of network diagram, the operations performed on each information unit. They are useful in defining the types of information units involved in office work, and the range of operations applied to each unit.

2. *Procedural Models.* This class of models is based on the view that office work is basically procedural in nature, involving the execution by office workers of predefined sequences of steps, called procedures. Like

information flow models, procedural models involve operations (procedural steps) and operands (units of information). They emphasize the task-oriented nature of office work, in the sense that each procedure is designed to perform a particular task, and they identify some of the important and often unpredictable roles played by people in carrying out procedures. In these two respects they offer a more accurate model than information flow. However they may be less accurate in portraying interprocedure and interdepartmental flows and dependencies.

3. *Decision-Making Models.* These models relate to the decision-making activities of managers and other office personnel. The more traditional model of decision making treats it as a fairly objective process of information gathering and analysis; recent studies have suggested that there is a large element of unpredictability in organizational decision making, which may be difficult to capture with this model.

4. *Database Models.* Office work can be modeled in terms of databases; these contain information records that are created and manipulated by means of transactions, and that can be viewed by generating reports of the databases contents. Business accounting and control methods are nowadays largely based on this model.

5. *Behavioral Models.* It is possible to view office work as a social activity, involving situations and encounters into which are woven the information-processing tasks of the office. Organizational and sociological studies provide a considerable amount of information to support this approach to office modeling.

Another classification[77] of office models arises from the involvement of different disciplines in the study of offices:

1. Organizational communication models;
2. Functional models;
3. Information systems models;
4. Quality of working life models;
5. Decision support systems models;
6. Operations research models.

These categories are not mutually exclusive or all inclusive; rather they show that a problem can be approached from different conceptual viewpoints.

5.4. Uses of Models

It can be debated whether current simplistic models are adequate for study purposes. It appears that this depends upon the use which is to be made of the model. There are a number of ways in which models are inherently inadequate (particularly modeling human behavior), but also there are significant ways in which models may be used:

1. For description and specification of a system;
2. For dynamic simulation of office activity;
3. For comparison of systems;
4. For prediction of system performance under change of input conditions or under variation of model structure;
5. As an aid to communication;
6. To assist in system design and implementation;
7. For purposes of evaluation;
8. For purposes of understanding by workers or other stakeholders.

6. Office Framework

Previous discussion of facilities and tools such as document editors, spreadsheets, and electronic mail was rather random. It suggests that we need a design model that is comprehensive enough to serve as a framework for discussion. This model should show the relationship of these and other facilities to each other. A model around which we will structure much of the discussion in this section is called the "Interconnect Model." It is a layered framework for discussing the pieces of an office information system and for explaining how these pieces fit together. One part of the model is concerned with the structuring of facilities and tools within a personal environment, and the other part of the model is concerned with structuring within a communal environment. For each of these environments we postulate that functional work gets done via an implicit or explicit hierarchy of processes. A diagram of this framework is shown in Figure 1.

The layering concept, as we will use it within the interconnect model, is a technique to structure complex systems. It has been used successfully to aid in the conceptualization and in the construction of complex systems of many types. A good example of the use of layering is within the international standards organization's OSI model of communication protocols.[87] This model has seven layers ranging from a wires-and-bits description at layer one to application protocol descriptions at layer seven. In general, a given layer implements a set of functions for use by the layer above the given layer. Each function is constructed from combinations of more primitive functions provided by the layer below the given layer. Note that there is simplification in a well-structured system, because a given layer only needs to be aware of and deal with the layer immediately below

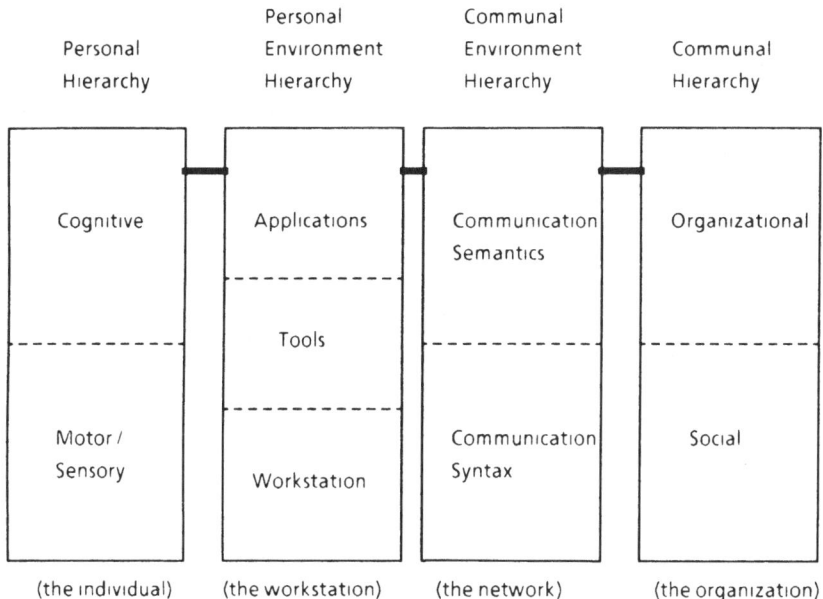

FIGURE 1. The interconnect model.

it. We define interfaces to be the functions, procedures, and data structures of a given layer which are known and usable by the next higher layer. Note that here is another opportunity for simplification, because a given layer can implement information hiding, and only make available needed procedures and data structures to the next higher layer, rather than all of its algorithms and complexity. For example, when I am driving my car from home to work, I really make use of layered functionality. I turn the steering wheel to the left when I want the car to turn to the left; and I turn the steering wheel to the right when I want the car to turn to the right. I do not know or particularly care exactly how the electromechanical subsystem carries out this function as long as it works. This electromechanical subsystem represents a lower layer which provides the driver with car-turning functionality while hiding the details of mechanical leverage and wheel torque suspension.

Another important concept is concerned with protocols. We define protocols to be agreed-upon sequences of actions by several entities (such as people or workstations) which they perform whenever they want to exchange information at the same layer in the hierarchy. For example, when answering the phone, it is customary to say a polite greeting such as "Hello." This "Hello" is not passing critical conversational information, but it is a useful action to let the caller know that a connection is complete.

In this sense, the "Hello" is a protocol. This example illustrates that protocols can be useful to gain a common understanding for communication without much overhead. Without protocols, mass confusion could ensue. To understand how protocols and multiple layers fit together, consider a three-layer example of offline conferencing within a multinational corporation. If the chief executive officer wants to make a statement to all of the executive officers, then he dictates this message in his native tongue (say German, if he is located in Germany) to his executive scribe who performs translation. When speaking, the chief executive officer considers that he is speaking directly to other executive officers using mutually understood protocols which we will call level 3 protocols. In reality, he passes his message to his scribe at layer 2 across the 3/2 interface. The layer 2 scribe has a task of using agreed upon protocols to communicate with all other scribes using layer 2 protocols.

For example, this scribe may translate the German message into English as the agreed upon common language among scribes. In reality, the scribe implements his task by passing the English message to the courier at layer 1 across the 2/1 interface. The actual means of transmission may be computer network, telegram, personal delivery, or other means depending upon the layer 1 protocol. When the courier working for the receiving executive officer receives the message (irrespective of what language it is in), he delivers it to his translator. If the receiving officer is based in Japan, then the translator may convert the text to Japanese and then pass it up to the officer at the top level. Note that the protocol at any layer can be changed without affecting other layers; this is great for system maintainability and alterability. Note that the only real transmission of information among different locations of the corporation takes place at the lowest layer. Since all communications at layer 3 must pass through all lower layers and, upon receipt, pass up all layers, one potential implementation disadvantage of this structuring scheme has to do with the overhead incurred by frequently moving down and up the hierarchy. The reader should bear in mind that the framework which we are introducing is for the purpose of conceptual modularization, and says nothing about how layers are actually implemented. Thus implementations may choose to shortcut some layers of the system.

It has been postulated that individuals get work done via a hierarchy of processes.[17] For example, a typical office task such as making a telephone call is accomplished by activities at many layers. At the cognitive layer, we must recall the phone number and fix in mind the goal of the communication (i.e., what we want to say). At the motor layer, our fingers must dial the correct digits and our vocal chords must formulate the utterances which compose the spoken conversation. Thus we again encounter layers. Also when a person is interacting with a workstation, there are high-level activities such as conceptualizing the layout of a document, and low-level activities such as keyboarding. Both are required for successful inter-action. This is also true on the receiving end to interpret stimuli from the personal environment. My low-level sensory perception notices such things as the alteration by the personal environment of the pixels on my display screen. Within my internal processing system, this may cause a signal and pictoral information to be sent up through various layers to a cognitive layer (my brain), which realizes that these bits form a new window which has just opened. At a higher semantic layer, I interpret this stimulus to mean that a high-priority message has arrived at my workstation which I should read immediately. Although it is not the primary focus of this paper, we note in passing that this user hierarchy which interacts with the workstation hierarchy should not be ignored in systems design. Likewise, in our discussion of the communal environment hierarchy, we must remember that the social and organizational hierarchies should not be ignored in system design. All of these hierarchies are present within the interconnect model (see Figure 1).

It is possible to divide the facilities provided by the personal environ-ment into three layers, each of which has two sublayers as illustrated in Figure 2. There is the workstation layer, the tools layer, and the applica-tions layer. Within the workstation layer, there is a basic workstation layer that provides facilities that are considered ordinary fundamental hardware and software. CPU, main memory, and "ordinary" input–output facilities are included here. An example of "ordinary" hardware in this category is the standard keyboard. The next higher sublayer in this hierarchy, called layer 2 in Figure 2, is the structured workstation layer. Above this layer users see a system whose memory is very large, and potentially different in size and nature from the actual main memory of the layer 1 raw machine. This layer may include other special instructions, special input or output facilities, and generally special hardware for special applications. Note that in all of the layering specification, the exact boundaries between layers are hazy. The particular time and setting may be partial determinants of whether a feature is considered special or ordinary. Layer 3 is the basic tools layer, which contains fundamental facilities useful for a wide variety of applications. Examples are compilers and window packages. Similarly, the facilities provided by the structured tools layer are application inde-pendent and used for a wide variety of applications, but they are typically more sophisticated tools usually implemented using or on top of the layer 3 tools. Examples are document editors and database facilities. These are

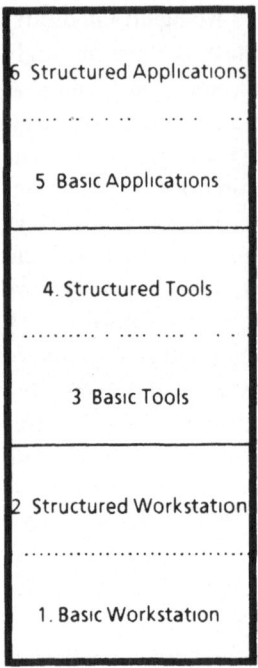

FIGURE 2. The interconnect model (personal environment).

typically built using a compiled language and they may occupy and run inside of a window, so we place them within layer 4. At layer 5 are basic applications packages such as the spreadsheet packages and the forms design systems. These are generic applications which are applicable to many domains of application. The structured applications layer (layer 6) contains domain-specific application packages and systems. Examples include financial planning systems, and income tax calculation packages. There are facilities for which it is unclear in which layer they should reside, such as calculators and calendars. Also, the exact nature, implementation, and sophistication of a package may determine in which layer it resides. For example a database management system may reside at level 4 as a tool, but if a graphical forms user interface is added or if it is customized for a particular order entry department with a very high-level interface, then it should be moved into the applications layer.

For layering of the various communal aids, we use the communications layering developed and standardized by the international standards organization (ISO model[87]). This has seven layers, as shown in Figure 3. The lowest layer is the physical layer, which is concerned with transmission of raw bit streams. The datalink layer (layer 2) is concerned with the reliable transmission of frames of data. The next layer is the network layer and is concerned with point-to-point routing of datagrams, and the avoidance of congestion. The fourth layer is the transport layer, which is concerned with reliable host-to-host communication for use by the session layer. The session layer is concerned with setting up and maintaining and tearing down process-to-process connections. The presentation layer (layer 6) performs transformations on the data to be sent, such as text compression. The seventh and highest layer is the applications layer, which contains protocols for specific applications so that generally available services can be communicated with and used by any cooperating customer. This includes generic applications such as file transfer protocols, and very domain specific protocols such as those used by the banking industry. Whereas the lower layers are typically covered in books and courses concerning communications, the higher layers receive much less study and are much less developed. The area of office information systems design is intimately concerned and interwoven with the higher layers of the personal and communal environments.

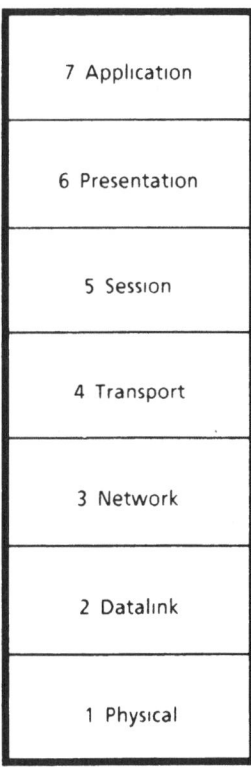

FIGURE 3. The interconnect model (communal environment).

7. Summary and Conclusions

We have tried to compose an introduction to office systems which presents some information and intuition about the office, its people, and its technology. Our emphasis has been on the human element and its interaction with technology. In keeping with this emphasis, we introduced a general framework which makes explicit the layered processes and protocols which occur between individuals and their personal environments; also between groups, organizations, and their communal environments. A framework such as this can be applied to a wide variety of problems[52] such as (1) explaining historical changes in the structure of American business organizations, (2) predicting changes in the structure of human organizations that may result from the widespread use of computers, and (3) analyzing the advantages and disadvantages of decentralized task scheduling in computer networks. It is noteworthy that the interdisciplinary nature of this field encourages techniques of one area (e.g., organizational design) to be applied to problems in seemingly unrelated areas (e.g., computer science). We conclude this paper with a few statements about integrated systems, current and future.

The ability to easily pass information between applications, and the ability to use the same interface, data structures, and commands for various applications has greatly increased the utility and ease of use of office information systems. One example is the table of data which can be displayed in a document, and also used to generate a bar chart and a pie chart and other data items in the same or another document. When an entry in the table is changed, the bar chart, the pie chart, and the other data items are all automatically updated correspondingly. Finally, we note that a number of conversion services are becoming available to convert documents from one format to another so that documents can be viewed and edited within heterogeneous environments. Also, gateway services are becoming available. These are computer systems which are connected to more than one network; they handle the translation and buffering needed in order to allow information to flow between heterogeneous networks. This progress with converters and integrators sets the scene for the next generation of knowledge-based office assistants which will be able to combine local user preference, global office knowledge, and inference to assist with many crucial functions. These computerized personal assistants knowing organizational and social structures and their relations to information and procedures can make sure that global office procedures are handled by the numerous parties involved in a timely fashion, and that crucial meetings are not forgotten. It is becoming increasingly apparent that office models that take into account *people in offices* and *technology in offices* are needed for the successful design and installation of office information systems.

References

1. *ACM Computing Surveys. Special Issue: Psychology and Computer Systems.* September (1980).
2. A. ADERET, *Menu-Driven and Command-Driven Word Processor Interfaces: Evaluation of Ease of Learning.* Proceedings of the 1982 Office Automation Conference, AFIPS Press, San Francisco 1982, pp. 875–884.
3. J. ALLEN, Synthesis of speech from unrestricted text, *IEEE Trans.* **64**(4): 433–442 (1978).
4. R. B. ALLEN, Composition and editing of speech, *Bell Syst. Tech. J.* (1981).
5. T. ALLEN, R. NIX, and A. PERLIS, PEN: A hierarchical document editor. ACM SIG-PLAN SIGOA symposium text manipulation, *ACM SIGPLAN Not.* **16**(6): 74–81 (1981).
6. S. ALTER, *Decision Support Systems: Current Practice and Continuing Challenges,* Addison-Wesley, Reading, Massachusetts, 1980.
7. C. ARGYRIS, Management information systems: The challenge to rationality and emotionality, *Manage. Sci.* **17**(6): 275–292 (1971).
8. J. H. BAIR, Communication in the office of the future: Where the real payoff may be, *Proceedings of the International Computer Communications Conference,* Kyoto, Japan, August, 1978.
9. J. H. BAIR, Productivity assessment of office automation systems (two volumes), Report for the National Archives and Records Service, SRI project 7823, Menlo Park, California, March 1979.
10. J. H. BAIR, J. FARBER, and R. UHLIG, *The Office of the Future,* North-Holland, Amsterdam, 1979, p. 379.
11. G. BARBER, Reasoning about change in knowledgable office systems, Workshop on Research in Office Semantics, Chatham, Massachusetts, June, 1980.
12. R. A. BOLT, Put-that-there: Voice and gesture at the graphics interface, *SIGGRAPH '80 Conference Proceedings,* July, 1980, pp. 262–270.
13. BOOZ-ALLEN and HAMILTON. *The Booz Hamilton Multiclient Study on Office Productivity,* New York, 1980.
14. L. BRACKER, and B. R. KONSYNSKI, A model of the automated office, MIS Technical Report, Department of Management Information Systems, University of Arizona, 1980.
15. *Business Magazine, Special issue on Ill-Structured Problem Solving,* Jan–Feb (1979).
16. S. CARD *et al.,* The keystroke level model for user performance times with interactive systems, *CACM* **23**(7), (1980).
17. S. CARD, and T. MORAN, *The Psychology of Human-Computer Interaction,* Lawrence Erlbaum Associates, New York, 1983.
18. J. M. CHANG and S. K. CHANG, Database alerting techniques for office activities management, *IEEE Trans. Commun.* **30**(1), 74–81 (1982).
19. D. COHEN, A protocol for packet-switching voice communication, *Comput. Networks,* **2**(4/5), 320–331 (1978).
20. R. UHLIG, ed., *Computer Message Systems,* North-Holland, Amsterdam, IFIP, 1981.
21. D. W. CONRATH, Communications environment and its relationship to organizational structure, *Manage. Sci.* **20**(4), 586–603 (1973).
22. D. W. CONRATH and J. H. BAIR, The computer as an interpersonal communication device: A study of augmentation technology and its apparent impact on organizational communication, *Proceedings of the 2nd International Conference on Computer Communications,* Stockholm, Sweden, August, 1974.
23. G. CORTESE and F. SIROVICH, A daemon-based programming system for office procedures, *Proceedings of the Second ACM-SIGOA Conference on office information systems,* ACM, New York, 1984, Vol. 2, p. 203.

24. I. W. COTTON, Technologies for local area computer networks, *Proceedings of the LACN Symposium, May, 1979, pp. 24–45.*
25. J. COUTAZ, Towards friendly systems, design concepts for interactive interfaces, *Proceedings of the 1984 Southeast Regional ACM Conference,* Atlanta, April, 1984, pp. 56–61.
26. B. CZEDO and D. EMBLEY, Office form definition and processing using a relational data model, *Proceedings of the Second ACM-SIGOA Conference on Office Information Systems,* ACM, New York, 1984, Vol. 2, p. 123.
27. S. P. DE JONG, The system for business automation (SBA): A unified application/development system, *Proceedings of IFIP 80,* October, 1980.
28. C. A. ELLIS, Information control nets: A mathematical model of office information flow, *Proceedings of the 1979 ACM Conference on Simulation, Modeling and Measurement of Computer Systems,* August, 1979, pp. 225–240.
29. C. A. ELLIS and G. J. NUTT, Computer science and office information systems, *Comput. Surv.,* **12**(1), 26–60 (1980) (also available as Technical Report, Xerox Palo Alto Research Center, Palo Alto, California, June 1979.)
30. D. W. EMBLEY, A natural forms query language—An introduction to basic retrieval and update operations in *Improving Database Usability and Responsiveness,* P. Scheuermann, Ed., Academic, New York, 1982, pp. 121–145.
31. D. C. ENGELBART, Augmenting human intellect: A conceptual framework, Summary Report, SRI, Menlo Park, California, 1962.
32. D. C. ENGELBART and W. K. ENGLISH, A research center for augmenting human intellect, *Proceedings of the Fall Joint Computing Conference,* AFIPS Press, New York, 1968, pp. 395–410.
33. S. GIBBS, Office information models and the representation of office objects, *Proceedings of the ACM Conference on Office Information Systems,* 1982.
34. M. GOOD, Etude and folklore of user interface design, *Proceedings of the ACM SIGPLAN SIGOA Symposium on Text Manipulation,* Portland, Oregon, June, 1981.
35. IRENE GREIF, The user interface of a personal calendar program, *Proceedings of the NYU Symposium on User Interfaces,* New York University, May, 1982.
36. K. D. GUNTHER, PLOP—A predicative programming language for office procedure automation, *IEEE Workshop on Languages for Automation,* Chicago, November, 1983.
37. M. HAMMER and M. D. ZISMAN, Design and implementation of office information systems, *Office Automation,* Infotech State of the Art Report. Series 8, No. 3. Infotech Limited, Maidenhead, Berkshire, England 1980. (Also in *Proceedings of the NYU Symposium on Automated Office Systems,* New York University Graduate School of Business Administration, May, 1979, pp. 13–24.)
38. L. HARRIS AND ASSOCIATES, INC. *The Steelcase National Study of Office Environments: Do They Work?* available in Canada from Steelcase Canada Ltd. P.O. Box 9, Don Mills, Ontario M3C 2R7, 1978.
39. C. F. HEROT, Spatial management of data, *ACM Trans. Database Syst.* **5**(4), 17–32 (1980).
40. R. S. HILTZ, Computer conferencing: Assessing the social impact of a new communications medium, *Technol. Forecast. Soc. Change* **10**, 225–238 (1977).
41. S. HILTZ and M. TUROFF, The evolution of user behavior in a computerized conferencing system, *Commun. ACM* **24**(11), 739–751 (1981).
42. INTERNATIONAL STANDARDS ORGANIZATION *ISO/TC97/SC16 Reference Model of Open System Interconnection ISO/DIS 7498,* International Organization for Standardization, December, 1980.
43. J. E. ISRAEL, and T. LINDEN Authentication in office system internetworks, *ACM Trans. Office Inf. Syst.* **1**(3), 193–210 (1983).
44. R. J. K. JACOB, Using formal specifications in the design of a human–computer interface, *Human Factors in Computer Systems Proceedings,* March, 1982, pp. 315, 321.
45. A. C. KAY, Microelectronics and the personal computer, *Sci. Am.* **237**(3), 231–244 (1977).
46. J. S. KE *et al.* SLOS: A specification language for office systems, *IEEE Workshop on Languages for Automation,* Chicago, November, 1983.

47. W. J. KEASE, Towards an in-house integrated publishing service, *Proceedings of the Second ACM-SIGOA Conference on Office Information Systems,* ACM, New York, 1984, Vol. 2, p. 113.
48. P. G. W. KEEN and M. S. SCOTT MORTON, *Decision Support Systems: An Organizational Perspective,* Addison-Wesley, Reading, Massachusetts, 1978.
49. G. C. KENNEY et al. An optical disk replaces 25 mg tapes, *IEEE Spectrum* **16**(2), 33–38 (1979).
50. G. LE LANN, Control of concurrent accesses to resources in integrated office systems, *Integrated Office Systems—Burotics,* N. Naffah, Ed., North-Holland, Amsterdam, 1980.
51. J. O. LIMB, Integration of media for office services, *Office Automation Conference Digest,* Houston, March, 1981, pp. 353–355.
52. T. W. MALONE, and S. A. SMITH, Tradeoffs in designing organizations: Implications for new forms of human organizations and computer systems, MIT Center for Information Systems Research Report SLOAN WP #1541-84, March, 1984.
53. A. H. MASLOW, *Motivation and Personality,* Harper, New York, 1954.
54. N. F. MAXEMCHUK, An experimental speech storage and editing facility, *Bell Systems Tech. J.* **59**(8), 1383–1395 (1980).
55. R. J. MCLEAN, Organizational structure as viewed by intraorganizational communication patterns, Ph.D. thesis, University of Waterloo Management Sciences Department, Waterloo, 1979.
56. J. M. MCQUILLAN, Local network technology and the lessons of history, *Proceedings of the LACN Symposium,* May, 1979, pp. 191–197.
57. N. B. MEISNER, The information bus in the automated office, *Integrated Office Systems—Burotics,* N. Naffah, Ed., North-Holland, Amsterdam, 1980.
58. V. MELNYK, Man–machine interface: Frustration, *J. Am. Soc. Inf. Sci.* **23**(6), 392–401 (1972).
59. H. MINTZBERG, *The Nature of Managerial Work,* Harper & Row, New York, 1973.
60. H. L. MORGAN, The interconnected future: Data processing, office automation, personal computing, *Proceedings of the LACN Symposium,* May, 1979, pp. 291–300.
61. H. L. MORGAN and O. P. BUNEMAN, Implementing alerting techniques in database systems, *Proceedings of the AFIPS National Computer Conference,* 1977.
62. H. L. MORGAN and J. U. SODEN, Understanding MIS failures, *Proceedings of the Wharton Conference on Research on Computers in Organizations, Database,* ACM, New York, 1973, Vol. 5:2, 3, 4, pp. 157–171.
63. W. NEWMAN, Office models and systems design, in *Integrated Office Systems—Burotics,* N. Naffah, Ed., North-Holland, Amsterdam, 1980, pp. 3–10.
64. W. OUCHI, A conceptual framework for the design of organizational control mechanisms, in *Organization Theory, A Design Perspective,* P. Connor, Ed., SRA Inc., 1979.
65. W. OUCHI, *Theory Z.* Addison-Wesley, Reading, Massachusetts, 1981.
66. M. PILOTE, The design of user interfaces for interactive information systems, Ph.D. thesis, Department of Computer Science, University of Toronto, 1982.
67. M. Y. PORAT, *The Information Economy: Definition and Measurement* (Nine Volumes), U.S. Government Printing Office, Washington, D.C., 1977.
68. R. PROPST, *The Office: A Facility Based on Change,* Herman Miller, Inc., Zeeland, Michigan, 1968.
69. R. REICHWALD, Cooperation in the office—Office communication systems as a management tool, *Proceedings of the Second ACM-SIGOA Conference on Office Information Systems,* ACM, New York, Vol. 2, p. 212.
70. I. RICHER, Voice, data and the computerized PABX: An electronic office, *Integrated Office Systems—Burotics,* N. Naffah, ed., North-Holland, Amsterdam, 1980.
71. B. SCHEURER, Office workstation design, *Workshop on OIS,* St. Maximin, October 1981, Republished in N. Naffah et al., *Office Information System,* North-Holland, Amsterdam, 1982, pp. 105–116.
72. M. SIRBU, Understanding managerial group work, *Proceedings of the 3rd AFIPS Office Automation Conference,* April, 1982.

73. M. SIRBU *et al.* OAM: An office analysis methodology, MIT Laboratory for Computer Science, Office Automation Group Memo, Cambridge, Massachusetts, 1981.
74. D. C. SMITH, E. HARSLEM, C. IRBY, and R. KIMBALL, The star user interface: An overview, *Proceedings of the National Computer Conference*, 1982, pp. 515–528.
75. S. A. SMITH and R. BENJAMIN, Projecting demand for electronic communications in automated offices, *ACM Trans. Office Inf. Syst.* **1**(1), (1983).
76. L. SUCHMAN, Office procedure as practical action: Models of work and system design, *ACM Trans. Office Inf. Syst.* **1**(4), (1983).
77. D. TAPSCOTT, *Office Automation: A User-Driven Method*, Plenum Press, New York, 1982.
78. K. J. THURBER, A survey of local network hardware, *Proceedings of COMPCON Fall 80*, September, 1980, pp. 273–275.
79. D. TILBROOK, Information systems primitives, *Proceedings of the International Workshop on Integrated Office Systems*, November, 1979.
80. D. TSICHRITZIS, F. RABITTI, S. GIBBS, O. NIERSTRASZ, and J. HOGG, A system for managing structured messages, *IEEE Trans. Commun.* **Com-30**(1), 66–73 (1982).
81. D. C. TSICHRITZIS, Integrating data base and message systems, *Proceedings of the 7th International Conference on Very Large Data Bases*, 1981, pp. 356–362.
82. D. C. TSICHRITZIS and S. CHRISTODOULAKIS, Message files, *ACM Trans. Office Inf. Syst.* **1**(1), 88–98 (1983); also presented at the ACM SIGOA Conference on Office Information Systems, Philadelphia, 1982.
83. M. TUROFF, Information, value, and the marketplace, New Jersey Institute of Technology Technical Report, September, 1983.
84. A. WEGMANN, VITRAIL; A window manager for an office information system, *Proceedings of the Second ACM-SIGOA Conference on Office Information Systems*, ACM, New York, 1984, Vol. 2, p. 1.
85. G. WILLIAMS, The Lisa computer system, *Byte* **8**(2), 33–50 (1983).
86. S. B. ZDONIK, Object management system concepts: Supporting integrated office workstation applications, Ph.D. thesis, Massachusetts Institute of Technology, May, 1983.
87. H. ZIMMERMANN, OSI reference model—The ISO model of architecture for open systems interconnection, *IEEE Trans. Commun.* April, 425–432 (1980).
88. M. ZISMAN, Representation, specification and automation of office procedures, Ph.D. thesis, Department of Decision Science, The Wharton School, University of Pennsylvania, Philadelphia, 1977.
89. M. M. ZLOOF, QBE/OBE: A language for office and business automation, *IEEE Trans. Comp.* **14**(5), 13–22 (1981).
90. M. M. ZLOOF, Office by example: A business language that unifies data and word processing and electronic mail, *IBM Syst. J.* **21**(3), 272–304 (1982).

2

LOGIC PROGRAMMING TAILORED FOR OFFICE PROCEDURE AUTOMATION

KLAUS D. GÜNTHER

1. Introduction

The design of the experimental logic programming language PLOP (*p*redicative *l*anguage for *o*ffice *p*rocedure automation) and an associated application-oriented programming environment aims at a proper integration of office databases and office procedure programming in a distributed environment, without being principally confined to office applications.

Unlike PROLOG, PLOP is not based on Horn clauses but on a restricted version of higher-order predicate logic enriched by function expressions, equations, and a genuine concept of "closed predicate definition." The PLOP database facilities comprise automatic access synchronization without explicit transaction delimiters or locks, and nearly unrestricted use of quantifiers and negation in queries. Implementation of PLOP programs will be based primarily on translation into distributed communicating sequential processes, each of which may access a "local" database and include recursive local computations.

The promising features of logic programming (to provide high-level, nonprocedural, descriptive expressive means) are well known (cf. Refs. 13–15, 6, 3), and there is some evidence that higher-order predicate logic has important additional advantages (cf. e.g., Ref. 16).

Chiefly for two reasons we decided to take a "logic programming" approach, i.e., an approach based on predicate logic:

(1) The application domain "office automation" requires proper support of database applications, and the most natural, currently known lan-

KLAUS D. GÜNTHER • Gesellschaft für Mathematik und Datenverarbeitung, Institut für Systemtechnik, Rheinstrasse 75, D-6100 Darmstadt, Federal Republic of Germany

guages for expressing database queries (for instance SQL) seem to be non-procedural and closely related to the expressive means of predicate logic.

(2) After several years of studying conventional procedural (including object-oriented) as well as net-oriented approaches to distributed applications programming we had come to the conclusion that nonprocedural, applicative programming languages, and in particular logic programming languages, were more likely to provide a safe and perspicuous way of distributed applications programming, at least in the long run, since they promise to free the applications programmer from the necessity of taking care of the correct synchronization of communicating parallel processes as well as of concurrent database access operations.

We felt, however, that current logic programming languages have to be improved and supplemented in several respects in order to become practical tools in office applications and related application domains. Our analysis of the drawbacks of current approaches to logic programming resulted in the design of PLOP, which attempts to overcome some of those drawbacks.

Section 2 presents our major office-related as well as office-independent design goals and requirements. Section 3 defines some particularly characteristic expressive means of PLOP in a BNF-like notation. In Section 4 we try to explain our respective approach to the solution of the problems identified in Section 2 and to illustrate our approach by small PLOP programs. Section 5 outlines the major implementation issues associated with our approach. Finally, Section 6 summarizes the most relevant points where PLOP differs from PROLOG, the currently most familiar approach to logic programming. The Appendix contains a listing of the basic subset of the PLOP syntax presented in Section 3.

2. Major Design Goals and Requirements

2.1. Office-Independent Design Goals

1. Conceptual integration of databases and data-processing.

Since databases play a central role in office applications PLOP should provide a unified conceptual approach to database access and dataprocessing.

2. Automatic synchronization of database access operations and a proper substitute for the transaction concept which is inapplicable in nonprocedural programming languages.

Concurrent access operations to multiuser office databases should be automatically synchronized, wherever possible, on the basis of a likewise automatic analysis of the data dependencies inherent in the respective office procedures. Just treating the entire office procedure as a single transaction which locks all data items to which it requires read or write access will generally not be acceptable since office procedures may run for days, weeks, or even months.

3. Application programs should be configuration independent.

The applications programmer should not be forced to think in terms of parallel processes and associated configuration, communication, and synchronization problems. It should be possible to run the same program on centralized as well as on highly distributed hardware configurations.

4. Automatic recovery from failures of network nodes and lines.

Well-known automatic error-recovery techniques in databases, based on "logging" and a suitable transaction concept, have to be extended so as to support also the automatic retransmission of data and/or the correctly synchronized, automatic restart of processes that have been affected by a partial breakdown of computers or communication lines involved in a distributed office procedure.

2.2. Office-Related Requirements and Design Goals

1. Automatic generation and handling of "forms" and "reports."

The system should automatically generate forms from the "pure logic" of office procedures and control the interaction between forms and users. The production of complex reports should be facilitated in a similar manner. The applications programmer should not be forced to manually program the precise layout and the entire interaction handling related to forms, although he must be able to influence these things if required.

2. Proper communication support for the organizational units of the office environment.

The office environment will usually consist of a hierarchy of organizational units ("org-units") having various responsibilities. So PLOP should provide proper expressive means for the addressing of organizational units, for the naming of objects contained in the individual name spaces which belong to these units, and for the communication and interaction between different organizational units.

3. It should be possible to reroute or copy incoming messages and to

reroute or delegate incoming input requests to other organizational units (manually, or as a triggered, automatic reaction; cf. point 6 below).

4. It should be possible to compute (at run time) the name of an organizational unit addressed by an input request or by an output.

5. It should be possible to program the automatic (instead of manual) production of responses to selected input requests of selected office procedures.

6. Sufficiently flexible alerter facilities for the triggered invocation of office procedures should be provided.

These alerter facilities should support the automatic starting of specific office procedures whenever certain application-oriented conditions occur (like arrival of specific new facts in the database, or arrival of a message or input request at an organizational unit, or arrival of a certain time or date). Cf. also Ref. 4.

7. An utmost degree of parallelism with respect to the work of persons involved in office procedures should be provided.

A rigid sequential mode of executing office procedures will be undesirable in general.

A lot of further important problems which call for more adequate solutions than present office systems would provide are currently more or less outside the scope of our project.

3. Characteristic Expressive Means of PLOP

Before we explain in Section 4 how the PLOP language and programming evironment tries to meet the design goals and requirements that have been identified in Section 2, we first outline the most characteristic and basic constructs of PLOP. We use a BNF-like notation, which is explained in the Appendix. The Appendix contains a condensed listing of the syntactic productions presented in this section.

The syntax of "expressions" and the special cases thereof are not presented on the first occurrence of the "expr" notion but are deferred to the end of this section, in order to avoid a lengthy interruption of the statement syntax at that place.

```
plop_program   ::=   statements
statements     ::=   statement {; statement} ...
```

A PLOP program is a sequence of "statements" separated by ";" which denotes the logical AND-connective.

statement   :: =   pred_def_stm | envmt_ctl_stm

A most important distinction is made in PLOP between "predicate definition statements," which may occur within predicate definitions, but also outside predicate definitions "in the main program," and "environment control statements," which serve to control the program execution environment. The latter do not make sense within predicate definitions and therefore must not occur there, since they do not express constraints for "local" data but specify modifications to the global memory of the system. (They have nonlocal "side-effects," without enforcing a deviation from the "logical" semantics of PLOP, however.)

pred_def_stm :: =   condition | definition | declaration

A predicate definition statement expresses either a "condition", or a "definition", or a "declaration." "Conditions" are the actual building blocks of PLOP programs. They describe relations between input, intermediate, and desired result objects (data objects and real world entities). "Definitions" are one way of introducing new predicates and serve also for introducing named, reusable office procedures ("macros"). "Declarations" do not influence the results of PLOP programs but may be used in order

- to optimize the computation process or the internal representation of data objects or
- to "beautify" the external representation of results or input requests or
- to simplify programs by default conventions.

condition   :: =   elem_cond | complex_cond

A condition is either elementary, or complex, i.e., composed from elementary conditions by application of logical connectives and quantifiers.

elem_cond   :: =   standard_cond |
                               infix_cond |
                               result_cond |
                               query_cond |
                               macro_call

An elementary condition has either the standard format known from predicate logic, or it makes use of a built-in predicate in infix notation, or it specifies "immediate results" by making explicit which of its parameters are considered to be "given" and which are to be determined as "results" from the former ones, or it is a special query condition, or a macro call.

$$\text{standard\_cond} ::= \text{pred\_ref (expr \{, expr\} ...)}$$

pred_ref	::=	pred_name [. org_unit]
pred_name	::=	identifier
org_unit	::=	sub_unit {. sub_unit} ...
sub_unit	::=	constant_su | computed_su
constant_su	::=	identifier
computed_su	::=	([expr])

A standard condition has the format: "predicate is true of parameter list," where a parameter may be an arbitrary expression (for "expr" see below). If the predicate is not defined but only characterized by facts stored in the database, then a standard condition expresses a database query, otherwise an algorithmically defined relation between its parameters.

The name of the predicate may be "qualified" by appending the name of an organizational unit of the application environment whose name space is expected to contain a definition or characterization of the predicate (by facts stored in the database associated with this org-unit and its name space). If "org-unit" is omitted the org-unit of the task initiator (who has started the program) is addressed. The individual components of the org-unit name may be constant as well as computed identifiers (dynamic addressing of org-units at run-time!).

infix_cond	::=	expr infix_pred expr
infix_pred	::=	< | <= | = | > | >=

The infix predicates denote equality/inequality as well as the usual ordering relations between numbers and the lexicographic orderings of strings of numbers or characters.

result_cond	::=	result = expr |
		pred_ref ? (expr {, expr} ...)
result	::=	var_name | (var_name|*} {,
		{var_name|*} } ...)
var_name	::=	identifier

Some typical examples of result conditions:
 Example 1.

$$z = (x + y)/2$$
$$z = f(2*x,y)$$
$$(x,y,z) = \text{coordinates (point\_1)}$$
$$(\text{year, month, day}) = \text{date? (business\_trip)}$$
$$\text{ok.boss? } (a,b,c)$$

The result parameters on the left-hand side of an equation "result = expr" are actually the trailing parameters of the "outermost" predicate of the expression on the right-hand side which have been ommitted there. In addition, such an equation implies that the "result parameters" are uniquely determined by the "given" parameters on the right-hand side and can be effectively computed from these (cf. "function_expr" below.) A result parameter "*" means "ignore" or "does not matter here."

The outermost predicate may be a built-in predicate, as "/" is in the first example. It may also be user-defined, or merely characterized by stored facts, as "*f*" and "coordinates" in the following two examples. Or it may be an "input predicate" (marked by "?") which presents its "given" parameters to its owning org-unit and expects the latter to provide suitable result parameters such that the input predicate (which is formally undefined) is true of the entire, manually completed parameter tuple (in the opinion of the owning org-unit, which is supposed to know the meaning of this "informal" predicate).

The last example illustrates the second kind of result conditions, where all parameters of the input predicate are "given" and the owner of the predicate has just to decide whether the predicate is true of this completely known parameter tuple. So he can only answer "yes" or "no" in this case.

$$\text{query\_cond} \quad ::= \quad [\text{FIND}]\{$$
$$\text{standard\_cond} \mid$$
$$\text{current\_cond} \mid$$
$$\text{is\_in\_cond} \mid \ldots \ldots\}$$

Besides the standard form of a condition there are several special and principally redundant constructs, which, however, greatly facilitate the formulation (and implementation) of certain frequently occurring queries asking for the "current" value of time-dependent entity attributes or "current" memberships or similar relations between entities. (For the entity concept see Section 4.1.) We have listed here only the most important of these constructs and only simplified versions thereof.

The FIND particle is in some cases required in order to distinguish query conditions from new facts that are to be stored (cf. "db_modif").

$$\text{current\_cond} \quad ::= \quad \{\text{CURRENT} | \text{OLDEST}\} \, [(\text{expr})]$$
$$\text{pred\_ref OF pred\_ref (expr) : expr}$$

Time-dependent attributes of real objects ("entities"), for instance the successive balances of an account, are described in PLOP not directly as attributes of these objects but as attributes of the individual update events which are treated as separate entities related to some fixed owner entity (cf. the "new_attr_val" construct below and Section 4.1).

$$\text{CURRENT(3) balance OF account}(a) : b$$

specifies b to be the third youngest balance of account a. Additionally, a specific time of reference different from the "current" time could have been specified by a preceding declaration (cf. the "ref_time" declaration below). Then the third youngest balance with respect to that time of reference would be determined.

$$\text{is\_in\_cond} \quad ::= \quad \text{pred\_ref (expr) IS IN pred\_ref (expr)}$$

Time-dependent relationships between real objects, for instance membership relations, are described by explicit "entry and exit events (entities)" which are associated with the former objects (cf. the "enters_leaves" construct below).

$$\text{employee}(e) \text{ IS IN project}(p)$$

specifies employee e to be in project p presently, i.e., the most recent entry/exit entity referring to employee e and project p should be an entry entity. Again a specific time of reference could have been declared before.

macro_call	::=	/macro_ref [macro_parms]
macro_ref	::=	macro_name [. org_unit]
macro_name	::=	identifier
macro_parms	::=	var_name = expr {, var_name = expr} ...

The PLOP programmer may assign a name to some complex condition and later refer to it by using this name prefixed by "/". He may store such a "macro" permanently and use also macros owned by other org-units. The most important application are reusable office procedures, which may be defined and "called" as macros.

The programmer may prevent a macro from issuing input requests for independent input parameters (cf. "expr" below) by assigning values to these parameters when calling the macro.

```
complex_cond  ::=   NOT log_operand |
                    log_operand log_infix_op log_operand |
                    if |
                    quant_cond


log_operand   ::=   condition | ( condition )
log_infix_op  ::=   {AND|&|;} | {OR|"|"} | {XOR|"||"} |
                    {IMPLIES|==>} | {IFF|<==>}
```

"XOR" denotes the exclusive OR, "IFF" the logical equivalence ("if and only if").

```
if            ::=   IF condition THEN statements
                    {ELIF condition THEN statements}...
                    [ELSE statements]
                    FI
```

The successive "ELIF" clauses can be considered to be nested "ELSE IF" branches of the outer "IF" statement.

```
quant_cond    ::=   {FOR EVERY | EXISTS [UNIQUE] | UNIQUE}
                    quant_base {, quant_base}...
                    statements
                    END


quant_base    ::=   standard_cond | enum_qbase | ......
enum_qbase    ::=   var_tuple IN grid
var_tuple     ::=   var_name | ( var_name {, var_name}...)
grid          ::=   ( number_seq {, number_seq}...)
number_seq    ::=   expr [TO expr [BY expr]] |
                    ( number_seq {, number_seq}...)
```

The purpose of the "quant_bases" is to restrict the quantified variables to finite base sets, for instance sets of entities known to the database or cross products of number sequences, in order to ensure that the verification of the respective "quant_cond" can be accomplished in finitely many steps and thus can be implemented.

Example 2.

```
FOR EVERY customer (c)
  IF NOT EXISTS UNIQUE account_of (c,a) END
  THEN
    name(c,!);
    FOR EVERY account_of (c,a)
      account_number (a,!);
      CURRENT balance OF account (a): !
    END
  FI
END
```

For every customer (of a bank, say) who has no or more than one account this program outputs those predicate parameters for which the output symbol "!" has been substituted (cf. "expr"), i.e., the name of the customer and the numbers and current balances of his accounts.

Example 3.

```
FOR EVERY (i,k) IN ((1,3,10 TO 100 BY 2, 200), −10 TO 10)
  f(i,k)!
END
```

This program outputs the value of the expression $f(i,k)$ for every argument tuple (i,k) from the cross product of the two number sequences "(1,3,10 TO 100 BY 2,200)" and "−10 TO 10".

3.1. Implicit Existence Quantification

If an identifier is neither bound by an explicit quantifier nor qualified by an org-unit name (and thus marked as being a stored constant) nor occurs as a formal parameter of a predicate definition, then the identifier is assumed to be subject to an implicit existence quantification.

Relative to the first occurrence of that identifier, the scope of such an implicit existence quantifier covers all subsequent occurrences of that identifier within the same explicit quantifier or definition scope and ends immediately before the first succeeding explicitly closing scope parenthesis. It starts at the last preceding place in the program where a matching "begin of scope" can be placed.

Implicit existence quantification is allowed only under certain further preconditions which are required as a substitute for the "quant_bases" of the explicit quantifiers in order to ensure implementability. Occurrence of the respective variable as "result" on the left-hand side of a result equation (see above) is, for instance, a sufficient precondition, since its value can then be computed by evaluating the expression on the right-hand side.

$$\text{definition} \quad ::= \quad \text{pred\_def} \mid \text{macro\_def} \mid \ldots \ldots$$

We shall only treat predicate and macro definitions here. In order to avoid overloading this first introduction to PLOP we have decided to omit the definition of several types of "complex axiomatic objects," in spite of their relevance to the office and other applications and to the envisaged PLOP implementation procedure. They can be used to represent complex data objects which in conventional programming languages would require pointers and/or named components, like "records" or directed graphs with attributed nodes and edges, or syntactic structures of formal texts, or the "forms" of the office environment, or partial "snapshots" of the database.

$$
\begin{array}{lll}
\text{pred\_def} & ::= & \text{def\_head def\_body END} \\[1em]
\text{def\_head} & ::= & \text{pred\_name [lifespan\_ind]} \\
& & (\text{ formal\_parm }\{, \text{formal\_parm}\} \ldots) \\
& & [\text{FOR }\{* \mid \text{org\_unit }\{, \text{org\_unit}\} \ldots \}] \\
& & == \\[1em]
\text{lifespan\_ind} & ::= & : \{S \mid L\} \\
\text{formal\_parm} & ::= & \text{identifier} \\[1em]
\text{def\_body} & ::= & \text{pred\_def\_stm }\{; \text{pred\_def\_stm}\} \ldots \mid \\
& & \text{expr\_tuple }\{; \text{expr\_tuple}\} \ldots \\[1em]
\text{expr\_tuple} & ::= & \text{expr }\{, \text{expr}\} \ldots
\end{array}
$$

If the life-span indicator is omitted the predicate definition will not survive the current program run (= "task"). S means that it is kept till the end of the current terminal session ("logoff"), L means that its life-span is "long-term", i.e., it is kept until it is deleted explicitly.

The FOR clause allows the programmer to provide automatic answers to input requests caused by (formally undefined!) input predicates (cf. "result_cond" above) that have been issued by programs owned by one of the enumerated org-units ("*" = by any org-unit). Input requests issued by different org-units which accidentally use the same name for the respective input predicates though these have different meanings can in this way be automated individually.

The first form of the definition body covers also arbitrarily complex recursive references to the defined predicate. In practice we will, of course, have to restrict the expressive means in such a way as to allow of a reasonable implementation.

The second form of the definition body provides a purely enumerative definition by listing all and only those parameter tuples of which the predicate is true by definition.

macro_def   ::=   macro_name [lifespan_ind] = = statements END

Cf. "macro_call" above.

declaration   ::=   input | output | ref_time | query_sep |
                        . . . (many others)

Declarations control the interpretation of the subsequent program text till the occurrence of a new, conflicting declaration, maximally till the end of the current quantifier or definition scope.

The automatic generation and handling of "forms" according to the input requests originating from the undefined input predicates and independent input parameters of a PLOP program (distinguished by "?" or by an input declaration) is a central goal of the envisaged PLOP implementation. A great variety of input declarations allow the programmer to control this process and to deviate from the automatically generated forms layout and interaction handling. Moreover he may provide default values and help information for every input parameter and assign attributes like "confidential" or "urgent" to an input request.

Similarly the programmer may influence the presentation of result data (marked by "!" or requested by an output declaration) in many ways.

ref_time   ::=   REF {TIME | ENTITY} expr|
                        REF TIME -

The "ref_time" declaration allows the programmer to specify a fixed time of reference or a reference entity for subsequent query conditions asking for "current" information, e.g., for the current balance of an account, or for current project memberships of employees. In Section 4.1 we shall explain how events of the real world are described by "entities" in the database which are always associated with a unique time-stamp reflecting the time when these entities have become known to the database.

If a reference time or entity has been declared explicitly queries take only such entities into account that have become known before or at that point of time.

"REF TIME -" resets the last such declaration, i.e., every currently known entity is taken into account thereafter.

$$\text{query\_sep} \quad ::= \quad \{\text{NEW} \mid \text{FREE}\}\ \text{QUERY}$$

Normally a complex query condition must be true in toto at some appropriate instant ("snapshot semantics" of complex queries, cf. Section 4.1.2). NEW QUERY allows a deviation from this default rule by allowing the system to determine a new, separate snapshot instant for the following query conditions within the same or nested quantifier or definition scope(s). FREE QUERY completely abandons the normal snapshot semantics. It rather allows every elementary query condition to have its own "time of validity."

$$\text{envmt\_ctl\_stm} \quad ::= \quad \text{db\_modif} \mid \text{alerter\_actvn}$$

Environment control statements may modify the current contents of the database or activate an "alerter" which waits for certain alert conditions to become true and then reacts to this occurrence in a specific way.

$$\text{db\_modif} \quad ::= \quad \text{insert} \mid \text{delete}$$

$$\text{insert} \quad ::= \quad [\text{STORE}]\ \text{elem\_fact}\ [\text{store\_opts}]$$

The STORE particle is generally required in order to distinguish elementary forms of queries from insertions of new facts. The store options "store_opts" may be used to associate additional trace information or remarks with the newly stored facts, e.g., information about date and time of this fact storage operation, or about the provider (org-unit) of the data parameters of the new fact.

$$\text{elem\_fact} \quad ::= \quad \text{standard\_cond} \mid$$
$$\text{new\_attr\_val} \mid$$
$$\text{enters\_leaves}$$

new_attr_val   ::=   NEW pred_ref OF pred_ref (expr) : expr

enters_leaves   ::=   pred_ref (expr) {ENTERS|LEAVES} pred_ref
                        (expr)
                        [EVENT entity_name]
entity_name   ::=   identifier

Example 4.

employee(*e*);
name(*e*,'Brocklehurst');
project(*p*);
name(*p*,'ABCD');
STORE
married(*e*);
NEW salary OF employee(*e*): 12345.67;
employee(*e*) ENTERS project(*p*);

The first four statements express a query which attempts to retrieve a certain employee and a certain project from the database. If its result is unique then the subsequent facts (which should be self-explanatory) are stored about this employee and this project.

The NEW fact implicitly introduces a new salary update entity "su", say, and stores the fact "salary(su,12345.67)". It stores also a binary relation between "*e*" and "*su*" which expresses that this salary update belongs to this employee and which has an internal name (unknown and inaccessible to the users). (Cf. the CURRENT construct above.)

The ENTERS fact likewise introduces a new entity which is characterized by internal predicates as representing an entry of an employee (viz., *e*) and into a project (viz., *p*). (Cf. the IS IN construct above.)

delete   ::=   DEL[ETE] del_item {,del_item} . . .
del_item   ::=   expr | fact_pattern | . . . . . .
fact_pattern   ::=   pred_ref (expr {, expr} . . .)

The DELETE construct allows the programmer to delete named data objects of life-span *S* (session) and *L* (long-term), as well as stored facts with given fact predicate and partly given fact parameters, but also entire entities (by deleting all stored facts referring to these entities).

alerter_actvn   ::=   WHEN[EVER] alert_conds [start] [stop]
                        THEN alert_stm {; alert_stm} . . . END

alert_conds	::=	alert_cond {{"	"	&	AND	;} alert_cond} ...
alert_stm	::=	statement | alert_action
alert_action	::=	copy | re-route | delegate | stop | ...

"WHENEVER" is, in terms of predicate logic, a universal quantification of the form "for every occurrence of the alert conditions (within the specified time interval specified by the "start" and "stop" clauses) the following alert_stm's are to be rendered true." We have omitted here the detailed syntax of "start," "stop," "alert_cond," and "alert_action." The WHEN construct causes the "alerter" to react only once, viz., upon the first occurrence of the alert conditions.

Alert conditions supported by the present draft of PLOP include arrival of specified clock times (also on a periodical basis), arrival of outputs or input requests at the current org-unit, and insertion/deletion of specific facts in the database. More complex alert conditions may be composed from these elementary alert conditions by logical OR and AND connectives. Here AND means: One of the ANDed conditions has occurred, and the others turn out to be also true at that instant.

The THEN clause may be a (fairly normal) complex PLOP program and may make reference to certain parameters which are mentioned in the alert condition and related to this alerter activation (e.g., the sender of the message, the time or date of this activation, etc.).

Typical reactions to the arrival of outputs are to copy or reroute them to other org-units. Input requests may be rerouted or delegated to some other org-unit. Delegation means that the input request is automatically rerouted to the delegating org-unit after the second org-unit to which it has been delegated has answered the request. The former may then reexamine and revise this answer or delegate it again before confirming it finally.

We can also cover event processing tasks by an extended version of this WHEN[EVER] construct which is typical of "systems programming," by causing the operating system to store a corresponding fact in the name space of a special "system org-unit" whenever a system event occurs that is to trigger an alert condition (like interrupts, exceptions, external signals, arrival of an input byte or block on some I/O device). "Systems programming" would additionally require a special version of an existence quantification requesting the realization of typical system events, like output of characters or blocks on various output devices. A further, frequent requirement will be to produce these output events in the same order in which the causing alert events have occurred in the respective alerter.

expr   ::= * | ?[!] | ! |
                     constant [!] |
                     object_ref [?] [!] |
                     function_expr |
                     infix_op_expr |
                     current_expr | . . . . . .

constant   ::= integer | real | char_string | bit_string |
                     NULL | . . . . . .

object_ref   ::= object_name [. org_unit] [lifespan_ind]
object_name   ::= identifier

function_expr   ::= pred_ref [?] ([expr] {, [expr]} . . .)

infix_op_expr   ::= operand infix_op operand
operand   ::= expr | (expr)
infix_op   ::= + | − | * | / | . . . . . .

current_expr   ::= CURRENT [(expr)] pred_ref OF pred_ref (expr)

"*" means "existent, but having an arbitrary value."

"?" either marks an "independent input parameter" whose value has to be provided by the user, without requiring the presentation of previous results or inputs, or, if occurring after the "pred_ref" in "function_expr," it marks an input predicate (cf. "result_cond" above).

"!" marks expressions whose value(s) are to be presented to the user.

The representation of constants is the same as in PL/1.

Function expressions are derived from predicates by omitting one or several "output parameters" of the predicate (but retaining the preceding comma, unless it is a "trailing" output parameter). The expression stands for every possible tuple of (omitted) output parameters which, together with the remaining ("input") parameters fulfills that predicate.

For example, if p denotes a predicate of four parameters and q a predicate of three parameters then the condition $p(3,q(,4),5)$ is equivalent to

> FOR EVERY x, y
>   IF $q(x,4,y)$ THEN $p(3,x,y,5)$ FI
> END ,

and the condition $(a, b) = q(,4)$ is equivalent to

$$
\begin{aligned}
&\text{FOR EVERY } x, y \\
&\quad \text{IF } q(x,4,y) \text{ THEN } a = x \ \& \ b = y \text{ FI} \\
&\text{END}
\end{aligned}
$$

Note that the latter assertion implies that $q(x,4,y)$ has only one solution x,y, while the condition $q(a,4,b)$ could have more than one solution a,b. (Cf. "result_cond" above.)

The CURRENT expression corresponds closely to the CURRENT condition (see above). It yields the current value of a time-dependent attribute (first "pred_ref") of some type of entity (second "pred_ref"), e.g., "CURRENT salary OF employee(e)."

4. How We Try to Approach Our Goals

In the following we shall try to explain for each of our general and our office-oriented design goals mentioned in Sections 2.1 and 2.2 how we attempt to solve the related problems.

Design and implementation of PLOP are still in a very early stage. Many basic questions are still open and will require a more thorough consideration. Implementation will proceed in a more or less experimental style and in small steps only, corresponding to small subsets of PLOP. (This is also the reason why this chapter is not completely consistent with Refs. 8 and 9.

Nevertheless we found it desirable to present our general ideas to a broader public already in this early stage in order to stimulate the discussion about possible improvements of the current "logic programming" conceptions and since we hope to get valuable feedback in this way.

4.1. Approach to Our General Design Goals

4.1.1. Integration of Databases and Data Processing. The conventional interpretation of a database is that it contains descriptive data about the current state of the world. Whenever some relevant state transition occurs in the real world the pertinent descriptive data are replaced with more actual data.

In our approach this state-oriented database interpretation has, for two reasons, to be replaced with a more history-oriented interpretation:

- From the standpoint of predicate logic the facts about real world objects and events stored in a database should not depend on time but should remain true for ever. This is only possible if the stored facts refer explicitly to definite objects and associated events, i.e., if they explicitly reflect the history of state transitions.

- As a rule, office applications would anyway require a complete history of the data "produced" by any office procedures to be recorded in the database, while deletion and/or replacement of stored facts are rare and critical exceptions and are applied at most in order to correct erroneous or to abandon obsolete data.

So we let all stored facts refer explicitly to some real "entity" (an "object," "event," or "process") whose existence, whose attributes, and whose interrelations are asserted by those facts.

In contrast to so-called "logic databases," we exclude such more general constructs for expressing facts from PLOP which would require general automatic theorem proving or other artificial intelligence techniques to be included in the PLOP implementation.

A further important decision was to adopt the so-called "closed world assumption," again in contrast to logic databases. This means that the currently stored facts are assumed to be "complete," or in other words that logical negation and universal quantification are interpreted relatively to the currently stored facts.

For instance, "NOT $p(e)$" means "the fact p about entity e is currently not stored," "NOT EXISTS . . . " means "is currently not known," "FOR EVERY . . . " means "FOR EVERY currently known. . . ".

Internally every real entity is represented by a system-generated unique identifier comprising a time-stamp which fixes the time when the entity has become known to the database. Similar notions of "entities," internally represented by unique, system-generated identifiers, have been discussed in the literature under the name of "surrogates." Cf., e.g., Refs. 7 and 12. As to the semantics of time in databases, cf. Ref 5.

Example 5.

> STORE
> book(b);
> title(b,'Hamlet');
> writer(w);
> name(w,'Shakespeare');
> author(b,w)

Here ";" denotes the logical AND. Variables b and w are implicitly quantified by EXISTS. (This is the default quantification assumed by PLOP for simplicity). The predicates "book," "title," "writer," "name," "author" are used for storing facts. The particle STORE is necessary in order to distinguish between insertion and specification (in particular query) semantics of PLOP statements. Without STORE the same statements would express a query asking for a certain book b, etc.

This associated query is in fact executed internally in order to check whether entities b, w having all these properties are already known. In the latter case an error message would be issued. Otherwise new entities are established.

The internal representation of the stored facts no longer contains the original entity variables b, w of the program but rather two associated unique internal entity identifiers comprising a time-stamp (see above).

The entities occurring in the preceding example were of a more static, permanent nature ("objects"). The following example combines such a static entity (an account) with eventlike entities (updates of the account).

Example 6.

$$account(a);$$
$$number(a,123456);$$
STORE
$$acc\_update(u,a);$$
$$balance(u,6543.21)$$

Here the first two statements express a query asking for an account a having account number 123456. The following statements (after STORE) express the existence of an update event u belonging to a which is associated with a new balance of the account. Note that the new balance of the account is not viewed as an attribute of the account but of the individual update event! The (system-generated) internal entity identifiers associated with all update events then reflect the time order in which these updates have become known to the system.

PLOP provides special, principally redundant syntactic constructs which, however, greatly facilitate the treatment of frequently occurring combinations between static "owner objects" (like accounts) and time-dependent attributes of these (like updated account balances), as well as the treatment of time-dependent relations (like "memberships") between static objects.

Example 7.

FOR EVERY employee(e)
  IF FOR EVERY child(c,e)
    DATE( ) - date_of_birth(c) < 18

> END
> THEN
>   NEW salary OF employee(*e*): CURRENT salary OF employee(*e*)
>   + 100.00
> FI
> END

This example makes use of function expressions (the built-in function "DATE()", "date_of_birth(*c*)", and the CURRENT-expression) and of two of the just-mentioned additional constructs (viz., "NEW" and again the CURRENT expression) related to time-dependent attributes ("salary"). Like the new balance of the account in the preceding example, the salaries are viewed as attributes of the individual salary update entities belonging to the employee entity.

Without going into details, let us just explain the meaning of this example: It grants an increase of 100.00 to the salary of every employee all of whose children are currently less than 18 years old.

4.1.2. Automatic Database Access Synchronization. The "transaction" concept is inapplicable in logic programming. In PLOP, the concepts of "consistency," "integrity," and "correct synchronization of concurrent database access operations" have to be redefined from the perspective of logic programming, which differs also in this respect principally from that of conventional procedural programming languages.

The conventional transaction concept cannot be viewed as a final and universally applicable answer to the problems of consistency and integrity in database applications. It cannot be readily taken over by a truly nonprocedural, for example logic programming, approach since it relies essentially on the basic concepts of procedural programming languages: The control flow of a procedural program is subdivided into sections, each of which is viewed as an actually complex but conceptually atomic indivisible action.

Compared with this, a nonprocedural program does not exhibit any explicit flow of control which could be subdivided into conceptually atomic (trans)actions.

In PLOP, correct synchronization is rather attempted by combining the above-mentioned entity concept with a more discriminating view of "queries" and with an automatic analysis of dependencies between already stored and newly entered facts.

Reader–writer synchronization. The explicit reference to real entities in all stored facts guarantees that inconsistencies due to yet incompletely stored new facts are impossible (provided storing a single, atomic fact is an indivisible, atomic operation.) For example, it is always clear to which individual account update event the pertinent attributes (for instance "balance," "date," "update_type," etc.) refer.

What can happen at most is that a reader does not find all the required facts about an entity that are concurrently introduced to the database by some writer. This (usually transient) incompleteness of facts that are currently entered into the database is bridged in PLOP quite naturally, and without requiring specific expressive means, just by the fact that a query is interpreted as a specification. It demands the existence of certain entities and requires certain facts to be currently known about these entities, and it waits, if necessary, until such entities fulfilling these facts have become known to the system.

So, in contrast to other query languages, queries in PLOP are generally interpreted as "waiting queries" in a sense. For example, the first two statements of Example 6 require an account having number 123456 to exist. Otherwise execution of the procedure is suspended until such an account becomes known to the database.

There may, however, exist parts of a query which definitely need not wait but which can always be executed immediately. The typical representative of this type of queries ("immediate queries" or "tests") is an IF-condition, since it would just ask whether certain entities fulfilling certain conditions are *currently* known or not known and then execute the THEN-clause or the ELSE-clause, respectively.

Example 8.

IF EXISTS account(*a*)
    CURRENT balance OF account(*a*) > 1 000 000.00
    END
THEN 'There exists an account with balance > 1 000 000'!
ELSE 'There does not exist an account with balance > 1 000 000'!
FI

The IF-clause tests immediately whether an account is currently known containing more than one million, while the THEN- and ELSE-clauses report the result of the test.

Besides the basic specification semantics of PLOP queries there is a second provision as to the semantics of queries which is related to the correct synchronization of readers and writers: We demand that complex query conditions must be true in toto at some instance after their evaluation has been started.

So, for example, if a query is of the form "cond1 AND cond2" it is not sufficient to find cond1 to be true at some time $t1$ and to find cond2 to be true at some later time $t2$, while cond1 is no longer true at $t2$, but this query requests cond1 and cond2 to be simultaneously true at some appropriate time t after it has been passed to the system. ("Snapshot semantics" of complex queries.)

Using the "NEW/FREE QUERY" declaration (cf. "query_sep" in Section 2), the programmer may, however, reset and thus advance this snapshot instant. This is sometimes desirable in the course of an office procedure or other application which takes days or weeks till termination but does not require a common "time of validity" to exist for all elementary query conditions occurring in the procedure.

Writer–writer synchronization. Sometimes new facts have to be stored or deleted only if certain other facts are known to be fulfilled, or newly stored attribute values depend on old, already stored values. In these cases we have to guarantee that these preconditions for the insertion/deletion remain valid until the insertion/deletion has been completed.

More precisely: We have to ensure the existence of an instant at which the new ("dependent")/obsolete facts are already completely stored/ deleted while the condition on which they depend would still be true if the new/deleted facts were considered to be not yet/still visible. ("Snapshot semantics" of a read and "dependent" insert/delete access.)

Note that actually a database modification becomes visible only after a new checkpoint has been reached (see below), and in many cases newly stored facts will terminate the validity of their own preconditions. For example, the current balance of an account will no longer be the current balance as soon as a new balance has been stored.

In the implementation of Example 7 above we thus have to ensure (for instance by causing the PLOP compiler to include appropriate locking and unlocking primitives in the generated object-program)

- that the current value of the salary of the respective employee really remains the current value (rather than being replaced with some other value by some other, concurrently executing PLOP program), and
- that no new child (perhaps due to a marriage or adoption) whose age is 18 or greater is entered into the database before the new salary is stored.

Prevention of "phantoms." The above-mentioned "snapshot semantics" of complex PLOP queries guarantees that so-called "phantoms" cannot occur, i.e., entities for which the parts of a complex query condition have never been true simultaneously but only successively at slightly different times during the evaluation of the entire complex query.

"Phantoms" must not be confused, however, with certain cases of misinterpreted queries and/or inappropriately structured entities/facts:

Misinterpretations may arise, e.g., if our general "closed world assumption" is ignored. For instance, for some person p, "NOT married(p)" is not equivalent to "unmarried(p)". If we find the former condition to hold true at some instant we can only conclude that p is not known to be married at that instant, i.e., the fact "married(p)" is not stored in the database currently, while "unmarried(p)" means that this fact "unmarried(p)" is effectively stored in the database. So if we need to know definitely whether p is married or unmarried we have to provide, e.g., a separate predicate "unmarried" besides the predicate "married," or we could provide a two-placed predicate "personal_status" and store the fact

$$personal\_status(p, \text{'married'})$$

if p is married and

$$personal\_status(p, \text{'unmarried'})$$

if p is unmarried.

Another type of potential misinterpretations can be illustrated by considering a transfer of money from one account to a second account, both accounts having the same owner. Now assume a query running concurrently to this transfer process attempts to sum up the "current" balances of all accounts of this owner in order to determine the total amount of his current capital. This query could catch an instant of time when the transferred amount has disappeared from the first account but has not yet been entered on the second account. The amount is thus temporarily reflected by neither of the two balances though it is owned by this person without interruption.

We do not consider this to be a violation of consistency and the computed sum to be a "phantom," since even in such transient "snapshot" states the database does not contain wrong facts. (But, of course, it does not yet contain all facts of which some user already knows that they soon will be stored.) Rather we consider this query to be an inappropriate way of keeping track of the current total capital of a person.

In the PLOP environment a correct result could, e.g., be achieved

- by keeping an additional pseudoaccount which continuously reflects the current total capital of that person and which, of course, will not be updated in the case of a purely "internal" money transfer between two "regular" accounts of that person, or

- by first stopping all internal transfers by appropriate (but surely exceptional) organizational measures (outside the regular data processing system) and then summing up the current balances. (Analogy: A firm is closed while it takes an inventory.)

The conventional solution, to establish the sum-up process as a (pure read) transaction, may appear to be somewhat simpler, admittedly, as it does not require additional data to be kept or manual organizational measures to be taken.

On the other hand, the fact that in PLOP a reader is allowed to see such intermediate, still "incomplete" (in some sense, but not really "inconsistent") states of the database promises to considerably reduce the problems inherent in "long transactions," which are a serious handicap in office as well as engineering applications.

External visibility of newly stored facts. Our present design provides that a task (= individual run of an application) is decomposed into sequential processes corresponding to the involved org-units (cf. Section 4.1.3). Insertions/deletions applied to the database by such a component process become visible to other tasks only after the process has completed a new checkpoint. They are backed out if automatic error recovery causes a restart before reaching that checkpoint.

A programmable roll-back of tasks is not provided in PLOP. If a PLOP task fails for internal reasons ("exceptions") then all facts stored so far become visible to other tasks. It is considered to be a matter of applications programming and particularly of the application-specific fact and entity structuring to avoid the premature storage of still uncertain facts, to replace out-of-date facts with new facts, and/or to provide explicit indicators in the stored facts which allow other tasks to assess the current status, the reliability, and completeness of the facts stored so far by a task.

4.1.3. Configuration–Independent Programs. Logic programs do not reflect the underlying distributed hardware configuration (in contrast to conventional communicating sequential programs). The problem remains to be solved, however, how to implement such programs in a distributed environment.

We have decided to attempt this by translating PLOP programs into communicating sequential processes. Each of these corresponds to an individual organizational unit involved in the office procedure.

For a restricted class of recursive programs such a compiler has already been developed[1] (written in SPITBOL, target language: SIMULA67). The extension to a more complete and up-to-date version of PLOP is planned, of course, and seems to be feasible at least for nonrecursive PLOP programs and simple types of recursion.

4.1.4. Automatic Error Recovery for Distributed Office Procedures. We try to achieve this goal by a separate "object-related communications and upkeep service (ORCUS)" for general distributed applications (not necessarily written in PLOP).

The component processes of a distributed application may submit complex data objects to ORCUS, together with application-oriented object identifiers which are unique within the distributed task. The objects are then transferred within ORCUS to those sites where potential consumer processes (belonging to this distributed task) reside. ORCUS is particularly responsible for the secure temporary storage, life-span management, and (possibly repeated) delivery of objects. Repeated reference to an object (via its identifier) may in particular be caused by (normally isolated!) restarts of components processes.

Crash recovery and restart of component processes are also controlled by ORCUS and include "undoing" all changes to the local component database performed by the restarted process since the respective checkpoint. (Cf. also Ref. 2, which is somewhat outdated, however.)

4.2. Approach to the Office-Related Goals

Owing to the lack of space we can only outline here some of our ideas concerning some of our office-related design goals.

4.2.1. Automatic Generation of Forms and Reports. Application-oriented input is based on formally undefined predicates which are evaluated "manually" by the user at run-time, and on "independent input parameters" (cf. "result_cond" and "expr" in Section 3, respectively).

We intend to include an automatic forms and interaction management in our PLOP compiler and the associated run-time system which generates an appropriate form from the undefined input predicates and independent input parameters occurring in the PLOP program and which controls the interaction with the users involved in the office procedure. (Cf. also the following subsections.)

So in our approach forms are not viewed as explicitly predefined data structures but are derived partly at compile-time, partly at run-time from the structure of the PLOP application program and additional input declarations.

A special role in this context is played by so-called "formal" and "informal annotations" that can be affixed to the input predicates and input parameters of a PLOP program (as to every other syntactic unit) by the programmer. Such annotations may be used to provide additional help information to the user at run-time or to cause the layout of the generated form to deviate from its standard format, or to modify the interaction between user and generated form. The syntax of these annotations has not yet been fixed in detail, however, whereas the annotation feature is already supported by our (language-independent) syntax-directed editor (cf. Section 5.4 below).

Our ideas concerning report generation are still in a very early state, likewise.

4.2.2. Supporting Office Communication. The organizational units of the office environment can communicate with one another in several different ways:

- An office procedure invoked by org-unit *A* can store facts in the database of org-unit *B* (proper authorization provided).

To this end, the predicates that are used for expressing those facts just have to be qualified by the name of org-unit *B*. The procedure cannot produce, however, a new named data_object, e.g., "new_obj.*B* = 123.45", in *B*'s name space, nor can it create there a new predicate which could then be used for storing a new kind of facts in *B*'s part of the database.

- The office procedure (invoked by *A*) can request input data from org-unit *B* (via independent input parameters or input predicates).
- The office procedure (invoked by *A*) can direct a message to *B*.

For instance "SHOW *e* TO *B*" would cause all currently known facts about a real entity "*e*" to be presented to *B* (as soon as *B* decides to inspect this message).

- *B* can copy or reroute the message to some org-unit *C*.
- In the case of example 1 org-unit "boss" may delegate the evaluation of the "ok" predicate to some other org-unit.
- Informal annotations may be affixed to stored facts, to rerouted, delegated, copied messages and input requests.

4.2.3. Rerouting, Copying, Delegation. This is supported by corresponding expressive means available on the task control level of the programming environment as well as by actual PLOP constructs that can be used within the THEN clause of the WHENEVER statement. (Cf. "alerter_actvn" in Section 3.)

4.2.4. Dynamic Addressing of Org-Units. It is quite clear that it must be possible to determine the participants in an office procedure at runtime. (For example, the manager of an employee who applies for a business trip can be determined and asked for his consent only after the employee is known). Therefore, besides constant org-unit identifiers, PLOP provides also computed org-unit identifiers.

The actual problem in this context is, however, the dynamic integration of additional processes (corresponding to those dynamically addressed org-units) and communication links into the already running distributed application.

4.2.5. Automation of Manual Inputs. In PLOP an input request is either due to an input predicate or an independent input parameter. The addressed org-unit may give a formal definition for an input predicate and in this way avoid its manual evaluation in the future, and it may also preassign values to independent input parameters in several ways.

A frequent way of automating manual inputs will be to refer to the value that had been provided "the last time" in response to the respective request.

4.2.6. Alerter Facilities. Alerter functions are supported in PLOP by a general alerter construct. The alert conditions provided in PLOP (cf. "alerter_actvn" in Section 3) are flexible enough to cover the cases mentioned in requirement 6 of Section 2.2.

In PLOP the triggered invocation of office procedures is not intended for enforcing application-related database integrity constraints, nor does PLOP support "assertions" for that purpose. In our approach, application-dependent integrity constraints are enforced by allowing only quite specific, particularly authorized office procedures to insert or delete such critical facts subject to an integrity constraint. We suppose that this will suffice in practice and that more independent integrity checks supported by triggers and assertions are not really required in this application domain.

4.2.7. Utmost Degree of Parallelism. This goal is pursued in PLOP by assigning an individual sequential process to each org-unit involved in an office procedure, even if these processes run within the same local computer. As has been mentioned already in Section 4.1.3, we implement a PLOP program by translating it into a system of communicating sequential processes, where a process corresponds to exactly one org-unit of the office environment. The necessary synchronization between these processes is based exclusively on the natural dependencies due to "data-flow," and thus the maximum parallelism with respect to the work of the persons representing these org-units is achieved.

5. Implementation Issues

As a consequence of its enhanced expressive power (covering, e.g., explicit quantifiers, negation, predicate definitions, alerter functions, various office support functions) and its distributed implementation environment it is clear that PLOP cannot be implemented on the basis of the same uniform pattern matching, resolution, and backtracking mechanism as PROLOG, but that more complex program transformations have to be performed to this end and that more specific computation procedures have to be applied to the major and essentially different subsets of PLOP.

Three major subsets have to be distinguished in this respect, whose expressive means may be mingled more or less arbitrarily, however, in concrete PLOP programs:

- the database subset (in particular complex "queries"),
- the subset of nonrecursive, nonquery (main) programs and predicate definitions ("nonrecursive algorithms"),
- the subset of recursive predicate definitions ("recursive algorithms").

5.1. Implementation of the Database Functions

Our first implementation attempt will be confined to the case of a single, centralized database. We found it desirable to base our implementation upon the applications programmer's interface of an existing relational database. In doing so, we have to take a number of special requirements of the PLOP database concept into account, and in particular of the concept of internal entity identifiers on the basis of time-stamps. We need, among other things,

- an appropriate, time-stamp-like attribute type,
- quick access to the "youngest" entity having some characteristic property,
- quick access to "all currently known" facts concerning some given entity (which is actually a most desirable service of a "data dictionary"),
- the capability of affixing optional trace information about the providers of newly stored facts, about date and time of their insertion, etc., to the respective facts (= tuples),
- the capability of replacing the conventional access synchronization of the underlying relational database with our completely different, automatic access synchronization concepts outlined in Section 4.1.2 above.

More details on the database requirements of computer-aided office procedures (from our point of view) can be found in Ref. 10.

We expect that the solution of several of the above-mentioned and further problems will be facilitated by the experimental "nonstandard" relational database system which is currently developed by the AIM (Advanced Information Management) group of the IBM Scientific Research Center in Heidelberg, with whom we have a cooperation contract.

The "specification semantics" of PLOP queries will be simply implemented by repeated execution of queries until all conditions are fulfilled. The time between two trials could perhaps be automatically extended, depending on the number of unsuccessful trials experienced so far. Very volatile database states cannot be "caught" in this way, of course. But such problems of type "management of limited resources" would anyway be better solved by a dedicated "resource management program" which would wait for and sequentially process incoming requests via the WHEN-EVER construct. The requests would be issued by storing corresponding facts in the database of the respective resource management org-unit and would be answered in the same way.

The "snapshot semantics" of complex queries and of "dependent" insertions/deletions requires any insertion/deletion to be delayed if it could change the truth value of a concurrently evaluated query condition from true to false.

We expect that the required automatic detection of potential collisions between insertions/deletions and query conditions is feasible at least for PLOP programs that do not involve recursive predicate definitions. In many frequently occurring cases this can be decided essentially at compile-time using fairly simple syntactic criteria. For example, if a query condition contains neither negations nor universal quantifiers and has been found to be true at some instant, then it cannot be rendered false thereafter by concurrent tasks which store new facts only. (It may be rendered false, of course, by the deletion of facts.)

In more complicated cases a potential collision cannot be definitely excluded by such a syntactic analysis already and thus must be assumed and prevented.

In any case the compiler decomposes the entire program into "coherent insert/delete sections" and "critical query sections" to which a simple and provably deadlock-free scheduling strategy is then applied at run-time by the access synchronization component of our (modified relational) database.

The "critical query sections" are either the "insert/delete preconditions" mentioned in the subsection on "writer—writer synchronization" of Section 4.1.2, or they are (parts of) pure queries (taking the "NEW / FREE QUERY" declarations mentioned in Section 3 into account).

5.2. Implementation of Nonrecursive Algorithms

The nested structure of universal and existence quantifiers of a non-recursive PLOP program determines a basic nested loop structure of the sequential computation process which is derived from the PLOP program in a first compilation phase. Owing to the associated "quant_base" every quantifier ranges over a finite set of admissible values only which have to be successively tried out by the corresponding loop.

Further determinants of this sequentialization are the explicit "result equations" (cf. "result_cond" in Section 3) occurring in the program since the variables on the left-hand side of such an equation are determined only after the assignment of values to the variables occurring on the right-hand side.

Finally the occurrence of inputs (marked by "?") and outputs (marked by "!") as well as of requests for long-term storage of facts and data objects determines the direction of the computation process (from input towards result data).

The sequential computation process derived in the first compilation phase is subsequently decomposed into communicating, parallel, sequential processes. A component process corresponds to that org-unit which is responsible for the production of those data objects whose names have been qualified by the name of the org-unit. The control flow of each sequential component process is induced by the control flow of the single sequential process produced by the first compilation phase. The compiler includes the required communication primitives in each component process and makes the necessary provisions for the envisaged automatic, distributed crash recovery procedure.

5.3. Implementation of Recursive Algorithms

About this subset of PLOP we have currently the least definite ideas, although its investigation has actually been the starting point of our work.

As an exploratory exercise we had developed a compiler which translates a simple class of primitive recursive function definitions into communicating sequential processes (expressed in SIMULA67); cf. Ref. 1.

Since office applications began to attract our attention we have concentrated on the integration of databases, communication between organizational units, alerter functions, and other office-related functions into our conception of "logic programming," the more so as we felt that recursive algorithms would not play a central role in this application domain.

Generally we would prefer an implementation strategy that tries to transform recursion into iteration wherever possible (as we did in the above-mentioned exploratory activities), rather than applying a more direct, straightforward, recursive computation procedure.

We would prefer this not only for efficiency reasons but also since we feel that automatic error recovery (e.g., by checkpoint/restart procedures) in a distributed implementation can be achieved in a more natural way if a recursive algorithm is translated into communicating iterative processes.

Since we are particularly interested in such automatic, distributed error recovery procedures, we shall probably concentrate on types of recursion which facilitate translation into iteration, viz., special types of primitive recursive functions of integer-valued arguments, and "repetitive" types of recursive function definitions.

So it will not be our primary effort to support recursive predicates by Horn-clause-like expressive means as in PROLOG, with "expert systems" and similar artificial intelligence applications in mind. Such essentially "noniterative" and "exploding" types of recursion will be supported by our approach only in the long run, and then only "locally," in a quite conventional and nondistributed manner.

5.4. Major Components of the Envisaged PLOP Programming Environment

The PLOP programming environment will presumably consist of the following major components:

1. A combination of a syntax-directed editor and an incremental parser.

These do not depend on the PLOP syntax but allow us to define the context-free syntax and the layout of any formal language text.

Their output is an internal tree representation (essentially a parse tree) of the respective formal language text. This can then serve as input to a subsequent compilation phase. The general tree-analysis and tree-synthesis procedures which are needed anyway by the editor can also be applied by the subsequent compilation phases. In fact, the actual editor is used as an input procedure for PLOP programs by the central monitor of the PLOP programming environment.

A first version of the editor has been developed on an IBM 4341 under VM/CMS. The incremental parser has been realized as a quite straightforward and inexpensive (but certainly not very efficient) "top down parser" which interprets the syntax definition tables also used by the editor. We hope that the lacking efficiency will not matter since the incremental parser will usually be applied only to single statements or even smaller syntactic units.

2. A compiler which takes the parse tree representation of PLOP programs as input.

The PLOP compiler produces a machine-independent, procedural like intermediate code for a number of communicating sequential processes corresponding to the individual organizational units that are involved in the respective office procedure.

3. An intermediate code interpreter (one per computer type involved in the respective distributed programming environment).
4. A run-time system which itself is essentially programmed in PLOP and translated into intermediate code.

Its most important purpose is to support the automatic generation of forms and formatted reports mentioned in previous sections and to control the associated interaction with the user, as well as the communication with the database and the other processes belonging to this same office procedure.

5. One management process per local computer system belonging to the respective distributed programming environment.

Its purpose is to initialize and to terminate distributed applications in an orderly way, and to provide the ORCUS service outlined in Section 4.1.4.

The components mentioned so far are more or less required in this or some other appearance. The following two supplementary features would be also highly desirable, however, in order to enhance the attractiveness of the system for nonprofessional applications programmers.

6. For certain simpler types of algorithmic applications frequently occurring in the commercial field a spreadsheetlike programmer's interface should be provided.
7. For (more or less) pure database applications a QBE-like interface would be desirable. (cf. Ref. 17.)

Corresponding translation functions from spreadhsheet and QBE format to PLOP (in internal parse tree representation) would be required then, too, of course.

6. Contrasting PLOP with PROLOG

Our analysis of the difficulties with the PROLOG approach to logic programming yielded three main conclusions:

I. The programmer should not be forced to fit everything into the expressive framework of Horn clauses.

II. There should be no "pseudopredicates" with memory or side-effects or similar features typical of a procedural or operational semantics. Particularly input/output and database applications should not rely on such predicates.

III. A logic programming language should provide a genuine concept of "closed predicate definition" as a basis for program modularization and in view of a proper separation of recursive and non-recursive expressive means.

These three basic conclusions have been taken into account in many ways during the detailed design of PLOP:

1. New predicates can be introduced by "closed definitions" (in contrast to the always extensible and thus preliminary "characterization" of predicates by an open-ended set of facts and rules). Defined predicates may represent program modules or composite data-objects (like "tuples," "strings," "arrays").

2. Such a module or composite data-object may occur as an argument of another predicate (i.e., PLOP supports "higher-order" predicates to some extent).

3. Defined predicates can be used to form function expressions (that can also be nested) and equations between expressions. Equality is expressed by " = " in any case; a distinction between " = " and "is" as in PROLOG is not required since function expressions just stand for their result rather than being special literal constants.

4. PLOP programs may contain existence and universal quantifiers (confined to some finite base set), negation, arbitrary logical connectives (including nested IF ... THEN ... ELSE ... FI-constructs), and equations between nested function expressions.

5. The PROLOG notion of a "question" is replaced with the more flexible notion of a "specification." This is an arbitrarily complex, composite statement ("main program") which describes the relation between inputs and the desired results.

6. In contrast to PROLOG, the textual order of PLOP statements does not matter. "Control primitives," like the "cut" in PROLOG, do not exist in PLOP.

7. In PLOP, there are no pseudopredicates, like "get," "put," etc. in PROLOG, having memory or side-effects. Input and output is effectuated by metalevel annotations to variables, expressions, and predicates (cf. "expr" and "result_cond" in Section 3).

8. PLOP offers special support for database applications without dispensing with the nonprocedural, purely logical semantics of the language.

   In particular, PLOP provides an approach to an automatic database access synchronization promising a high degree of parallelism between different applications, including "long transactions," without requiring explicit, user-programmed locks or transaction delimiters.

9. Our implementation, which is still in a very early state, was from the beginning on directed towards a distributed implementation of PLOP programs, including an automatic error recovery procedure after temporary failures of components of the distributed system. This has in turn influenced the language design in several respects.

10. PLOP is especially suited for office applications, due to a number of syntactic constructs and other provisions, some of which have been mentioned above while others have been omitted in this first introduction:

- The naming conventions for variables and predicates support a hierarchy of name spaces, each of which contains the named data-objects and stored facts of an individual organizational unit ("org-unit") of the office environment. An org-unit (or "role") may be authorized to play the role of another org-unit. Two or more concurrent activations of the same "role" are prohibited, however.

- We intend to automatically generate (interactively processed) forms reflecting the input requests of a PLOP program.

- The programmer can express complex side conditions ("checks") that are to be imposed on input data, the violation of which leads to an automatically generated error message and a subsequent input correction request.

- The reaction to input requests can be automated by the addressed org-unit, for instance by giving a formal definition for the formally undefined predicate that caused the request.

- Input requests can be rerouted or delegated by the addressed org-unit to some other org-unit; outputs can be rerouted or copied to several other org-units. "Delegation" of an input request, in contrast to "rerouting" it, has the consequence that the delegating org-unit is asked to acknowledge or revise the answer produced by the second org-unit.
- PLOP provides "alerters" which initiate the execution of specific office procedures on various application-oriented events; cf. the "alerter_actvn" construct in Section 3.
- The PLOP programmer may request trace information to be affixed to stored user data in the database. This tracing facility allows the user to record and retrieve the production date and time of stored data and/or the producing org-unit. This is an important requirement in office applications.
- PLOP provides facilities for expressing authorization constraints. Access rights may be granted or denied not only to org-units but also to individual defined predicates (owned by some org-unit).

The last feature allows the applications programmer to enforce integrity constraints applying to certain types of facts stored in the database. To this end write access with respect to these facts is granted only to certain "privileged" (fact storage) modules. The latter are represented as defined predicates or macros owned perhaps by an org-unit that is "responsible" for those types of stored facts. The execution right for these macros or predicates may be granted to many org-units or even to "everybody."

Appendix: Syntax of a Basic Subset of PLOP

Notation

:: =   denotes a syntactic rule (production)
| separates syntactic alternatives
{}   are the parentheses on the syntactic level
[]   enclose optional items
. . . marks the preceding syntactic item as being repetitive; it may be repeated 0, 1, 2, . . . , *n* times
" "   enclose literal strings (only required if one of the above special symbols is to be used in PLOP, e.g., "|").

The names of syntactic variables consist of lower case letters, digits, and underlines and begin with a letter. A single blank stands for an arbitrary number of blanks.

```
plop_program  ::=  statements
statements    ::=  statement {; statement} ...

statement     ::=  pred_def_stm | envmt_ct1_stm

pred_def_stm  ::=  condition | definition | declaration

condition     ::=  elem_cond | complex_cond

elem_cond     ::=  standard_cond |
                   infix_cond |
                   result_cond |
                   query_cond |
                   macro_call

standard_cond ::=  pred_ref ( expr {, expr} ... )

pred_ref      ::=  pred_name [. org_unit]
pred_name     ::=  identifier
org_unit      ::=  sub_unit {. sub_unit} ...
sub_unit      ::=  constant_su | computed_su
constant_su   ::=  identifier
computed_su   ::=  ( [expr] )

infix_cond    ::=  expr infix_pred expr
infix_pred    ::=  < | <= | = | > | >=

result_cond   ::=  result = expr |
                   pred_ref ? ( expr {, expr} ... )

result        ::=  var_name | ( {var_name|*} {, {var_name|*} } ... )
var_name      ::=  identifier

query_cond    ::=  [FIND]{
                   standard_cond |
                   current_cond |
                   is_in_cond | ... }

current_cond  ::=  {CURRENT|OLDEST} [( expr )]
                   pred_ref OF pred_ref ( expr ) : expr

is_in_cond    ::=  pred_ref ( expr ) IS IN pred_ref ( expr )
```

```
macro_call    ::=  /macro_ref [macro_parms]
macro_ref     ::=  macro_name [. org_unit]
macro_name    ::=  identifier
macro_parms   ::=  var_name = expr {, var_name = expr} . . .

complex_cond  ::=  NOT log_operand |
                   log_operand log_infix_op log_operand |
                   if |
                   quant_cond

log_operand   ::=  condition | ( condition )
log_infix_op  ::=  {AND|&|;} | {OR|"|"} | {XOR|"||"} |
                   {IMPLIES|==>} | {IFF|<==>}

if            ::=  IF condition THEN statements
                   {ELIF condition THEN statements} . . .
                   [ELSE statements]
                   FI

quant_cond    ::=  {FOR EVERY | EXISTS [UNIQUE] | UNIQUE}
                   quant_base {, quant_base} . . .
                   statements
                   END

quant_base    ::=  standard_cond | enum_qbase | . . . . . .
enum_qbase    ::=  var_tuple IN grid
var_tuple     ::=  var_name | ( var_name {, var_name} . . . )
grid          ::=  ( number_seq {, number_seq} . . . )
number_seq    ::=  expr [TO expr [BY expr]] |
                   ( number_seq {, number_seq} . . . )

definition    ::=  pred_def | macro_def | . . . . . .

pred_def      ::=  def_head def_body END

def_head      ::=  pred_name [lifespan_ind]
                   ( formal_parm {, formal_parm} . . . )
                   [FOR {* | org_unit {, org_unit} . . . }]
                   ==

lifespan_ind  ::=  : {S|L}
formal_parm   ::=  identifier
```

def_body	::=	pred_def_stm {; pred_def_stm} ... | expr_tuple {; expr_tuple} ...
expr_tuple	::=	expr {, expr} ...
macro_def	::=	macro_name [lifespan_ind] == statements END
declaration	::=	input | output | ref_time | query_sep | ...... (many others)
ref_time	::=	REF {TIME | ENTITY} expr | REF TIME -
query_sep	::=	{NEW | FREE} QUERY
envmt_ctl_stm	::=	db_modif | alerter_actvn
db_modif	::=	insert | delete
insert	::=	[STORE] elem_fact [store_opts]
elem_fact	::=	standard_cond | new_attr_val | enters_leaves
new_attr_val	::=	NEW pred_ref OF pred_ref (expr) : expr
enters_leaves	::=	pred_ref (expr) {ENTERS|LEAVES} pred_ref (expr) [EVENT entity_name]
entity_name	::=	identifier
delete	::=	DEL[ETE] del_item {,del_item} ...
del_item	::=	expr | fact_pattern | ......
fact_pattern	::=	pred_ref (expr {, expr } ...)
alerter_actvn	::=	WHEN[EVER] alert_conds [start][stop] THEN alert_stm {; alert_stm} ... END
alert_conds	::=	alert_cond {{"|"|&|AND|;} alert_cond} ...
alert_stm	::=	statement | alert_action
alert_action	::=	copy | re-route | delegate | stop | ......

expr   ::=   * | ?[!] | ! |
                      constant [!] |
                      object_ref [?] [!] |
                      function_expr |
                      infix_op_expr |
                      current_expr | . . . . . .

constant   ::=   integer | real | char_string | bit_string |
                      NULL | . . . . . .

object_ref   ::=   object_name [. org_unit] [lifespan_ind]
object_name   ::=   identifier

function_expr   ::=   pred_ref [?] ([expr] {, [expr]} . . .)

infix_op_expr   ::=   operand infix_op operand
operand   ::=   expr | (expr)
infix_op   ::=   + | − | * | / | . . . . . .

current_expr   ::=   CURRENT [(expr)] pred_ref OF pred_ref (expr)

References

1. B. BAUMGARTEN, F. LORENZ, and P. OCHSENSCHLÄGER, DISTSYS—Ein System zur Übersetzung funktionaler Programme in kommunizierende parallele Prozesse, Internal Report GMD/IFV, 1982, 43 pages.
2. B. BAUMGARTEN and P. OCHSENSCHLÄGER, Checkpoint/Restart-Verfahren gegen Komponentenausfall und Leitungsfehler in verteilten Systemen, Arbeitspapiere der GMD, No. 13, 1983, 26 pages.
3. M. VAN CANEGHAN, Ed., *Proc. First International Logic Programming Conference,* Marseille, France, 1982; see also "Logic Programming Conferences" of subsequent years.
4. J.-M. CHANG and S.-K. CHANG, Database alerting techniques for office activities management, *IEEE Trans. Commun.,* **COM-30**(1), 74–81 (1982).
5. J. CLIFFORD and D. S. WARREN, Formal semantics for time in databases, *ACM TODS,* **8**(2), 214–254 (1983).
6. W. F. CLOCKSIN and C. S. MELLISH, *Programming in* PROLOG, Springer-Verlag, Berlin, 1981.
7. E. F. CODD, Extending the database relational model to capture more meaning, *ACM TODS,* **4**(4), 397–434 (1979).
8. K. D. GÜNTHER, Basic concepts of PLOP—A predicative programming language for office procedure automation, Arbeitspapiere der GMD, No. 52, 1983, 37 pages.
9. K. D. GÜNTHER, Syntax der prädikatenlogischen Programmiersprache PLOP, Arbeitspapiere der GMD, No. 53, 1983, 71 pages; and more recent versions (English version in preparation).
10. K. D. GÜNTHER, Database requirements of computer-aided office procedures, Arbeitspapiere der GMD, No. 54, 1983, 6 pages.

11. K. D. GÜNTHER, PLOP—A predicative programming language for office procedure automation, *Proceedings IEEE Workshop on Languages for Automation,* IEEE Computer Society Press, New York, 1983, 94–101.
12. W. KENT, *Data and Reality,* North-Holland, Amsterdam, 1978.
13. R. A. KOWALSKI, *Predicate Logic as Programming Language,* Information Processing 74, North-Holland, Amsterdam, 1974, pp. 569–574.
14. R. A. KOWALSKI, Algorithm = logic + control, *Commun. ACM,* **22**(7), 424–436 (1979).
15. R. A. KOWALSKI, *Logic for Problem Solving,* Elsevier North-Holland, New York, 1979.
16. P. MATERNA and J. POKORNY, Applying simple theory of types to databases, *Inf. Syst.* **6**(4), 283–300 (1981).
17. M. M. ZLOOF, Query by Example, in: *AFIPS Conference Proceedings 1975 NCC,* 431–437.

3

OPAL: AN OFFICE PROCEDURE AUTOMATION LANGUAGE FOR LOCAL AREA NETWORK ENVIRONMENTS VIA ACTIVE MAILING AND PROGRAM DISPATCHING

K. KISHIMOTO, K. ONAGA, AND H. UTSUNOMIYA

1. Introduction

LANs(local area network) have been recognized now as an indispensable means for realizing the integration of information resources (such as microcomputers, data files, printers) widely dispersed around buildings and areas. Because of recent remarkable developments of LAN technology and its cost effectiveness, LANs are now beginning to be put to practical use in offices (see, for example, the Aladdin office system of NEC, 1983.[1]) Present passive usage of LANs in file transfer, electronic mailing, and simple device sharing is, however, often far below its potential, and much greater utilization of its capabilities should be explored. A main cause of this situation seems to be the lack of suitable high-level programming languages for LAN environments.

From this point of view, we should note the recent development of concurrent languages[2-10] based on message-passing mechanisms such as CSP,[11] DP,[12] Rendezvous[8] and PLITS.[6] Contrasted with interprocess com-

K. KISHIMOTO, K. ONAGA, and H. UTSUNOMIYA • Department of Circuits and Systems, Faculty of Engineering, Hiroshima University, Higashi-Hiroshima, 724 Japan
NOTE: A revised version of the paper was presented at the IEEE Workshop on Language for Automation held on November 7–9, 1983, Chicago, Illinois.

munication mechanisms using event variables,[13] semaphores,[14] or monitors,[15-17] the message-passing mechanism does not require shared variables whose access control is critical in concurrent executions of programs in LAN environments.

Unfortunately, however, this work is still not adequate for developing practical application programs in offices, because heavy emphasis is placed on developing operating systems, as in Hansen,[12] whose characteristic is different from that of OA (*office automation*), particularly in the nature of event occurrences, data consistency, and human interactions.

In particular, the following defects are serious from the viewpoint of a high-level OA language:

(i) In executing application programs in an office, effective detection of and recovery from failures are of vital importance. However, conventional message-passing mechanisms are inadequate to provide practical remedies for solving these problems.

(ii) Concurrency control is one of the most troublesome parts in distributed systems. Since many languages proposed for LAN environments aim at system description, data locking is the user's responsibility and must be explicitly coded even in developing application programs.

(iii) In CSP and PLITS, the interprocess communication is described through "send" and "receive" commands, which are similar to the "goto" statement. Improper use of these communication commands may result in notorious "spaghetti" programs. This might sometimes be unavoidable and overlooked in low-level coding of operating systems, but it should not occur in higher-level application programs in an office.

(iv) In DP, communication is effected through a procedure call so that the defect stated in (iii) is reduced somewhat. Note that once a communication is initiated, execution of the main program is halted until the reply is received, and hence only a signal piece of mail can be exchanged at a time, even when there is concurrency.

This chapter endeavors to develop a LAN system, with an aim at OA, by proposing a programming language called OPAL (*office procedure automation language*) together with its man–machine communication environment, called an OPA-system (*office procedure automation system*), which is capable of running concurrently batch-type programs networkwide.

To make active use of a LAN as well as to cope with the above-mentioned difficulties, OPAL and OPA-system are provided with several new devices in LAN execution of OA programs and in its interprocess communication mechanisms among distributed tasks.

(1) Two distinctive features of office procedure are the human elements of authorization and managerial decision-making which are rather

difficult to automate and must be regulated by the personnel themselves. In OPAL, such an interaction is treated as a local subprogram of the OP-program which enables the main station to demand responses to its query from others and then to perform some processing or manipulation activated by those returned responses. (We will show in a later section some examples of how to use this mechanism.) To realize this mechanism, OPAL adopts the concept of a *program dispatching scheme,* instead of the conventional *data dispatching scheme,* which gives OPAL a concurrent language capability. Thus, OPAL aims at coding not real time programs, but batch-type ones.

(2) To simplify detection and recovery from a failure, some restriction is made in control flow patterns, which also contributes to good readability of OPAL programs.

(3) In OPAL, file sharing among all the work stations is presumed, so that concurrency control, such as file locking, is automatically executed by the system.

(4) Following DP, the message passing is always executed though a procedure call, but designed execution of the main program is continued until a specified point in the program. For clarity, such a specification is described in a special part called the communicator. (The name communicator is taken from CDP,[10] which seems to have been the first to propose this idea.)

(5) To avoid interference with other programs, a program that comes to a halt for receiving the return message is rolled back in the disk and stays there until a condition becomes fulfilled to resume the task again.

Based on these communication mechanisms, OPAL and OPA-system exhibit the following characteristics:

(a) OPAL is a high-level language for execution in LAN environments. While it is designed to assist efficient programming for experienced or semiexperienced programmers, it might not be easy for novices to program in OPAL. Thus, in a real office OP-programs are assumed to be already loaded into the network by system programmers, so that neither managers nor clerks are needed to code OP-programs in their daily work.

(b) OPA-system is to be implemented on a federation of work stations connected to a LAN and controlled by conventional single-CPU OS's such as CP/M with some enhancements. This enables the user to separately run global OPAL prorams as well as local non-OPAL programs on his work station. And yet, the OPA-system behaves as if it were controlled with a global LAN operating system.

(c) OPAL provides a facility of automated I/O formatting as a communication interface.

(d) To keep simplicity of control and internodal uniformity of the programming, we assume a so-called personal database residing in the work-station where stored data accessible to others are organized as two-dimensional tables, and the data processing is written in a table manipulation language.

We summarize comparisons between OPAL and other languages in Tables 1 and 2. We see that OPA-system together with OPAL is equipped with sufficient capability of building an OA system in LAN environments. We should note that such a system is indispensable for realizing true OA and that implementation in LAN system is technologically imminent. In an earlier stage, OA simply meant introduction of single OA devices such as word processors, facsimiles, passive electronic mailing systems, and so on. However, this naïve type of automation is now under criticism, and automation of office procedures (i.e., sequences of components office tasks such as documentation, data retrieval, issuing purchase-order and managerial decisions, for instance) has become a new target of OA.

Indeed, Hammer *et al.*[18] has proposed a very high-level language, BDL, which aims at analyzing office procedures; Zloof,[19] Lum *et al.*,[20] Shu *et al.*,[21] and Zloof,[22] have proposed, respectively, QBE, OPAS, an enhanced OPALS, and OBE, as user-friendly interfaces for relational database systems, which, at the same time, aim at office procedure automation.

Hammer *et al.*'s language is, however, at a very high level without mention of any actual implementation. The other four are rather revolutionary languages/systems, but each of them works under the support of a single relational database system residing in a host machine that is more expensive and more limited in accommodation of work stations than LANs. Furthermore, since intercommunication between separate database

TABLE 1
Comparison of Various Message-Passing Mechanisms

Communication primitive	Description of message exchanges	Necessity of synchronization between sender and receiver process	
CSP[11]	By send-and-receive statements	Required	
PLITS[6]	Same as above	Not required	(messages are temporarily stored in a buffer)
DP[12]	By procedure calls	Required	
Rendezvous in Ada[8]	Same as above	Same as above	
OPAL (our proposal)	Same as above	Not required	(messages are temporarily stored in a buffer)

TABLE 2.
Tabulation of Concurrent Languages with Message-Passing Mechanisms

Language	Communication primitive	Recovery from failures	Specifications in case of timeout, etc.	Broadcasting capability	Data abstraction in communication	Communicator declaration
OCCAM[2]	CSP	Difficult	Indescribable	X	X	X
CONCURRENT C[3]	CSP (modified)	Difficult	Describable	X	X	X
PASCAL-m[4]	CSP	Difficult	Indescribable	X	O	X
PASCAL-PLITS[6]	PLITS	Difficult	Indescribable	O	O	X
SOMA[7]	PLITS	Difficult	Indescribable	O	O	X
Distributed-Path PASCAL[5]	DP	Difficult	Indescribable	X	X	X
Ada[8]	Rendezvous	Difficult	Indescribable	X	X	X
CP[9]	Rendezvous	Difficult	Indescribable	X	X	X
CDP[10]	Choice of CSP, DP and PLITS	Difficult	Abort and retry are describable	X	X	O
OPAL (our proposal)	OPAL	Easier	Describable	O	X	O

systems is very costly and ineffective, their integration is limited (see Table 3.)

In the following, we give some target office procedures to be automated by OPAL; in Section 2, an outline of the OPA-system; in Section 3, informal descriptions of the OPAL specification; in Section 4, by showing sample programs for target office procedures; in Section 6, a detailed behavior of OPAL program execution, and in Section 7, further discussion.

2. Some Target Office Procedures

In this section, we describe programming of some target office procedures for an OPA-system, which will explain certain features of OPAL specifications and will help the reader understand the OPA-system better, where one office procedure is described by one OP-program. In Section 5, we will show how these procedures are implemented in the OPA-system by using OPAL.

Example 1: Procurement Procedure

(1) A section clerk proposes to his section manager procuring some equipment, say a word processor.

(2) If the section manager agrees to this proposal as contributing to the productivity, he sends it to the procurement manager. If not, he sends a refusal to the section clerk, and the procedure terminates.

(3) The procurement manager assesses whether the proposal is in accord with the management policy of the company and sends the approval to the procurement clerk if the proposal is justifiable. Otherwise he sends the refusal to the originating section and the procedure terminates.

(4) The procurement clerk checks whether the equipment is really marketed. He will invite several vendors to bid for the equipment and issue the purchase order to the winner. He will inform the section manager and the procurement manager of his results, and the procurement comes to an end.

Example 2: Meeting Arrangement Procedure

(1) An organizer of a meeting sends to the participants several, say five, plans of the dates and times of the meeting and asks them to select plans that fit their schedules.

(2) The participants send back plans of the dates and times they can attend.

(3) If there exists at least one plan that fits all the participants, the organizer decides to hold the meeting at this time. If it turns out that no

TABLE 3

Tabulation of Various Office Procedure Description Languages

Language	Language type	Target users	Target environment	Communications among end users	Triggering events	The support of (relational) database system
BDL[18]	Nonprocedural two-dimensional graphic language	End users	Implementation is not aimed at	Describable	Arrival of data; arrival at prespecified time	—
QBE[19]	Nonprocedural table-oriented language	End users	Under TSS environment of a single host machine	Indescribable	Impossible	Required
OBE[22]	Nonprocedural table-oriented language	End users	Under TSS environment of a single host machine	Through passive electronic mailing	Arrival of data; arrival at prespecified time; renewal of stored data	Required
OPAS[20]	Nonprocedural table-oriented language	End users	Under TSS environment of a single host machine	Through passive electronic mailing	Arrival of data; arrival at prespecified time	Required
Modified OPAS[21]	Nonprocedural table-oriented language	End users	Under TSS environment of a single host machine	Semiprogrammable	Arrival of data arrival at prespecified time	Required
OPAL (our proposal)	Procedural	Experienced or semiexperienced programmers	Under LAN environment which connects work stations	Programmable	Arrival of data	Not required

plan satisfies all the participants, the organizer decides the date and time by a certain prescribed algorithm which assures attendance by key participants.

Example 3: Order Acceptance Procedure

(1) A clerk receives a purchase order from a customer.

(2) If the customer's name is not found in the customer list, he/she registers the customer information (customer's name, address, telephone, credit limit, etc.) in the customer list. Here, the credit limit is set at a certain lowest value. This value may be altered in step (3) when appropriate.

(3) The clerk registers the order on the order list, and proceeds to the task CREDIT-CHECK: the clerk calculates the total accumulation of the charges, and checks whether the total charge exceeds the customer's credit limit. If an excess is found, he asks and waits manager's decision on how to process this transaction. The manager, if necessary, issues an investigation of the customer's credence to the credit section and sends back his decision to the clerk. If the answer is "go ahead," the clerk proceeds to the next step. If not, he sends a letter of sales refusal to the customer, and the procedure terminates.

(4) The task INVENTORY is executed: the clerk at the sales section checks the amount of stock of the relevant commodity; if the amount is sufficient, he subtracts from the stock the amount of the order, issues a bill to the customer, and sends the instruction to the shipping section. If not, he/she informs the inventory section of the stock depletion, and informs the customer of an expected delay of the delivery.

3. Outline of the OPA-System

This section gives a rough sketch of the OPA-system which will help the reader understand more easily the specification of OPAL in the next section. Detailed behaviors of the OPA-system will be given in Section 6, after the introduction of OPAL.

In the OPA-system it is assumed that several dozen workstations are connected by a LAN. Each workstation is provided with a disk to store data common to other stations whose data structure is of a two-dimensional table. The hardware of each workstation may differ from station to station, but each one is equipped with the same system software of the OPA-system.

The OPA-system provides two ways of accessing nonlocal data residing in other stations: direct access, via one's own editor, to files that are transferred from others' into one's own station, and indirect access to files residing in others', via execution of a program written by OPAL, the OP-program. The OPA-system is partly protected with a synchronization scheme for keeping the data consistency free from disturbances caused by interlacing of OP-program executions, with the exception of the file transfer. Hence, writing on a file in other stations via the file transfer has to be prohibited in principle, but reserved data are allowed to be accessed through dispatched OP-programs.

The OPA-system operates as a federation of communicating workstations, each of which is under a single or multitask OS—enhanced CP/M, for instance. However, it behaves as if it were under a global LAN-OS. This implementation enables the user to run OP-programs as well as his own local non-OPAL programs separately. The reader should notice how this mechanism is naturally realized in the OPA-system in Section 6.

In the OPA-system, an office procedure is described by an OP-program whose subprograms are to be executed on several workstations concurrently with each other as well as with other OP-programs. Once an OP-program is coded at a workstation, and is compiled into intermediate codes there, it is subdivided into subprograms, and each is dispatched to the designated workstation where the intermediate code is further compiled into the machine code. It is this two-step compilation that enables the OPA-system to absorb hardware diversity at each workstation. We call this two-step process the "loading of a program into the network."

After completing the loading of a program into the network, the program is triggered at the *home station* by the arrival of signals or the occurrence of some events, and the execution starts there. According to the specification of the OP-program, executions of the OP-program in other stations are triggered by communication from the home station. An important feature in the OPA-system is that when execution of a program comes to a halt while waiting for an answer from another station, the program is rolled back into the disk until the answer actually arrives. The top OP-program in the waiting queue is then picked up for the restart.

Since the OPA-system is composed of several dozen work stations, it is capable of executing dozens of OP-programs concurrently. The automatic rollback mechanism of waiting or interrupted programs diminishes the necessity of global control of the LAN. Indeed, several parts of the same program are simultaneously at execution on several stations when a document flow program is running, for instance (see Section 5.)

4. Specification of OPAL

An OP-program is composed of two parts: global declaration (Fig. 1, lines 1–8) and local sub-programs (Fig. 1, lines 10–51). The global declaration part defines data types, procedures, etc., which are valid throughout

```
 1: PROGRAM <program name>;
 2:
 3: CONST     <identifier> = <constant>;···;···;
 4: TYPE      <identifier> = <data type>;···;···;
 5: DSTATION <identifier>,< >,···,< >;
 6: ESTATION <identifier>,< >,···,< >;
 7: DISPLAY  <display procedure name>(<parameter list>);···;···;
 8: PRINT    <print procedure name>(<parameter list>);···;···;
 9:
10: TERMINAL <estation name> IN <dstation name>;
11:
12:   PROCESS <process name>(<parameter list>);
13:
14:     { local declaration of labels, constants, data types,
15:        variables, display procedures, print procedures,
16:        procedures, functions }
17:
18:     BEGIN
19:        { execution part }
20:     END;
21:
22:   PROCESS <process name>(<parameter list>);
23:
24:     { local declaration }
25:
26:     BEGIN
27:        { execution part }
28:     END;
29:                           .
30:                           .
31: TERMINAL <estation name> IN <dstation name>;
32:
33:                           .
34:
35: MPROCESS <main process name> IN <dstation name>;
36:
37:     { local declaration of labels, constants, data types,
38:        variables, edef procedures, display procedures,
39:        print procedures, procedures, functions,
40:        communicatiors }
41:
42:     EDEF <edef procedure name>(<parameter list>);
43:
44:     COMMUNICATOR <communicator name>
45:        SEND    : <send specification>    |
46:        RECEIVE : <receive specification>|
47:     END;
48:
49:     BEGIN
50:        { execution part }
51:     END.
```

FIGURE 1. Typical constitution of an OP-program.

the program. A local OP-program describes a local declaration and a local specification of processing at individual stations.

An OP-program begins with the reserved word PROGRAM followed by the program name, and then come constant definitions.

Since local subprograms of an OP-program must be dispatched to designated workstations, the compiler/dispatcher must be informed of their station names. Such station names are defined at just after the constant definition by the reserved word DSTATION, followed by several formal parameters for the respective station set name. Preceding the compilation their contents must be prescribed in the DSTATION file such that the compiler/dispatcher is able to refer to them.

The stations at which a particular OP-program is to be executed form a subset of DSTATION. The identifiers of such stations are declared just after the DSTATION definition by the reserved word ESTATION, followed by identifiers. Their contents are either described directly in the program or are input at the home station (i.e., the station that triggers the OP-program) from the I/O-screen via the EDEF-procedure, with which the execution of the OP-program starts.

The type declaration is given in the same manner as in PASCAL.[23] The only difference is that DATE, TIME, and some other additional types are provided as a standard type in OPAL, which frequently appears in office procedures.

The global declaration part may contain a procedure that is to be called only by a local subprogram at a certain station, and is copied and sent to the destination together with the associated subprogram.

A local program is composed of the declaration part and the execution part, while there is no such division in the global program. Specification of the declaration part of the local program is almost the same as that of the global declaration part, but the execution part requires some explanation.

The local subprogram to be executed at the home station is called the *main process,* which plays the central role in executing the program. Communication among the remote local programs always takes place via the main process and no direct communication facility is provided. A remote local program is always written as a procedure to be called from the main process, and communication is always written as a procedure call to them from the communication part of the main process. Such local procedure is written in a body, and must specify the input terminal statement in the form of

TERMINAL [terminal name] IN [dstation set name].

A resource in a station is always accessed by a procedure call of the local subprogram residing in the station. We should note that this approach partially realizes data abstraction (see, e.g., Hansen[15]) in OPAL.

A control statement affecting other stations as well as specification of home executions must be described in the main process in the OPAL program. Usually the execution part of the home subprogram begins with the call to EDEF procedure, a special I/O screen procedure, which passes the actual parameter of the ESTATION name to the program. There are two ways in the EDEF procedure for passing the station name to the program: either by indirectly pointing to the route specification file or by directly supplying the station name.

A single procedure-call in the home subprogram may trigger local subprograms in several workstations simultaneously, and hence must specify when and how to receive answers from others. Such specifications are written in the communicator part of the program. The communicator name follows the reserved word COMMUNICATOR, and the send-and-receive specifications follow the reserved words SEND and RECEIVE, respectively. There are several rules which are valid only in the communication part: the procedure call is primarily specified by the reserved word CALL; the standard function SENDER is available for the default, to denote the station now in communication; the receive part may contain a standard TIMEOUT function, whose facility is often required in office procedures.

The point when requests are to be sent to other stations is indicated by the reserved word COMMUNICATE, followed by the communicator name. Completion of execution of this communication statement, however, does not necessarily mean the halt of the execution at the home station. (Let us remember that OPAL allows concurrent executions of local subprograms.) The point where the execution of a program comes to a halt is indicated by the reserved word WAIT, followed by the communicator name. Since the specification of WAIT has been described in the communication part, no further description is given at this WAIT statement.

The I/O-specification is always a troublesome part of interactive systems. To diminish this burden, the OPA-system is equipped with a translator which aids a programmer in writing the I/O part of a program. What a programmer has to do is to edit I/O-screen images. The translator then converts it into the intermediate code. In the OPAL program, such a task is declared just as DISPLAY, followed by its identifier without any specification in it. The EDEF procedure is a special type of this DISPLAY procedure. Similarly, the output format at a printer is declared as PRINTER, followed by its identifier.

Other details of the OPAL specification are PASCAL-like, so that the reader familiar with PASCAL[23] will understand it quite readily.

5. Sample Programs

In this section, we describe how the sample office procedures given in section 2 are automated in the OPA-system. We will at the same time help the reader to understand the details of the specifications of the OP-program.

5.1. Procurement Procedure

Figure 2 gives an OP-program for the procurement procedure given in Section 2. When this program is triggered at a station belonging to a station set "triggers," a screen specified by a EDEF procedure "disp_in" at line 61 appears on the triggered station (the home station). This display procedure is specified separately, as in Figure 3, on a special editor that includes the I/O-image translator mentioned earlier.

When the station identifier and the routing information of the trigger signal are passed to the OP-program by the EDEF procedure, the section clerk fills in the document format by the display procedure "disp_source_reply," which is specified on the editor as in Figure 4.

The section manager can read this document by taking it out of his own mailbox and filling his/her response in the designated space. The return key press automatically sends the document to a station designated in the OP-program.

5.2. Scheduling of a Meeting

Figure 5 gives an OP-program for a scheduling of a meeting. When this program is triggered, the meeting organizer defines the station name of the parameter parts on the screen specified by the EDEF procedure "stn_in." He then fills in several plans on the screen specified by the display procedure "plan_in."

An important fact to be stressed here is that the organizer will no longer be troubled with housekeeping in this scheduling and may be able to leave the station entirely, for the OP-program takes care of itself to the end.

5.3. Order Acceptance Procedure

This procedure is a rather popular order acceptance procedure often cited by QBE,[19] OPAS,[20] OBE,[22] and others. The OP-program for this procedure becomes slightly longer, and we do not give an example. The reader will easily understand it from the former two examples.

6. Mechanisms of the OPA System

This section describes the communication mechanisms in the OPA-system.

```
 1: PROGRAM procurement ;
 2:
 3: CONST    max_station   = 10 ;
 4:
 5: TYPE     stationnum    = 0..max_station ;
 6:          acknowledge   = (yes,no) ;
 7:          table_type    = RECORD
 8:                            station_name : ARRAY[1..20] OF CHAR ;
 9:                            ans          : acknowledge ;
10:                          END ;
11:          table         = ARRAY[stationnum] OF table_type ;
12:          heading_type  = RECORD
13:                            title   : ARRAY[1..30] OF CHAR ;
14:                            dating  : DATE ;
15:                            name    : ARRAY[1..20] OF CHAR ;
16:                          END ;
17:          screen type   = RECORD
18:                            equipname  : ARRAY[1..20] OF CHAR ;
19:                            equipcode  : ARRAY[1..5] OF CHAR ;
20:                            equipnum   : INTEGER ;
21:                            equipprice : INTEGER ;
22:                            reason     : ARRAY[1..400] OF CHAR ;
23:                            comment    : ARRAY[1..160] OF CHAR ;
24:                          END ;
25:
26: DSTATION triggers , managers ;
27:
28: ESTATION ws[stationnum] ;
29:
30: DISPLAY  disp_reply(i,last:IN stationnum; heading:
31:             IN OUT heading_type; sign:IN OUT table;
32:             screen:IN OUT screen_type; ack:OUT acknowledge);
33: DISPLAY  disp_out(heading:IN heading_type; sign:IN table;
34:             screen:IN screen_type) ;
35:
36: TERMINAL ws[stationnum] IN managers ;
37:
38: PROCESS  permission(i,last:stationnum; heading:heading_type;
39:             screen:screen_type; sign:table; #ack:acknowledge);
40:   BEGIN
41:     disp_reply(i,last,heading,sign,screen,ack) ;
42:   END ;
43:
44: PROCESS  resultproc(heading:heading_type; sign:table;
45:             screen: screen_type) ;
46:   BEGIN
47:     disp_out(heading,sign,screen) ;
48:   END ;
49:
```

```
50: MPROCESS mainproc IN triggers ;
51:
52:   LABEL   99 ;
53:
54:   VAR     i,j,last       : stationnum ;
55:           sign           : table ;
56:           heading        : heading_type ;
57:           timeout_period : DATE ;
58:           ack            : acknowledge ;
59:           screen         : screen_type ;
60:
61:   EDEF    disp_in(ws:OUT ESTATION; last:OUT stationnum;
62:                   timeout_period:OUT DATE) ;
63:   DISPLAY disp_timeout(heading:IN heading_type) ;
64:
65:   COMMUNICATOR com1
66:     SEND    : CALL ws[i].permission(i,last,heading,screen,sign;
67:                                     ack) ; |
68:     RECEIVE : RETURN(timeout_period) ;
69:               sign[i].ans:=ack ; |
70:   END ;
71:
72:   COMMUNICATOR com2
73:     SEND    : FOR j:=i-1 DOWNTO 1 DO
74:                 CALL ws[j].resultproc(heading,sign,screen) ;
75:               ENDFOR ; |
76:     RECEIVE : FOR j:=i-1 DOWNTO 1 DO
77:                 RETURN ;
78:               ENDFOR ; |
79:   END ;
80:
81:   BEGIN
82:     disp_in(ws,last,id,timeout_period) ;
83:     heading.dating:=TODAY ; i:=0 ;
84:     disp_reply(i,last,heading,screen,sign,ack) ;
85:     REPEAT
86:       i:=i+1 ;
87:       COMMUNICATE(com1) ;
88:       WAIT(com1) ;
89:     UNTIL (i=last) OR (sign[i].ans=no) OR TIMEOUT(com1) ;
90:     IF TIMEOUT(com1) THEN
91:       disp_timeout(heading) ;
92:     ELSE
93:       COMMUNICATE(com2) ;
94:       disp_out(heading,sign,screen) ;
95:       WAIT(com2) ;
96:     ENDIF ;
97: 99:END.
```

FIGURE 2. Illustrative program for automating the procurement procedure by OPAL.

<u>Procurement Procedure (EDEF screen)</u>

Please fill in the blank of the following table.

```
r----------------------------T-----------------1
I   total work stations      I@total         @I
I----------------------------I-----------------I
I   timeout period           I@period        @I
L----------------------------i-----------------J
```

Please input work station names on the route.

```
WS[1] -------------> @ws
WS[2] ------------->           r--------------------------------
WS[3] ------------->           I  <@total          @>
WS[4] ------------->           I
WS[5] ------------->           I   I/O ------> I
WS[6] ------------->           I   TYPE -----> I
WS[7] ------------->           I   RANGE ----> 1..64
WS[8] ------------->           I   REPEAT ---> N
                               I   SINK -----> last
                               I
```

FIGURE 3. Program for the EDEF procedure "disp_in."

6.1. *Outline of Network Communication Modules*

In the OPA-system, any communication is made through its (active) mailing system, where NCMs (network communication modules) residing in each station play the core role. NCMs are triggered for sending when

(S1) the dispatcher is ready for transmitting OP-subprograms;

(S2) an execution of a local subprogram comes to a certain point;

(S3) delivery of a urgent message is required.

NCMs are triggered for receiving when

(R1) a message is received;

(R2) the local subprogram execution is completed.

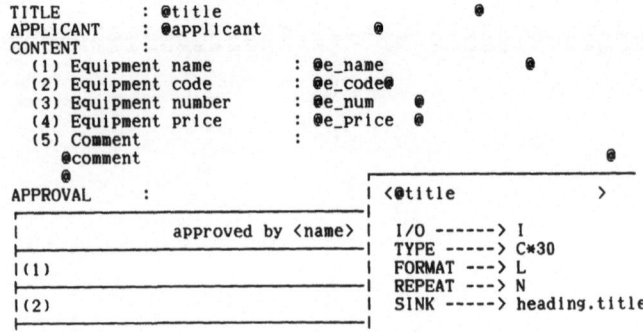

Purchase Request for Internal USE

FIGURE 4. Program for the display procedure "disp_reply."

The send part of NCMs is composed only of the SEND routine and is rather simple, while the receive part of NCMs is more complex. Each station has four mailboxes for receiving: second class, first class, command, and urgent messages, the last preemptively examined by NCM. The receive part of NCMs consists of several routines:

1. RECEIVE(and its subroutines SECOND-CLASS, FIRST-CLASS, COMMAND, URGENT-MESSAGES), BUFFER-CLEAR;
2. SEND, CONNECT, DISK-STORE, PROGRAM-TRIGGER.

The RECEIVE and BUFFER-CLEAR routines are, respectively, triggered by events (R1) and (R2), while the latter four are service routines which are called within the former two.

The RECEIVE has four subroutines, each of which handles the respective mailboxes. DATA transfer via the second-class mail is used for bulk data as well as for dispatching the OP-subprogram, and requires preestablishment of the connection, while data transfer via first-class mail requires no such preestablishment and is used only for passing control and actual parameters to dispatched subprograms. The other two mails also require no such preestablishment. The command mailbox is used for connection request prior to the transfer via second-class mail, lock and unlock requests for data files, and other lower-level communication control commands. The urgent messages box is designed to be used in an emergency, but it is not yet implemented in the present prototype.[24] The first-class and command mailboxes possess a reserve of common special buffer in the main memory, which are used as tentative storage of data received when the CPU is in a busy period and that explains why they do not require the preestablishment of the connection.

Figures 6 and 7 illustrate the OA-architecture of mailing, file organization, triggering, and dispatching at each station, and Figures 8–10 illustrate the behaviors of the routines in the NCMs which handle respective mail boxes.

6.2. Loading of an OP-Program into the Network

When an OP-program is coded and registered at a particular station, the OPAL-compiler compiles local subprograms into intermediate codes, by referencing for the DSTATION identifiers in the OP-program and the DSTATION file residing at the station. Display procedures and printer procedures are also compiled into other kinds of intermediate codes. Then the dispatcher starts by sending the connection request via the mailing system, to dispatch these codes to the command mailbox of designated work-

```
 1: PROGRAM scheduling_of_a_meeting ;
 2:
 3: TYPE    stationnum   = 1..63 ;
 4:         acknowledge  = (yes,no) ;
 5:         prpsl_type   = RECORD
 6:                          hold_day  : DATE ;
 7:                          hold_time : TIME ;
 8:                          ans       : acknowledge ;
 9:                        END ;
10:         array_prpsl  = ARRAY[1..5] OF prpsl_type ;
11:         heading_type = RECORD
12:                          title  : ARRAY[1..30] OF CHAR ;
13:                          dating : DATE ;
14:                          name   : ARRAY[1..20] OF CHAR ;
15:                          content: ARRAY[1..400] OF CHAR ;
16:                        END ;
17:         set_ws       = SET OF STATION ;
18:
19: DSTATION organizers , participants ;
20:
21: ESTATION parts[stationnum] ;
22:
23: PRINT    finalresult(final:IN prpsl_type;
24:                      heading:IN heading_type) ;
25:
26: TERMINAL parts[stationnum] IN participants ;
27:
28: PROCESS ansproc(heading:heading_type;proposal:array_prpsl;
29:                  #proposal:array_prpsl) ;
30:
31:         DISPLAY ansin(heading:IN heading_type;
32:                       proposal:IN OUT array_prpsl) ;
33:
34:         BEGIN
35:           ansin(heading,proposal) ;
36:         END ;
37:
38: PROCESS finalproc(heading:heading_type;final:prpsl_type) ;
39:         BEGIN
40:           finalresult(heading,final) ;
41:         END ;
42:
43: MPROCESS orgproc IN organizers ;
44:
45: VAR     proposal      : array_prpsl ;
46:         final         : prpsl_type ;
47:         vip,replied_st : set_ws ;
48:         heading       : heading_type ;
49:         timeout_period : DATE ;
50:         i,last        : stationnum ;
51:         ans_stack : ARRAY[stationnum,1..5] OF acknowledge;
52:
```

```
53:  EDEF     stn_in(parts:OUT ESTATION; last:OUT stationnum;
54:                  timeout_period:OUT DATE)
55:  DISPLAY  plan_in(proposal:OUT array_prpsl; vip:OUT set_ws;
56:                   heading:OUT heading_type) ;
57:
58:  FUNCTION  decision:BOOLEAN ;
59:
60:  PROCEDURE failure ;
61:
62:  PROCEDURE pushans(host:IN STATION) ;
63:  { store the response from the station host
64:    in the variable "ans_stack" }
65:
66:  PROCEDURE timeoutproc ;
67:
68:  COMMUNICATOR com1
69:  SEND    : FOR i:=1 TO last DO
70:              CALL parts[i].ansproc(heading,proposal;
71:                                    proposal) ;
72:            ENDFOR ; |
73:  RECEIVE : REPEAT
74:              RETURN(timeout_period) ;
75:              pushans(SENDER(com1)) ;
76:              replied_st:=replied_st+SENDER(com1) ;
77:            UNTIL (replied_st=parts) OR TIMEOUT(com1) ; |
78:  END ;
79:
80:  COMMUNICATOR com2
81:  SEND    : FOR i:=1 TO last DO
82:              call parts[i].finalproc(heading,proposal) ;
83:            ENDFOR ; |
84:  RECEIVE : FOR i:=1 TO last DO RETURN ; ENDFOR ; |
85:  END ;
86:
87:  BEGIN
88:  stn_in(parts,last,timeout_period) ;
89:  plan_in(proposal,vip,heading) ;
90:  replied_st:=[] ;
91:  COMMUNICATE(com1) ;
92:  WAIT(com1) ;
93:  IF TIMEOUT(com1) THEN timeoutproc ; ENDIF ;
94:  IF decision THEN
95:     COMMUNICATE(com2) ;
96:     finalresult(final,heading) ;
97:     WAIT(com2) ;
98:  ELSE
99:     failure ;
100: ENDIF ;
101: END.
```

FIGURE 5. Illustrative program for automating the scheduling-of-a-meeting procedure by OPAL.

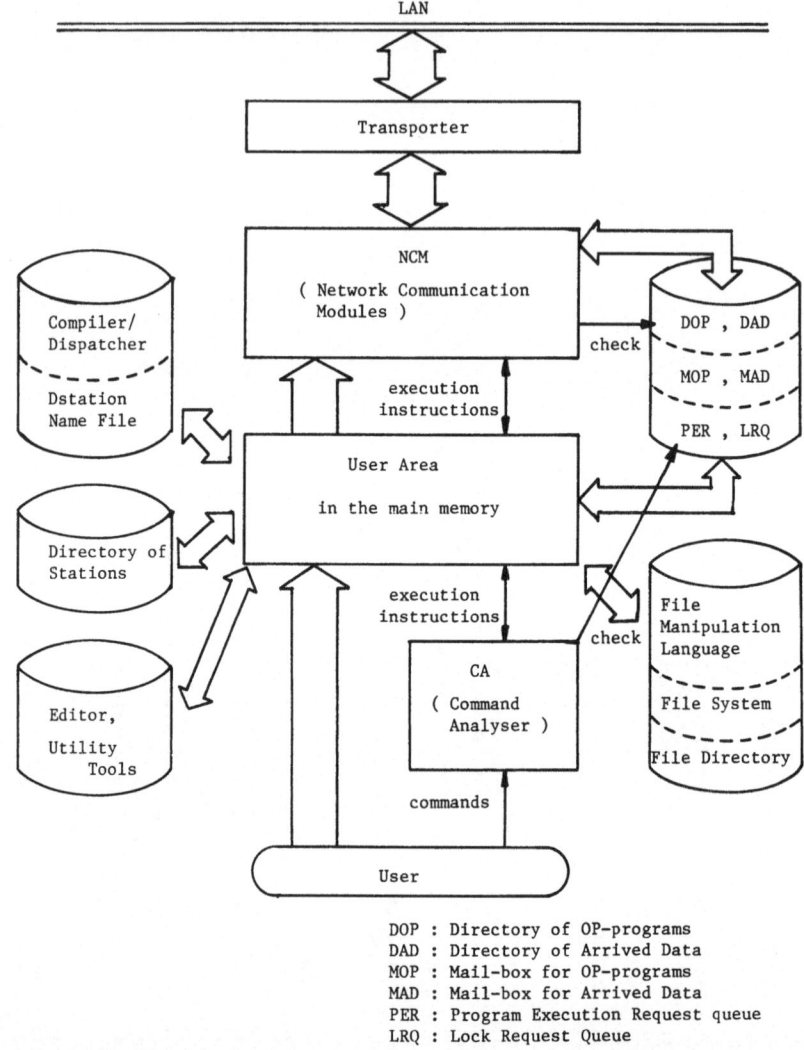

FIGURE 6. Relations and control flows among communication architecture components in a station.

stations together with program information such as

    a. the OP-program identifier of the following format:
       {Home station name of the program}. {Program name};
    b. whether this local subprogram requires interactive inputs and out-
       puts by personnel of this station. (A dispatched program may play
       the role of mail to the personnel who is often interested in seeing

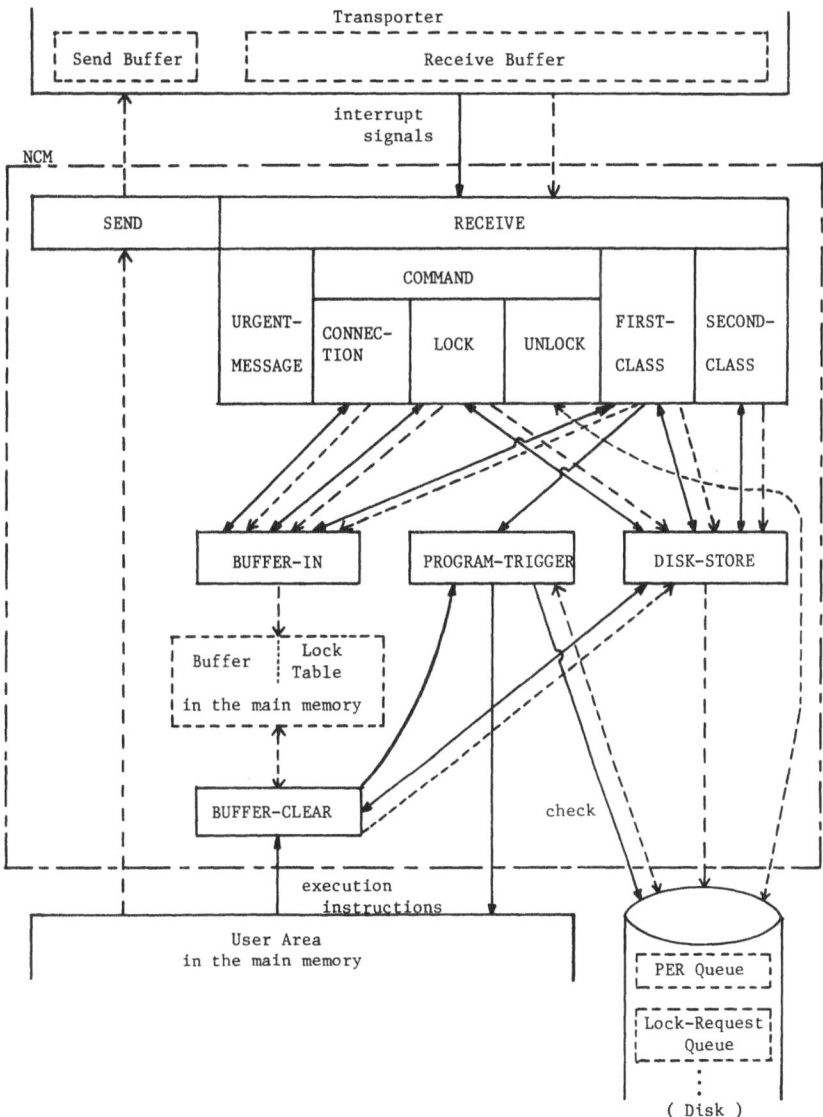

FIGURE 7. Relations and control flows among various routines in the NCM and their associated files and buffers. The solid line indicates a routine calling, while the dashed line indicates access to memories and files.

```
< SECOND-CLASS-MAIL >

{ preceding this execution, connection must be established }
BEGIN
  REPEAT
    store the received submail into the disk :
    send the request for the next submail
  UNTIL mail transfer is completed :
  release the connection
END ;
```

FIGURE 8. Algorithm of the SECOND-CLASS-MAIL routine.

a particular piece of mail prior to handling other mail. Thus, the order of such program executions may be arbitrarily arranged by the command from the editor.)

When a destination receives a connection request at its command mailbox, the COMMAND routine residing in the NCM (see Fig. 7) activates the service routine CONNECT. If the CPU of the station is idle, it establishes the connection by reserving the CPU, the second-class mailbox, and returns the acknowledgment to the sender. When the CPU is working at a local transaction, an interrupt occurs in order to return the busy signal to the sender, and to store the sender station name temporally in a buffer in the main memory. As soon as the transaction is completed, the retry-recommendation signal is emitted to this sender (see also Section 6.4).

Once a connection is established, the sender starts to execute file-transfer via the second-class mailbox and DISK-STORE routine at the receiver, called by the SECOND-CLASS-MAIL routine, and begins to receive the subprogram by creating a new file for it and by inserting program information into the OP-program directory. Then the OPAL compiler compiles the intermediate code of the subprogram into the object code, and the acknowledgment is returned to the sender at the end of the compilation. When acknowledgment is received from all the dispatched stations, the loading of the OP-program into the network has been completed.

```
< FIRST-CLASS-MAIL >

{ the first class mail is used only for PER }
IF CPU=idle THEN BEGIN
  store the PER in the execution queue file ;
  trigger the requested program
END ELSE BEGIN
  store the PER in the buffer
  { later, the stored PER is picked up by BUFFER-CLEAR }
END ;
```

FIGURE 9. Algorithm of the FIRST-CLASS-MAIL routine.

```
< COMMAND >

CASE request OF
  connection : IF CPU=idle THEN BEGIN
                    establish connection ;
                    send acknowledgment to the sender station
                 END ELSE BEGIN
                    send busy signal to the sender station ;
                    store the sender station name in the buffer
                    { when the task is completed, BUFFER-CLEAR,
                        one by one, emits retry-recommendation
                        to the senders. }
                 END ;
  lock       : { Lock table is held in the main memory for
                    realizing quick response, while lock-
                    request-queue file resides in the disk. }
                 IF already locked THEN BEGIN
                    send busy signal to the sender station ;
                    IF CPU=idle THEN
                       store the lock request in the lock-
                       request-queue file
                    ELSE
                       store the lock request in the buffer
                       { When the task is completed, BUFFER-CLEAR
                           stores it in the lock-request-queue
                           file. When the unlock command is
                           received retry-recommendation signal
                           is emitted to the sender station. }
                 END ELSE BEGIN
                    lock ;
                    send the lock acknowledgment
                       to the sender station
                 END ;
  unlock     : BEGIN
                    lock release ;
                    REPEAT
                       pick up the lock request
                           from the lock-request-queue file ;
                       send the retry-recommendation signal
                           to the sender station
                    UNTIL lock request arrives within
                               the prescribed time ;
                    delete the picked-up lock request
                        from the queue
                 END ;
END ;
```

FIGURE 10. Algorithm of the COMMAND routine.

6.3. Execution of an OP-Program

When an OP-program is triggered at the home station, it first sends a lock request to the command mailbox to try to seal data records, which are to be used by the program, from other accesses. If this locking succeeds at all relevant stations, the execution begins at the home station. If not, an unlock request is sent to all the stations where locking succeeded, and the program is rolled back to the disk for another trial at a later time when retry-recommendation signals have arrived from all the stations where the former request was unsucessful.

When the OP-program at the home station comes to execute a COM-MUNICATION procedure, the PER (*p*rogram *e*xecution *r*equest) is to be sent to relevant stations. no connection request is needed to be for PER, because the request uses the privileged first-class mailbox. The PER contains ID-information

a. program name;
b. host station name;
c. the number of times of execution of that program at the home station;

and an actual instance of the procedure parameters in it.

Upon the receipt of the PER, at the first-class mailbox the RECEIVE routine of the NCM activates the FIRST-CLASS-MAIL subroutine. If the station is idle, the routine immediately accommodates the PER into the execution-queue file, and then begins to execute the formerly accommodated OP-program by rolling it into the main memory. Otherwise, the routine tentatively stores the PER in the buffer, and then the BUFFER-CLEAR routine puts it at the tail of a waiting queue in the execution-queue file as soon as CPU is freed from the current task. Then the PER at the top of the queue is picked up by the CPU for the next execution.

Finally, at the completion of all the OP-subprogram executions, the home station broadcasts the unlock command to all command mailboxes for releasing the lock on data. The UNLOCK routine that is called by the COMMAND routine actually performs this unlocking operation.

6.4. Buffer-Clear

During the execution of the current OP-subprogram, command messages of other OP-programs are tentatively stored in the buffer for later processing: a PER is picked up first from the buffer by the BUFFER-CLEAR routine and is put into the execution-queue file. (Actual execution does not start at this point.) A connection request is handled next by issuing the acknowledgment to the sender from which a new connection request is expected. When the connection is established, the OP-program transfer takes place. This procedure for the connection and the program transfer is handled one-by-one until the queue in the file becomes empty. Then, the BUFFER-CLEAR activates the PROGRAM-TRIGGER routine which picks up the OP-program in the queue for the next execution. Figure 11 summarizes this algorithm.

7. Future Work

Our present research is aimed at developing a prototype which pays little attention to efficiency in speed and memory. For instance, NCMs and

```
< BUFFER-CLEAR >

BEGIN
  FOR every PER in the buffer DO
    BEGIN
      pick up a PER from the buffer ;
      store the PER in the PER queue file
    END ;
  FOR every lock-request in the buffer DO
    BEGIN
      pick up a lock-request ;
      store the lock-request in the lock-request-queue file
    END ;
  FOR every connection request in the buffer DO
    BEGIN
      pick up a connection request ;
      send the freed signal to the station ;
      wait for a prescribed time ;
      IF the request comes THEN BEGIN
        establish the connection ;
        send acknowledgment ;
        receive the second-class-mail
      END
    END ;
  execute the OP-subprogram at the top of the PER-queue
END ;
```

FIGURE 11. Algorithm of the BUFFER-CLEAR routine.

the compiler for the display procedure have been completed on the NEC PC-8801 microcomputer, whose CPU is Z80. The former is coded by an assembly language with 2800 lines, and the latter by PASCAL with 1600 lines. As the result of several tests, it turned out that 12 sec are required to trigger a program in another station if an 8-in. floppy disk is used, which is by no means very fast. We are now implementing NCMs on a faster machine, the NEC PC-9801E, whose CPU is 8086 with a hard disk, and are expecting more rapid responses.

In the present version of the NCM, as seen in Section 6, all the data required in an OP-program are automatically locked preceding the execution of its body, so that the existence of common data never becomes a cause of deadlock, except for a local one within a single station. Since individual operating systems are deadlock-free, the OPA-system is free from the deadlock. We notice here that the rollback of a halted program into a disk set the CPU free for the task at the top of the queue, and hence deadlock is prevented even if the operating system is single-tasked. This design, however, incurs a heavy cost in parallelism. Indeed, efficient implementation of assuring the deadlock-freeness in distributed systems is still an open issue.

Security and reliability are another great concern in the OPA-system, which tries to ease this trouble by deferring the actual writing of the file at the very end of the OP-program execution. This mechanism is easily realized in the OPA-system, since the home station has a complete grasp of the state of the execution.

There are many other problems to be solved before the OPA-system has any practical use:

1. Analysis and classification of structures of actual office procedures are prerequisite to any serious development of standard OP-program packages;
2. A compact set of OPAL-instructions must be designed for description of communications and manipulations in business transactions;
3. Human engineering assessment of man–machine interfaces, such as procedure decomposition, personnel intervention, authorization, and I/O responsiveness, is fundamental to an office worker's acceptance of the OPA-system;
4. Incorporation of data security and document authorization is essential to the user's confidence and acceptance;
5. The OPA-system is primarily designed to operate without support from a host, but such support is certainly welcome if the host is not overburdened.

Our prototype is a first step toward a further development of business procedure automation in LAN environments.

References

1. ACOS System 410 System Manual (in Japanese), CAZ05-1, NEC Co., 1983.
2. D. MAY, OCCAM, *SIGPLAN Not.* **18**, 69–79 (1983).
3. M. ANDO, Y. TSUJINO, T. ARAI, and N. TOKURA, Concurrent C: A programming language for distributed multiprocessor systems—Design and implementation (in Japanese), *Trans. Inf. Process. Soc. Jpn.* **24**, 30–39 (1983).
4. S. ABRAMSKY and R. BORNAT, PASCAL-m: A language for loosely coupled distributed systems, in *Distributed Computering Systems*, Y. PAKER and J. P. VERJUS, Eds., Academic, London, 1983, pp. 163–189.
5. R. H. CAMPBELL, Distributed-path PASCAL, in *Distributed Computing Systems*, Y. PAKER and J. P. VERJUS, Eds., Academic, London, 1983, pp. 191–223.
6. J. A. FELDMAN, High-level programming for distributed computing, *Commun. ACM* **22**, 353–368 (1979).
7. J. L. W. KESSELS, The soma: A programming concept for distributed processing, *IEEE Trans. Software Eng.* **SE-7**, 502–509 (1981).
8. U.S. DEPARTMENT OF DEFENSE, *Ada Reference Manual*, Springer-Verlag, Berlin, 1980.
9. T. W. MAO and R. T. YEH, Communication port: A language concept for concurrent programming, *IEEE Trans. Software Eng.* **SE-6**, 194–204 (1980).
10. M. T. LIU and C. M. LI, Communicating distributed process: A language concept for distributed programming in local area networks, in *Local Network for Computer Communication*, A. WEST and P. JANSON, eds., North-Holland, Amsterdam, 1980, pp. 375–406.
11. C. A. R. HOARE, Communicating sequential processes, *Commun. ACM* **21**, 666–667 (1978).
12. P. B. HANSEN, Distributed process: A concurrent programming concept, *Commun. ACM* **21**, 934–941 (1978).

13. HITAC Manual 8080-3-212-31, Optimizing PL/I and Checking type PL/I language (in Japanese), 1980.
14. A. VAN WIJNGAARDEN, B. J. MAILLOUX, J. E. L. PECK, C. H. A. KOSTER, M. SINTZOFF, C. H. LINDSEY, L. G. L. T. MEERTENS and R. G. FISKER, Revised report on the algorithmic language Algol 68, *Acta Informatica* **5**, 1–236 (1975).
15. P. B. HANSEN, The Programming language CONCURRENT PASCAL, *IEEE Trans. Software Eng.* **SE-1**, 199–207 (1975).
16. N. WIRTH, *Programming in Modula-2*, Springer-Verlag, Berlin, 1983.
17. J. G. MITCHELL, W. MAYBURUY, and R. SWEET, *Mesa Language Manual*, Xerox Corp., 1978.
18. M. HAMMER, W. G. HOWE, V. J. KRUAKAL, and I. WLADAWSKY, A very high level programming language for data processing applications, *Commun. ACM* **20**, 832–840 (1977).
19. M. M. ZLOOF, QUERY-BY-EXAMPLE: A database language, *IBM System J.* **16**, 324–343 (1977).
20. V. Y. LUM, D. M. CHOG, and N. C. SHU, OPAS: An office automation system, *IBM System J.* **21**, 327–350 (1982).
21. N. C. SHU, V. Y. LUM, F. C. TUNG, and C. L. CHANG, Specification of forms processing and business procedure for office automation, *IEEE Trans. Software Eng.* **SE-8**, 499–512 (1982).
22. M. M. ZLOOF, OFFICE-BY-EXAMPLE: A business language that unifies data and word processing and electronic mail, *IBM System J.* **21**, 272–304 (1982).
23. K. Jensen and N. Wirth. *PASCAL User Manual and Report,* Springer-Verlag, Berlin, 1978.
24. H. Utsunomiya, K. Kishimoto, and K. Onaga, Implementation of network communication modules for an office procedure automation system (in Japanese), Technical Report of IECEJ, EC83-56, 1984, pp. 19–28.

4

TOOLS FOR FORMS ADMINISTRATORS

JAMES A. LARSON AND JENNIFER B. WALLICK

1. Introduction

The area of forms interfaces to database management systems was pioneered by Moshe Zloof with his Query-By-Example (QBE) interface.[1] Recent research has extended the tabular interface of QBE to the more natural forms interface. The forms interface has been considered by the CODASYL End User Facilities Committee[2] as

> ... the most natural interface between an end user and data because a large number of end users employ forms (e.g., purchase order forms, expense report forms, etc.) ...

There have been many interesting approaches to developing forms interfaces, including Refs. 3–6. In Ref. 3 Tsichritzis explains three aspects of form management: (1) form creation, (2) specification of form procedures, and (3) analysis of office work flow. This paper only deals with the first of these topics, form creation.

The Electronic Forms System prototype under development at the Honeywell Computer Sciences Center bridges the gap between the office worker and database management systems. Using a visual language, the office worker manipulates electronic forms displayed on a terminal screen to file, modify, and retrieve information from a database. The system provides all the capabilities of traditional database query languages in a way that is easy to learn and use.

There are two types of Electronic Forms System users. There are those who use the system to input, manipulate, and view information in the database. We will refer to these people as office workers. In addition,

JAMES A. LARSON and JENNIFER B. WALLICK • Honeywell Computer Sciences Center, Bloomington, Minnesota 55420

there are people who define the electronic forms that office workers use. We will refer to these people as the forms administrators. The Electronic Forms System prototype satisfies both the office worker's needs and the form administrator's needs.

This chapter defines the requirements of a forms administrator and gives an overview of tools in the Electronic Forms System prototype that support these requirements. The main contribution of this paper is a description of an interactive interface for forms administrators to build and modify form templates which map to a database. This paper describes the Electronic Forms System prototype only from the form administrator's point of view. For the office worker, the Electronic Forms System prototype provides facilities to enter data, a pictorial language to formulate queries and updates, and methods to browse a database. For a more detailed description of the office worker requirements and tools, see Ref. 7.

Two visual representations are used in our prototype. An entity relationship diagram is used to display the contents of the database to the forms administrator. A form template representation is used to display the contents of the database to the forms user. Figure 1 illustrates a visual representation of an Entity Relationship (ER) schema.[8] Rectangles represent entity sets, diamonds represent relationship sets between entity sets, and ovals represent attributes of entity sets and relationship sets. The cardinalities of a relationship set are indicated by numbers at the points of the diamond representing the relationship.

The forms administrator manipulates the ER diagram to build a visual view or representation suitable to the office worker. The visual representation used to display this view of the database to the office worker is a form template. Figure 2 illustrates a *form template*. A form template is the office worker's view of part of the database. It is a two-dimensional layout of one or more database entities. When displayed on a video screen, all database entities conform to the layout of the form template. A form template contains background information that always appears on each occurrence of a form. Background information includes the form template name, the names of attributes, and instructions that always appear on each form occurrence. A form template also contains variable information whose values may vary from form occurrence to form occurrence. In all of the figures in this paper, variable information will be denoted by example values and will be underlined. As illustrated in Figure 2, form templates may be nested. The remainder of this chapter describes how a forms administrator may create a form template corresponding to a subgraph of an ER graph schema.

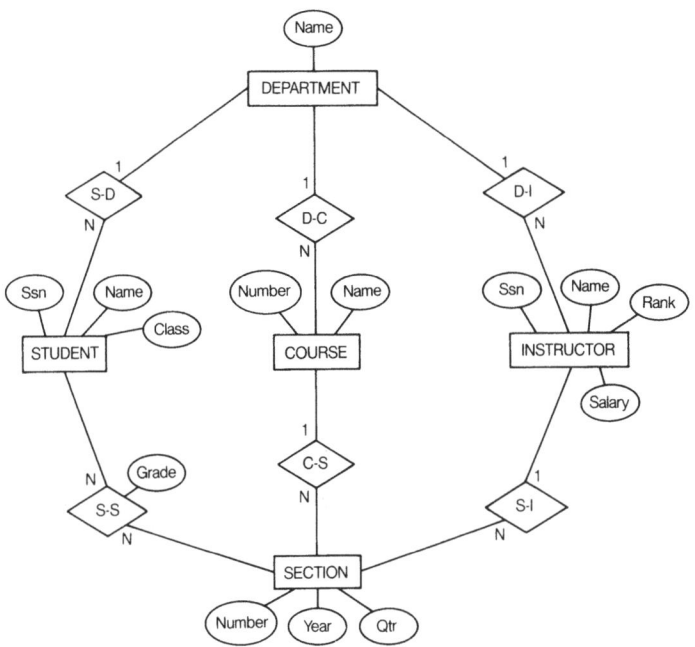

FIGURE 1. Graphical representation of schema.

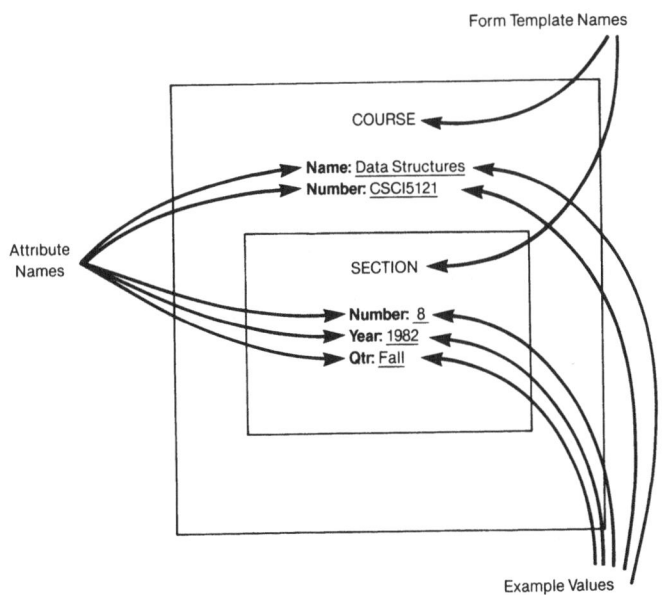

FIGURE 2. Form template.

2. Forms Administrator Requirements

A forms administrator defines the format and layout of an electronic form, called a form template, to be used by office workers. There may be several forms administrators in an enterprise. An experienced office worker may act as his/her own forms administrator if permitted by the policies set by the enterprise. A forms administrator must consider the office worker's needs, the requirements of the application, and database security constraints when designing form templates. A form template is the office worker's window to the database. It can be considered to be a view or subschema of a database and contains descriptions of data of interest to an office worker. When designing form templates the forms administrator must be aware of the types of information an office worker is going to need from the database. Through these form templates, a forms administrator controls who can access data in the database and how they can access it.

The Electronic Forms System Prototype will provide automated, flexible, and easy-to-use tools for building and modifying form templates so that the forms administrator can concentrate his/her efforts on designing form templates which meet the office worker's needs. The Electronic Forms System Prototype will provide two sets of tools for the forms administrator. The first set guides the forms administrator through a semiautomatic generation of form templates from data descriptions in a data dictionary. The second set allows the forms administrator to interactively modify existing form templates. The functional requirements of these tools are outlined as follows:

1. *Interactive Facility for Translating Data Descriptions into Form Templates.* The Electronic Forms System Prototype will provide semiautomated methods to access a data dictionary and design form templates using the data descriptions in the dictionary. The method should present information about the structure and contents of the database from the data dictionary and allow the forms administrator to interactively select which information belongs on the form template. The system should automatically generate a form template using the selected information.

2. *Interactive Facility for Modifying Form Templates.* Tools are needed that allow the forms administrator to view and interactively redesign any form template by editing background information and modifying the layout of the form template, adding and deleting variable information in the form template, and providing special

help instructions for the office worker who will use the form template to access a database.

To facilitate the forms administrator, tools to build and design form templates should be designed to meet the following additional criteria:

- The interface should be visually oriented so that forms administrators can see the results of each action.
- The interface should be interactive so that forms administrators can quickly analyze the results of each action.
- The interface should be as automated as possible so that the forms administrator can concentrate on making decisions and leave the bookkeeping details to the tool.
- Information about the existing database should be provided to the forms administrator so the forms administrator does not need to be intimately familiar with the database contents and structure.
- Forms administrators should be guided through the forms design choices so that the forms administrator needs little training and can easily use the system after long periods of nonuse.

The remainder of this paper presents a more detailed description and scenarios of the form template building and modifying tools in the Electronic Forms System Prototype.

3. Interactive Tool for Translating Data Description into Form Templates

A data description used to generate form templates may reside in a data dictionary, in the file description section of an application program, or it may be supplied by a user. Data descriptions of almost any database management system can theoretically be used to generate form templates. The particular environment we have used includes a data description based on the ER data model. The data description resides in a data dictionary. The ER model is a graphical model which lends itself nicely to a pictorial interface. We now present a scenario of a graphical interface which guides the forms administrator through six steps to generate form templates that correspond to portions of a database described using the ER data model. First we describe the approach and then present an example.

1. An ER graph is automatically generated on the form administrator's video screen from information in a data dictionary. To generate this graph, the system reads adjacency information about entities, relationship

sets, and attributes that reside in the database and then automatically displays the layout of the associated ER graph on a graphics terminal screen. This step relieves the forms administrator of the burden of remembering the exact contents of the database.

2. The forms administrator examines the ER graph, viewing the contents and structure of the database. The system supports various levels of visibility. For example, the system allows attribute visibility to be turned off so that the forms administrator can view only the entity sets and relationship sets without being distracted by their attributes.

3. The forms administrator now trims off irrelevant information from the ER graph. The system supports two modes for doing this. The forms administrator can either specify the relevant information or specify the irrelevant information. In the first mode the forms administrator examines the graph and points to entity sets, attributes, and relationship sets that are relevant for the form template, i.e., that are to appear on the form template. The forms administrator would probably want to use this mode if the graph contained a lot of irrelevant information. In the second mode, the forms administrator examines the graph and points to entity sets, attributes and relationship sets that are irrelevant to the form template, i.e., which are not to appear on the form template. The forms administrator probably wants to use this mode if there is not much trimming to be done on the graph. When all relevant and irrelevant information has been identified, the system erases irrelevant information from the displayed graph. When the forms administrator has completed this stage, the displayed graph contains only entity sets, relationship sets, and attributes to be contained on the form template.

4. The ER graph must now be converted from a graph structure to a tree-structured or form view of the database. To aid in the conversion, the forms administrator identifies an entity set of primary interest on the screen. The designated entity set will become the root entity of the tree structure and the top-level or outermost template in the nested form template.

5. The system automatically translates the ER graph with the designated entity into a nested form template. There are two stages necessary for this translation. The first stage uses an approach suggested by Elmasri and Larson[9] to convert the ER graph into a tree structure of entities. The second stage converts the entity tree into a form template. To convert the ER graph into an entity tree the attributes of relationships are migrated from the relationship downward (from the root) to the entity set below the relationship in the tree, and all relationships nodes are eliminated. In the hierarchical view, $M:N$ and $1:N$ relationships are viewed as $1:N$, while $N:1$ and $1:1$ relationships are viewed as $1:1$. Any two entity sets in the hierar-

chical view with a 1:1 relationship between them are merged to form a single entity set with the attributes of both entity sets.

The entity tree is then automatically converted into a collection of nested form templates as follows:

   a. The root of the entity tree becomes the top-level or outermost form template.
   b. While traversing the tree, all children nodes are translated into a form template nested inside the form template corresponding to the parent.
   c. Attribute names are positioned on the form template. Each attribute name appears on a separate line.
   d. An example value is a value from the domain of an attribute that is displayed on the form template as an example of a valid value that may be entered for that attribute. An example value for each attribute is selected from a default list of example values for the attribute. Example values are placed on the same line as and to the right of its attribute on the form template.
   e. Appropriate information is stored in the forms data dictionary. This includes information about entities that were merged during the hierarchical conversion and information about relationships connecting entities on the form. This information is used by the Electronic Forms System Prototype when translating queries formulated by office workers using the form template into equivalent queries suitable for processing by the underlying database management system.
   f. A form template name corresponding to the entity name is placed at the top of each form template.

The resulting form template may be used by the office worker to enter data, to formulate queries and updates to the database, or to browse through the database contents.[7]

The above approach can also be applied to databases described by the CODASYL data model. If the database is described by a hierarchical data model, then the displayed graph is already a tree which the forms administrator trims. The descripton of a relational database can also be displayed as a graph structure by defining a node of the graph for each relation and an edge of a graph for each corresponding foreign key and primary key pair.

An Example. As an example, suppose the schema of Figure 1 is displayed on the screen. Suppose that the forms administrator wants a form template consisting of entity sets COURSE, STUDENT, and SECTION entity sets. The forms administrator points to those three entities and their

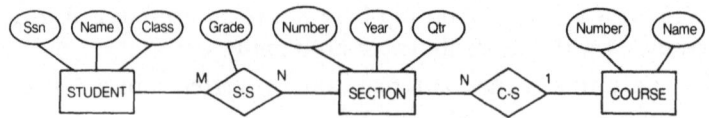

FIGURE 3. Schema subgraph.

relationships, S-S,C-S on the screen. The system then erases the remaining entities and relationships from the screen. The resulting subgraph is illustrated in Figure 3.

Now the forms administrator must choose a root entity set. Several choices can be made. We have extended the suggestions of Elmasri and Larson[9] to include the following criteria for choosing a root entity set:

1. The entity set with the most attributes upon which conditions will be specified by office workers.
2. The entity set with the most attributes to be included on the form template.
3. The entity set which seems most natural for the types of information retrieved or paper-form look-a-likes. For example, suppose that the office worker would be most likely to ask "Get me all of the courses taken by a particular student"; then the STUDENT should be the root entity set. On the other hand, if the office worker is more likely to ask "Get me all of the students who have taken a particular course," then COURSE should be the root entity.

Suppose the forms administrator chooses COURSE as the root. The ER subgraph of Figure 3 is converted into the hierarchy of Figure 4 and the corresponding form template for Figure 5.

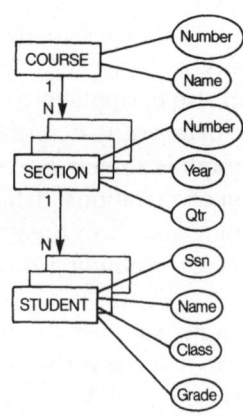

FIGURE 4. Schema hierarchy.

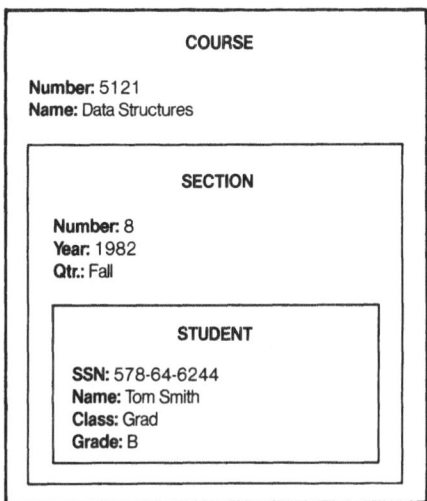

FIGURE 5. Form template.

Relationship attributes are migrated to the entity sets at the lower end (towards the leaves of the tree) of the relationship in the hierarchy. Thus the grade attribute is migrated from the S–S relationship to the STUDENT entity set, and gives the grade value of each student who took the section of a course.

4. Interactive Tools for Modifying Existing Form Templates

The forms administrator might want to modify the format and layout of the form template created above in order to

- Make the form template more closely resemble paper forms used in the office.
- Rearrange the order in which attributes appear on the form template to make data entry easier. Hide certain information in the database from the office worker.
- Change information on a form template when the structure of a database is changed. Create help messages and special instructions for office workers entering data.
- Improve the aesthetics of the form template's appearance.

A rich set of interactive tools for modifying a form template to conform to the styles and needs of office workers is provided by the Electronic Forms System Prototype.

1. Modify Tool. This tool allows the forms administrator to interactively edit and reposition all background information on a form template.

2. Delete Tool. The forms administrator can delete attributes or subform templates from a form template. The forms administrator must also supply a default value for a deleted attribute to be used when the office worker uses the revised form template for data entry.

3. Undelete Tool. The forms administrator can bring back attributes, example values, and subtemplates that have been deleted from a form template.

4. Move Tool. The forms administrator can interactively reposition information on the form template. Entire templates, background information, individual attributes, and individual example values can be relocated on the form template.

5. Enlarge/Shrink Tool. The forms administrator can insert or delete lines in the form template.

6. Help Instruction Tool. The forms administrator can define help instructions corresponding to an attribute name on the form template. This provides additional information to the office worker who is unclear about the type of information associated with an item on a form. An office worker entering information into a database who encounters an attribute on the form template about which they are confused can press a help key and the predefined instruction or message is displayed.

7. Special Editing and Validation Requirements. The forms administrator can specify a wide variety of validation constraints to be enforced during data entry and a wide variety of editing and formatting specifications to be used when data are displayed on a form template.

8. Add Derived Item Tool. The forms administrator can define attributes on the form template which are derived from other attributes on the form template.

The forms administrator may invoke any of these tools, which appear on a menu displayed in a window separate from the window displaying the form template. In effect, the user "repaints" the form template until he/she is satisfied with the resulting layout. Using these tools, the forms administrator is able to formulate a wide variety of form templates for office workers. The forms administrator can create a form template similar to the spread sheet of VISICALC, the added (derived) attributes being automatically calculated from nonadded attributes in the form template. The retrieval and storage features of a database management system and the capabilities of a spread sheet are combined in the Electronic Forms System Prototype.

5. Conclusions

The requirements of a forms administrator have been identified and tools that can be used to meet those requirements have been specified. The Electronic Forms System prototype implemented at the Honeywell Computer Sciences Center includes some of these tools. The system will be used to evaluate and refine our approach to supporting the tasks of the forms administrator. User studies need to be made to see how our system meets the needs of forms administrators and office workers. Future work in this area should include investigation in the other two areas of form management specified in Ref. 3: the specification and implementation of automatic form procedures and the analysis of office work flow.

References

1. M. ZLOOF, Query-By-Example, IBM Technical Report, RC-4917, July, 1974.
2. CODASYL EUFC, A status report on the activities of the CODASYL end user facilities committee (EUFC), *ACM SIGMOD Rec.,* **10** (2 and 3), 1978, 1–26.
3. D. TSICHRITIZIS, Form management, *Commun. ACM* **1984,** 453–478.
4. D. LUO and S. B. YAO, Form operation by example, *Proc. ACM Sigmod,* **1981,** 212–233.
5. L. A. ROWE and K. A. SHOENS, A form application development system, *Proc. ACM Sigmod,* **1982,** 28–38.
6. S. B. YAO, A. R. HEVNER, Z. SHI, and D. LUO, FORMANAGER: An office forms management system, *ACM Trans. Office Inf.* **2**(3) 235–262 (1984).
7. J. A. LARSON, The forms pattern language, *Proc. IEEE Conference on Data Engineering,* 1984, pp. 183–192.
8. P. P.-S. CHEN, The entity–relationship model: Towards a unified view of data, *ACM Trans. Database Syst.* **1**(1) 9–36. (1976).
9. R. A. ELMASRI, and J. A. LARSON, A user friendly interface for specifying hierarchical queries on an ER graph database, Honeywell Technical Report, Corporate Technology Center, HR-82-252: 17–38, March, 1982.

5

A FORM-BASED LANGUAGE FOR OFFICE AUTOMATION

S. Oyanagi, H. Sakai, T. Tanaka, S. Fujita,
and A. Tanaka

1. Introduction

One of the most important problems in developing an office information system is that each office environment has its own peculiar requirements for managing its task. Often, the knowledge about the task is not fully documented but is accumulated in the office worker's mind. Hence, one of the most effective solutions to developing and maintaining office information systems is that the office worker be provided with tools for programming his work by himself.

There are a number of end user oriented systems in commercial use. However, they are not powerful enough to describe a wide spectrum of office procedures.

We have developed a form-based language called LAMBDA for office automation. LAMBDA takes a form as an interface between an office worker and his tasks. A form is a generalized concept of business forms.[1,5,6] The major reasons for adopting this approach are that office workers are familiar with business forms in their daily work, and that most office work is carried out through forms. The concept of the form base can provide not only a natural interface for an end user but also powerful facilities to describe office procedures.

In addition, we are attaching more intellectual functions for managers. A manager is another kind of office worker who wants to get information without the precise knowledge of office tasks. Hence, the system must provide functions for resolving ambiguities and filling in all missing

S. OYANAGI, H. SAKAI, T. TANAKA, S. FUJITA, and A. TANAKA • Toshiba Research and Development Center, Komukai, Kawasaki, Kanagawa, 210 Japan

information. In this chapter, we describe the facilities for information retrieval and graph generation as the most effective functions to support managers.

2. *Concept of the Form Base*

The design philosophy of LAMBDA is to unify various business applications into a systematic approach based on the generalized form. It can provide a unified end user interface for a wide spectrum of business data processing.

There are many conventional systems that provide forms as a user interface. In these systems, forms play various roles as listed below:

- Input sheet;
- View of the database;
- Work sheet;
- Output report.

Most business systems use forms as an input sheet and as an output report. However, forms are designed for each application by a programmer. Hence, the end user cannot describe his own tasks in the system.

QBE[2,3] is an advanced end user oriented system which handles relational database on the screen. However, the screen image is restricted by its underlying data structure, which sacrifices the natural representation of a business form on the screen.

VISICALC[4] is another example which handles a form on the screen. It provides a simple two-dimensional work sheet and various functions on it. Although it is easy to use for an end user, it cannot be applied to a wide range of business applications using database.

LAMBDA aims to unify these roles of forms into the concept of the form base. Major characteristics of the form base are listed below:

1. It is based on the generalized form, which can represent most business forms naturally.
2. The process can be described on the form. Hence, a user can manipulate forms easily without any command or program.
3. Powerful form operations are prepared so that a user can handle the structure of a form easily.
4. Form operations can be combined into a form procedure. It makes it possible to execute a sequence of form operations automatically.

Prior to the implementation of LAMBDA, we have investigated practical business forms used in our office to clarify the specification of the generalized form. As the result, it was found that data structure is involved in a form. It is rather simple, and can be represented by an array of a set of fields.

The generalized form consists of the following elements:

Field: a slot to hold a value.
Array: a repetition of a set of fields.
Title: an arbitrary literal string which serves as a comment or a guidance for data manipulation.

An example form implemented on LAMBDA is shown in Figure 1.

A document is an instance of a form. Documents generated with the same form are bundled into a file. A user can retrieve and process each document through its form.

A user can design a form on the screen interactively. Then the file image to store the documents of the form is automatically defined by the screen image. The user need not know how they are stored in the database. On the contrary, a user can connect a form to a preexisting file in the database by describing the structure of the file.

An important characteristic of the form base is that the process can be described on the form. Namely, a user can calculate and retrieve doc-

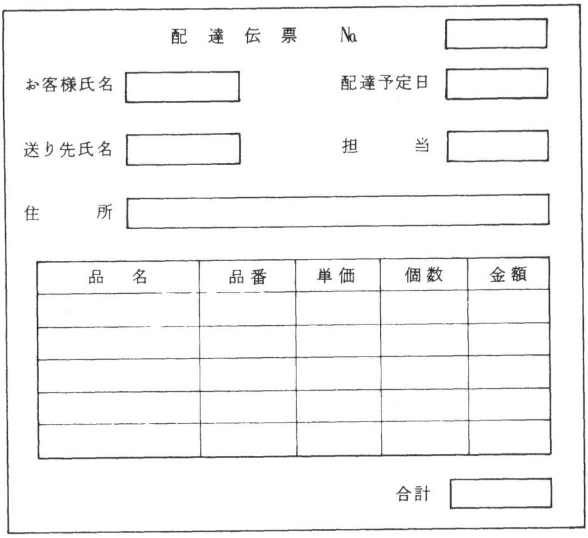
配 達 伝 票　№

お客様氏名　　配達予定日

送り先氏名　　担　当

住　所

品　名	品番	単価	個数	金額

合計

FIGURE 1. Example of a form.

uments on the form. The following sections discuss how the form operations are described on the form.

3. Form Operation

There are two kinds of form operations: intraform operations and interform operations.

Intraform operations are applied in the following situations:

1. Generating a document and storing it in the database.
2. Calculating expressions defined on the form.
3. Retrieving a document that satisfies retrieval conditions defined on the form.

Intraform operations are described by expressions defined on a field of a form. There are two kinds of expressions, a calculation expression and a condition expression with the same format. An example of the calculation expression is given in Figure 2.

Two expressions are defined on the INVOICE form:

$$\text{amount} : \_3 * \_4$$
$$\text{total} \quad : \text{sum}(\_5)$$

where $\_i$ denotes the ith field value in the form; that is, the expressions mean

$$\text{amount} : = \text{unit price: } * \text{ quantity}$$
$$\text{total} \quad : = \text{sum (amount)}$$

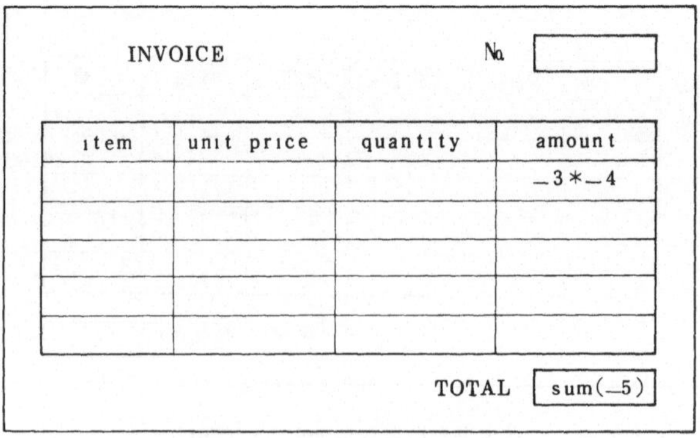

INVOICE			No.

item	unit price	quantity	amount
			$\_3 * \_4$

TOTAL   sum($\_5$)

FIGURE 2. Calculation expressions.

```
┌──────────────────────────────────────┐
│              REPORT CARD             │
│                                      │
│     NAME            ┌──────────┐     │
│                     └──────────┘     │
│     ENGLISH         │ >80      │     │
│                     └──────────┘     │
│     MATHEMATICS     │ >_ 4     │     │
│                     └──────────┘     │
│     SCIENCE         ┌──────────┐     │
│                     └──────────┘     │
│     TOTAL           ┌──────────┐     │
│                     └──────────┘     │
└──────────────────────────────────────┘
```

FIGURE 3. Condition expressions.

Expressions are evaluated as soon as possible. In the example, the value in the amount field is calculated as soon as both the unit price field and the quantity field values have been assigned. Then the value of the total field is calculated automatically.

Condition expressions can also be described on a field. An example is given in Figure 3. The conditions mean

ENGLISH > 80 and

MATHEMATICS > SCIENCE

Using these expressions, the user can generate, retrieve and modify each document of a form on the screen.

Interform operations are described by the mapping between forms. An example of the mapping is given in Figure 4. Figure 4 shows the operation to create an output form from an input form. Condition expressions are written on the input form which restrict the documents to be mapped. Assignment expressions are written on the output form which specify how to generate documents of the output form. Of course, it is possible to join multiple input forms into an output form.

Another example that aggregates a bundle of documents is shown in Figure 5. The difference between Figure 4 and Figure 5 is that the former generates multiple documents, while the latter generates only one docu-

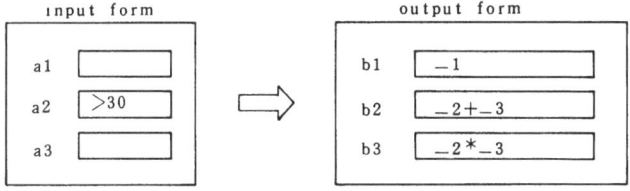

FIGURE 4. Mapping between forms.

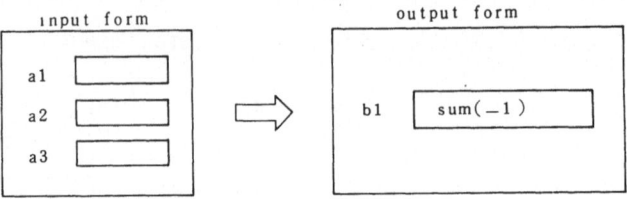

FIGURE 5. Aggregation.

shows that expressions written on the output form classify the types of interform operations.

As mentioned before, a form has an array structure, which requires more powerful interform operations to handle the data structure of a form. Some examples of them are given below.

Decomposition. It decomposes the array of the input form into distinct documents of the output form (Figure 6).

Composition. It composes distinct documents of the input form into an array of the output form (Figure 7).

Grouping. It classifies each document in the input form into groups by the value of the first field, and composes them into an array of the output form (Figure 8).

Summarization by grouping. It classifies each document of the input form into groups by the value of the first field, and summarizes each group into documents of the output form (Figure 9).

These operations are simple and powerful enough for an end user to describe his work. In addition, transferring information from one form to another form is a common practice in an office task. Hence, the concept of the form base can provide a familiar interface for an end user.

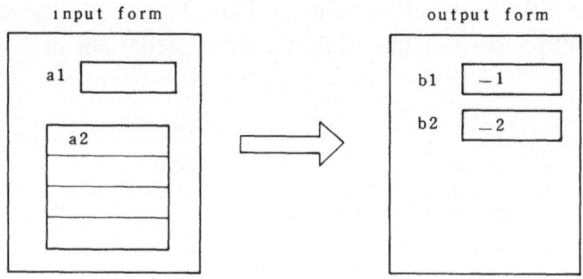

FIGURE 6. Decomposition.

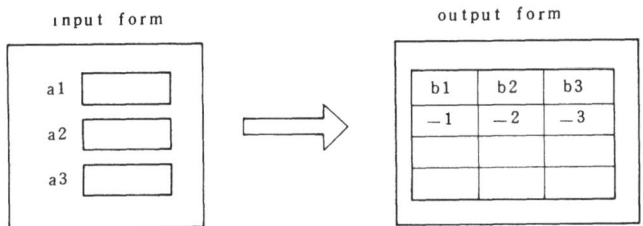

FIGURE 7. Composition.

4. Form Procedure

We have discussed the form operations on LAMBDA. They are the primitives to describe the office procedures, and most office tasks can be represented by combinations of these form operations.

LAMBDA provides a command file to realize the form procedure. The command file can remember the operations performed on the screen, and can play them back just by specifying the name of the command file. Using this playback function, the user need not describe the procedure in any format, but merely execute the sequence on the screen.

In this section, we will give a simplified example of a form procedure to explain the basic facilities of LAMBDA, and also discuss some problems about playback function.

4.1. Example of a Form Procedure

This example is a simplified version of a practical form procedure developed by an office clerical in our laboratory. It performs the management of patent proposals, and helps the managers to follow up the patent proposal of each section.

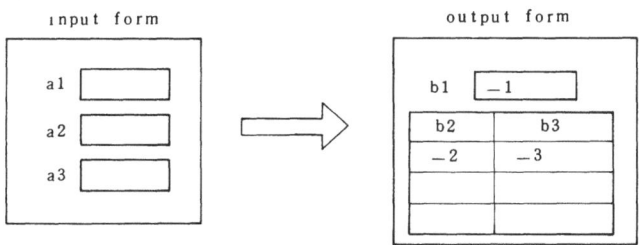

FIGURE 8. Grouping.

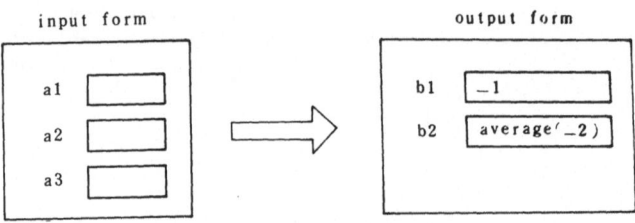

FIGURE 9. Aggregation by grouping.

The flow of the procedure is shown in Figure 10. Explanation of each step is as follows.

In step 1, a document of FORM-1 is generated whenever a researcher proposes a patent. FORM-1 is the master form in this example.

In step 2, the documents of FORM-1 are summarized into FORM-2 by grouping them into each section. The resulting documents of FORM-2 show the count of the proposals in each section.

In step 3, the summary of the patent proposals (FORM-2) and the target of the patent proposals of each section (FORM-3) are joined into FORM-4. The target of the patent proposals (FORM-3) is defined at the beginning of each year by the managers.

In step 4, the target and the result of each section (FORM-4) are composed into an array of FORM-5.

In step 5, a graph of the target and the result of each section is generated from FORM-5.

These steps are implemented in a form procedure by playing them back.

4.2. Augmentation of the Playback Function

The playback function is the simplest way to generate a form procedure for an end user; however, it has some drawbacks as follows:

- The user cannot give parameters to the command file.
- Only straight line programs can be generated.
- The user cannot refine the command file.

We augmented the playback function by providing a procedure editor as shown in Figure 11.

There are three operation modes in the procedure editor: teaching mode, editing mode, and executing mode.

In the teaching mode, the procedure editor receives data concerning all the operations performed on the screen, and records them in the command file.

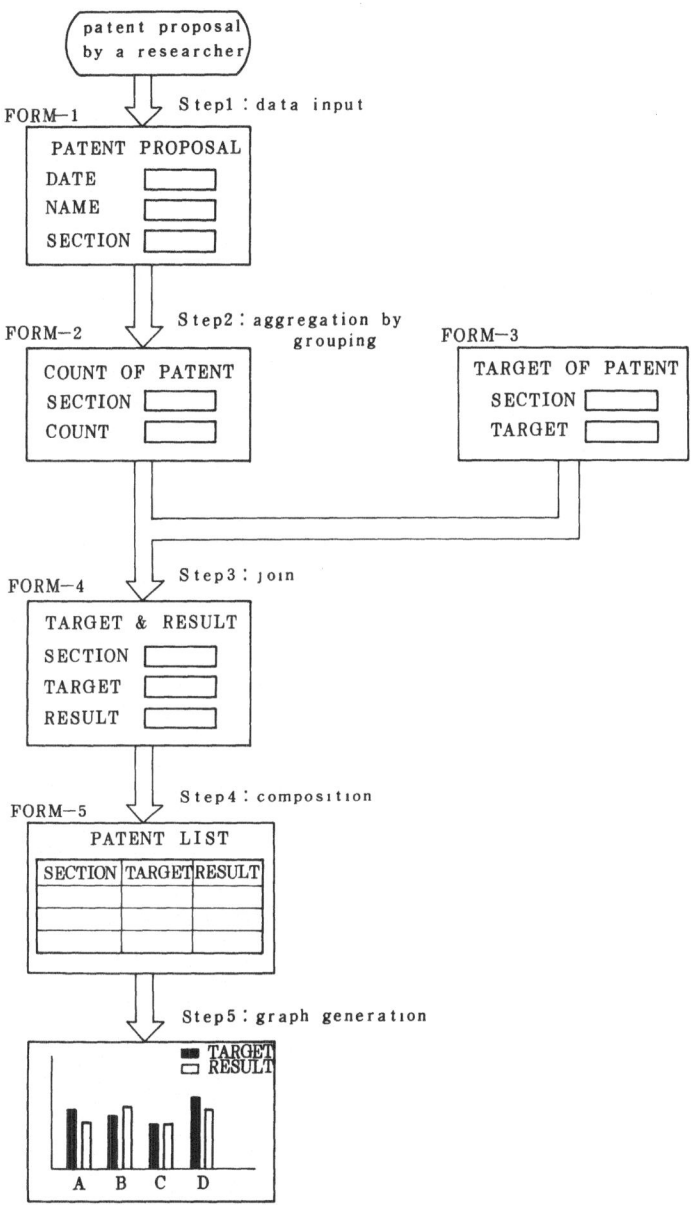

FIGURE 10. Flow of the procedure example.

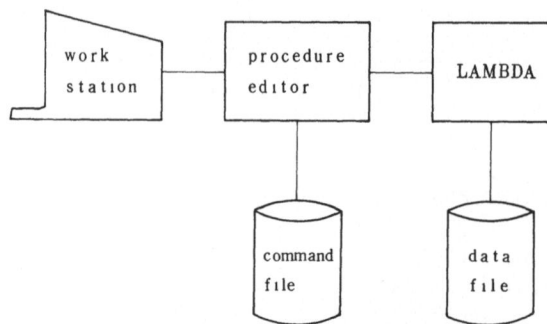

FIGURE 11. Procedure Editor and LAMBDA

In the editing mode, the user can refine the command file toward a more user-friendly format. Namely, the user can insert a message or a guidance for an operation into the command file, control the display mode, control the cursor movement, and so on.

In the executing mode, the editor recalls data concerning the operations from the command file, and sends them to LAMBDA. Then LAMBDA works just as well as receiving the data directly from the workstation. The editor can also manipulate some operations on the data of the command file, which augments the simple playback function as follows.

4.2.1. Parameters. Parameters play an important role in generalizing the function of the command file. To realize the handling of parameters, the following two functions are prepared.

Interactive assignment. This function allows a user to assign a parameter value at the required situation during the execution of a command file. For example, assume that a command file retrieves documents under some specific conditions, and transforms them into another form. The user wants to assign conditions during the execution of the command file. In this case, a stop code can be inserted into the command file. It enables stopping the execution after displaying the input form on the screen, and waiting for the user's operation. When the user finishes the assignment of the conditions, the procedure editor resumes the command file execution.

Parameter form. The interactive assignment of parameters is sometimes troublesome for a user, because he must wait at the work station until the command file stops. This function allows a user to assign the parameter value at the beginning of the execution of a command file. The procedure editor displays the parameter form at the beginning, and waits for the user's assignment. After completion of the user's operations, the editor begins to execute the command file using the assigned value of the parameter form. The user can design an arbitrary form as the parameter form.

In addition, this approach enables the procedure editor to pass parameters between form operations. The user can utilize an arbitrary field of the parameter form as a sytem variable. The procedure editor manages the data transfer between a field on the processing form and a field on the parameter form.

4.2.2. Representation of Control Flow. The most important drawback for the playback function is that it is limited to straight line programs. In order to overcome this drawback, LAMBDA adopts combining multiple straight line procedures into a command file. Each straight line procedure can be created by the playback function. To connect them, a concept of a link form is introduced. Each straight line procedure is associated with a link form which specifies the conditions to activate the procedure. An easy example of connecting procedures is shown in Figure 12.

Figure 12 shows the conditional branch depending on the value of the parameter form $1, and the loop structure until $2 equals zero. Each procedure can share the parameter form, which can be used as global variables.

5. Facilities to Support Managers

We have described fundamental facilities of LAMBDA. These facilities are suitable for office clerks who know their works in detail. However, there is another kind of office worker, i.e., the manager, who is not con-

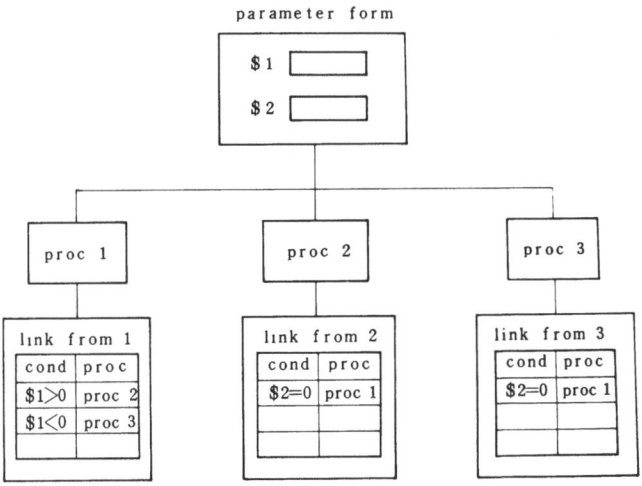

FIGURE 12. Connection of procedures.

cerned about how office tasks are carried out. To support managers, an office information system must behave intelligently without specifying their work in detail.

This section describes our approach toward extending LAMBDA to support the manager. Facilities for information retrieval and graph generation are considered as the most effective tools for the manager.

5.1. Japanese Language Interface for Information Retrieval

Users of office database systems have been forced to expend a great effort to learn the query language. LAMBDA is one solution to this problem for office clerks. However, managers still encounter difficulties regarding the following points:

1. They must know the form name to be retrieved;
2. They must know the data format written in the form;
3. They want more powerful retrieval functions using synonyms and thesaurus.

In order to solve these problems, a natural language interface is appropriate. There has been a good deal of research in this area. However, such interfaces may not be powerful enough for practical use in the office environment.

We took the approach of limiting the syntax of the Japanese language to a sentence of nouns and particles. The major reasons for this limitation are as follows:

1. In Japanese, a query sentence can be written without verbs and adjectives naturally. It is also clearer for a user to recognize the scope of acceptable representation for the system.
2. Information contained in the business forms are only nouns, hence it is natural to represent a query sentence with only nouns and particles.
3. Owing to the limited query syntax, the analysis is simple and fast enough for practical use.

Figure 13 shows an example of a query in this system. It means "List the names of customers located in Tokyo and whose delivery date is later than December first."

This query is translated into the usual expression of LAMBDA as shown in Figure 14, and the result of retrieval is given in the table format shown in Figure 15.

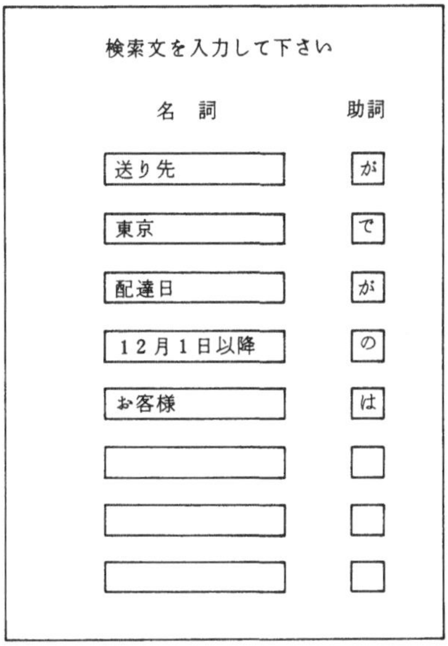

FIGURE 13. Query in Japanese.

5.2. *Automatic Graph Generation from a Form*

Managers usually want summary information represented in a graph. Conventional business graph package software can generate various kinds of graphs by specifying every parameter required. However, it is quite troublesome for managers to select the most suitable graph type, and to assign a number of parameter values.

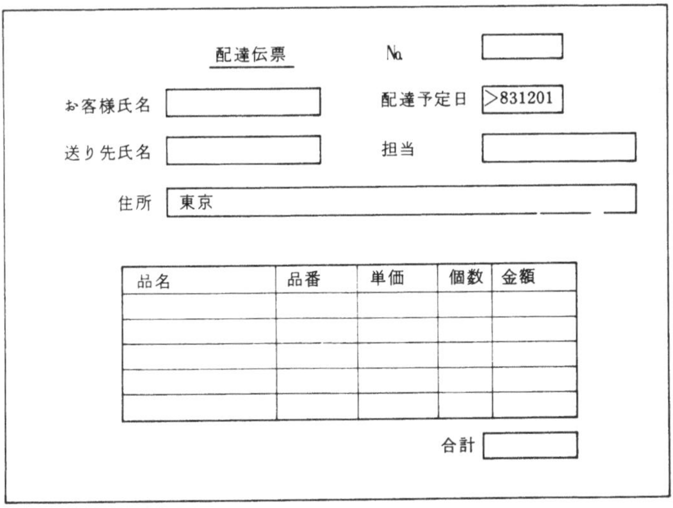

FIGURE 14. Condition expressions on a form.

検 索 結 果

№.	お客様氏名
103	須藤直樹
105	高橋豊
124	内山久雄
129	長谷川真
130	伊藤春男
146	菊地幸子
153	武藤達雄
169	大野和正
184	林弘子
201	磯部一男

FIGURE 15. Result of retrieval.

We aimed to generate a graph automatically from a form by considering information involved in the form. First, we clarify the problem by giving an example. Figure 16 shows a typical example of a form.

The title "Sales for Branches" explains the meaning of this form. The top line and the leftmost column explain each corresponding value in the table. We call them explanation terms.

The leftmost column (TOKYO, OSAKA, NAGOYA, and KYUSYU) shows the names of branches of the company, and the top line (video, audio, and others) shows the kinds of goods sold by the company. These lists of explanation terms are classified into the following four groups:

Element: elements of a set are listed.
Class: all the classes of a set are listed.
Time: periods of time are listed.
Attribute: different aspects of an object are listed.

Sales For Branches

Year   1984

Branch	Video	Audio	Others
TOKYO	13	14	10
OSAKA	10	12	6
NAGOYA	3	7	6
KYUSHU	4	3	5

FIGURE 16. Example of a form.

The element and the class are similar concepts. Namely, if a set consists of several elements, then each element corresponds to a class. Here we define that if a list of explanation terms covers all the elements of a set, then it is classified into a class; otherwise it is classified into an element.

Considering the example, the top line is classified into the class because all the elements are listed including "others," and the leftmost column is classified into the element.

Combination of the classification of the explanation terms can select a feasible graph format, and generate parameters automatically. For example, the graph shown in Figure 17 can be generated from the form in Figure 16.

6. Conclusion

LAMBDA has been implemented on the small business computer, the TOSBAC T-65. LAMBDA is presently used in more than 20 applications by about 20 users in our office. They can develop their own systems by themselves with only several hours' training. Through these practical applications, we are convinced that the form-based approach is well acceptable for office workers.

LAMBDA is also going to be combined with other OA tools such as the electronic mail system and the word processor. Combined with the electronic mail system, the user can send a document to another user, and invoke an execution of a command file by receiving mail. Combined with

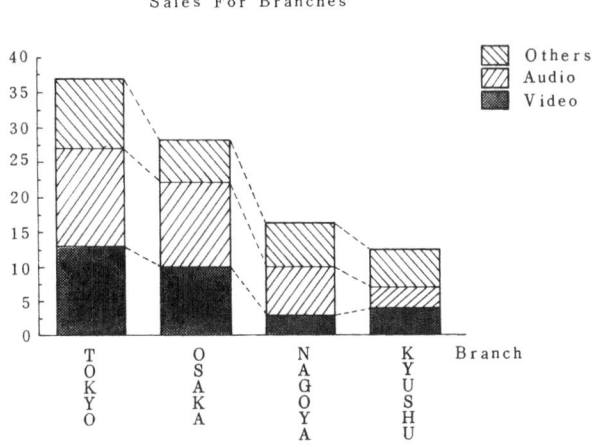

FIGURE 17. Example of a graph.

the word processor, the user can insert data of a document of LAMBDA into a text of the word processor. The user can also describe a text on the form of LAMBDA to generate an output report.

References

1. A status report on the activities of the CODASYL end user facilities committee (EUFC), *ACM SIGMOD,* **10**(2 and 3), (1979).
2. M. M. ZLOOF, QBE/OBE: A language for office and business automation, *IEEE Comput.* **14**(5), 13–22 (1981).
3. M. M. ZLOOF, Office-by-Example: A business language that unifies data and word processing and electronic mail, *IBM Syst. J.* **21**(3), 272–305 (1982).
4. PERSONAL SOFTWARE INC., VISICALC (1979)
5. D. TSICHRIZIS, Form management, *Commun. ACM* **25**(7),453–478 (1982).
6. V. Y. LUM, D. M. CHOY, and N. C. SHU, OPAS: An office procedure automation system, *IBM Syst. J.* **21**(3), 327–350 (1982).

QUERY LANGUAGES

6

REQUIREMENTS FOR DEVELOPING NATURAL LANGUAGE QUERY APPLICATIONS

Yannis Vassiliou, Matthias Jarke, Edward Stohr, Jon Turner, and Norman White

1. Introduction

The use of a "natural language" (for example, English) for direct interaction with databases, promises to broaden the market and the scope of computer utilization.[9,16,20,24] Technical feasibility has been demonstrated by the large number of experimental systems and the commercial availability of at least one such system. Yet there are many who voice concern as to the usability of natural languages for database querying by emphasizing the inherent dangers from its ambiguity.[17]

Natural language query systems (NLQSs) have been primarily looked at from the viewpoint of potential benefits/limitations for the database user.[11,19] Questions of whether they are easier to learn and more convenient to use in comparison to traditional database query languages are frequently raised. This debate has not been fully resolved. On the benefit side it is believed by supporters of natural language that an effective NLQS would

- reduce the amount of time required to train users;
- provide better memory retention;
- reduce total query execution time (including query formulation, query entry, and execution times);

YANNIS VASSILIOU, MATTHIAS JARKE, EDWARD STOHR, JON TURNER, and NORMAN WHITE • Computer Applications and Information Systems, Graduate School of Business Administration, New York University, New York, New York 10006

- broaden the base of users to include those who have application knowledge but neither the time nor the inclination to learn a formal query language.

Note that these all concern the cost-effectiveness of end-user computer interaction.[11] On the other hand, there are costs involved in implementing and maintaining an NLQS. These include the following:

- acquiring the system;
- processing and storage overheads;
- training systems analysts and data base administrators;
- analyzing the linguistic traits of users during systems analysis;
- developing a lexicon and grammar for the application;
- maintaining and tuning the system over time to accept new language variants and vocabulary;
- obtaining answers to the residue of queries that the NLQS cannot answer by other means—perhaps by trained personnel using a formal language.

Thus, while it seems that NLQS are becoming at least "usable," the more important question of their economic viability requires consideration of all of the factors listed above. In addition, other alternatives should be investigated including customizing a system (perhaps a database management system (DBMS) with a fourth-generation language) to produce a wide range of reports via a simple user-friendly interface or training end-users or staff in the use of a more formal DBMS query language.

In the remainder of this chapter we discuss these implementation issues. Section 2 provides some background on NLQS primarily by describing an existing prototype system. This is followed by a description of the application development process in Section 3. Section 4 raises some important unresolved issues related to the process of implementing NLQS, while Section 5 presents our conclusions.

2. Natural Language Query Systems

Using a natural language to query databases shifts onto the system the burden of mediating between two, sometimes very different, views of data: the way in which data are stored (records, files, etc.), and the way in which users perceive them. Structurally, NLQS are implemented as front-ends to general-purpose database management systems (DBMS)—see Figure 1. In order to make our discussion more concrete we discuss some general design issues and provide a fairly detailed description of a particular system.

2.1. Special-Purpose Versus General-Purpose Systems

There is an obvious conflict between the requirement for transportability of natural language database query systems to different application

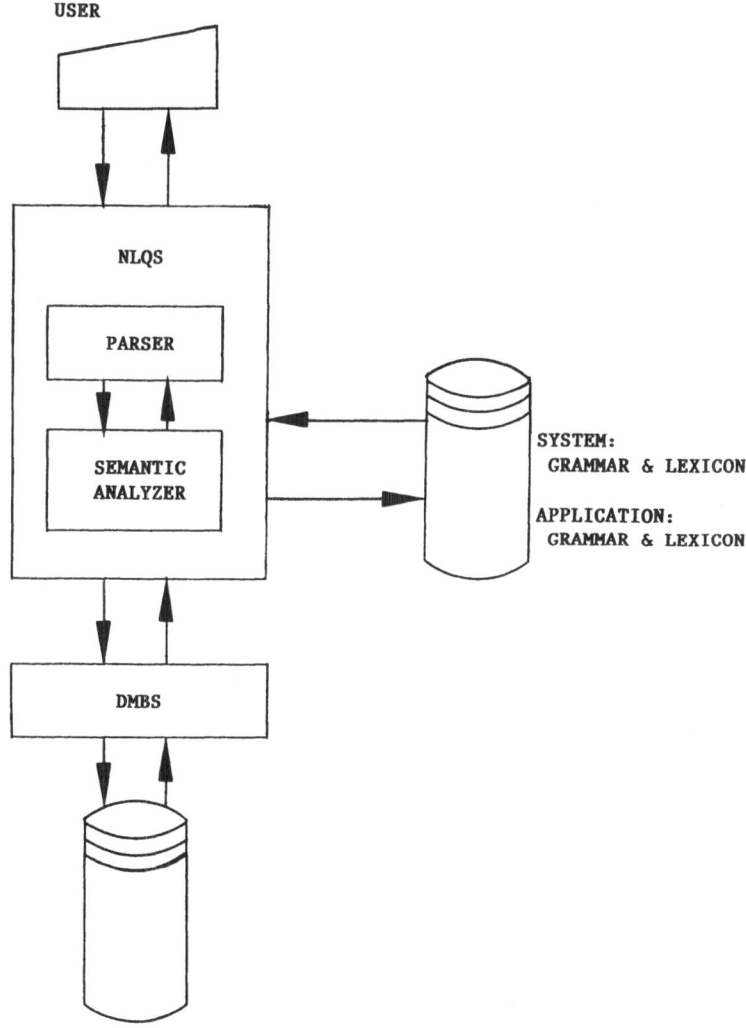

FIGURE 1. Structure of a general-purpose NLQS.

domains, and the large amount of knowledge that can more easily be represented in specialized systems. The emphasis on generality and transportability distinguishes general-purpose systems such as USL,[12] INTELLECT,[2] IRUS,[5] and TEAM,[8] from special-purpose systems such as LUNAR,[25] and the commercial products of Cognitive Systems Inc. of New Haven, Connecticut.[18]

Special-purpose (turn-key) systems mix the domain model, the database model, and the syntax and semantics of a particular domain to reach very high performance. This customization to a specific domain requires a substantial effort and experienced application developers. Moore[15] estimates that it takes between two months and five person years to complete the application for a small domain.

In contrast, for general-purpose systems, some critical design decisions have been made in order to adapt easily to new application domains:

- separation of information about language from information about the domain and the database;
- separation of the query's meaning from how to obtain the information requested;
- minimal user effort and as few and simple as possible new concepts introduced for each new application;
- selection of an optimal set of application-independent syntactic, semantic, and pragmatic concepts.

This paper only considers general-purpose NLQSs and examines application development for such systems. A representative NLQS is now described.

2.2. *Example of a General-Purpose NLQS*

The NLQS under study is a prototype.[12] Its aims can be summarized as economically allowing users to pose questions (queries) to a database in a natural language (e.g., German, English, Spanish), and to receive well-formated, meaningful responses within a short period of time. Figure 2 presents an example of the interaction between the NLQS and its users.

There are two major sub-systems driven by the NLQS: the generalized parser USAGE,[6] and a relational database management system. The process begins when a user enters an English query, such as: 'List the alumni who live in Seattle.' The NLQS first analyzes each query's syntactic structure, attempting to find the valid syntactic interpretations for the given query. This is accomplished through the coordination of three system components: the parser, the phrase-structure grammar, and the vocabulary.

USAGE accepts general phrase structure grammars written in a modi-

```
| ...                                                        |
| PLEASE ENTER YOUR QUERY:                                   |
| how many alumni have no donations?                         |
|                                                            |
|      ‾‾‾‾‾‾                                                 |
|      3679                                                  |
|                                                            |
| PLEASE ENTER YOUR QUERY:                                   |
| ...                                                        |
```

FIGURE 2. NLQS interaction.

fied BNF (Backus–Naur form). It produces all parses in parallel working bottom-up and from right to left. As in augmented transition networks,[25] arbitrary routines can be invoked with any rule (e.g., to perform arbitrary actions, to control the rule's application, to check and test syntactic features, etc.). The job of the parser is to build syntactic interpretations (parses) by calling on the rules specified in the grammar and the lexicon. The parser begins by trying to recognize each of the words in the query. Recognition of vocabulary occurs if a word (character string) (a) matches one of the application-independent words that form a permanent part of the NLQS (such as "list," "the," "who," and "in," whose meaning does not vary from application to application), (b) matches one of the application-specific vocabulary words defined upon installation of the system within the application context (such as "alumni" and "live"), or (c) is a value within the database whose field can be implicitly determined from the other words in the sentence (such as "Seattle"). The application-independent and application-specific words that explicitly defined (a and b above) comprise the lexicon, or vocabulary.

After all the words are recognized, the parser applies the grammar rules in order to encompass the entire query in a tree (a full parse). If the system cannot complete a full parse then the interpretation process is abandoned at the syntactic stage. However, if one or more full parses is achieved, the query proceeds to undergo semantic analysis.

The NLQS's semantic analyzer and executor receive as input the trees completed by the parser during syntactic analysis. Since many of the invalid interpretations of the query are "killed" during syntactic analysis, the scope of the relatively expensive semantic analysis is narrowed, thereby heightening the efficiency of the system.

The semantic analyzer and executor invoke the interpretation routines referred to in grammar rules. The intrepretation routines often call upon the database management system to access semantic information stored in the database. This information includes relation and attribute names, relational views, domain characteristics, etc. Various semantic interpretations are considered; however, most of them will "die" due to a semantic inconsistency (i.e., they will not "make sense" according to the semantic information available to the natural language processor).

SQL[3] is the formal query language into which the English query must be transformed. Once the SQL query has been produced, it is repeated back for the user to see (the echo capability), and is passed to the DBMS for retrieval of the data requested in the SQL statement.

The system's grammar assumes a close correspondence between language constituents and semantic relationships. Thus the design of the intrepretation routines is based on the assumption that syntactic structures carry meaning which is independent of the application domain. *Factual* knowledge is stored in the database and is not duplicated in a dictionary. *Conceptual* knowledge is encoded in relational *views* (virtual relations). We will pursue this issue in the following sections.

In summary, there are two major processors, the DBMS and the natural language processor, which must be connected for the overall system to operate. This connection is made during application development.

3. Application Development Process

The development of any system, exhibiting high standards of sophistication, usability, and effectiveness, is a tedious and complex process requiring careful analysis, planning, and design together with close cooperation with users. Natural language applications are no exception.

Generally, the development of an application for which natural language will be used as the query language includes three main stages:

1. Development of the application database;
2. Development of the natural language interface, involving
   • some work on the natural language system,
   • some work on the database system,
   • training of users;
3. Evolutionary Improvement of the System with respect to
   • database,
   • linguistics.

In the following sections these development stages are detailed. It will be demonstrated that the separation between the stages is not clear-cut and, therefore, iteration is necessary.

3.1. Database Development

The previous two sections have illustrated how the NLQS objectives of portability and flexibility are met by supplying an application-independent grammar and by coupling the NLQS to a general-purpose DBMS. It will often be important for the DBMS to simultaneously satisfy nonnatural language uses (application program and formal language interfaces). This being the case it would be undesirable if the database design requirements for the different types of interface were mutually incompatible. Fortunately this does not seem to be the case in general, although (as discussed below) the database design may sometimes need to be modified to accommodate an NLQS. Adding an NLQS interface to an existing database may therefore have expensive ramifications. For a new application the special requirements of the different database interfaces should be considered at the design stage and the appropriate trade-offs made. With this in mind the basic tasks of traditional database design practice should be followed. However, some additional tasks due to the NLQS are necessary as indicated below and discussed more fully in Section 4. Our discussion is based on the design process proposed in [Ref. 13].

Task 1: Corporate Requirements Analysis. This includes such things as defining the corporate general constraints (e.g,. how the organization does its business), the information requirements (e.g., how managers and users view data), and the processing requirements (e.g., planning, control, and operations on data). All the analysis is made independently of the information system structure, and requires extensive interviews and examination of existing documentation. Object-oriented requirements specification languages, e.g., Galileo,[1] Omega,[4] and RML,[7] appear to be best suited for the description of these requirements.

Additional tasks related to NLQS development include the collection and analysis of word usage patterns. Our suggested methodology is given in Section 4.3.

Task 2: Information Analysis and Definition. This task deals with conceptual data modeling and has also been termed "high-level logical design." The conceptual description, or schema, defines the enterprise's view of the database. In constructing such a global description, the views of different users and management, the determination of different user group needs, and the processing styles and needs play very important roles. Modeling the different user views and subsequently modifying them to reach a "compromise" enterprise view is the major problem in this task. Conceptual models, such as the entity-relationship data model, are usually employed at this stage.

When developing an NLQS many additional database views are necessary to provide links between the DBMS and NLQS as discussed in Sections 4.1 and 4.2.

Task 3: Implementation Design. For this task a particular database model and system are employed. Guided by such issues as integrity, consistency, privacy, recovery, security, and efficiency, DBMS-processible descriptions are generated. For instance, when the relational data model is used, then the base relation descriptions, virtual relation ("view") descriptions, and the design of programs to be run against these relations are developed.

Task 4: Physical Database Design. This is the final implementation of the database with emphasis on functioning and performance measures. According to data volume, process volume, DBMS characteristics, and hardware characteristics, the database storage and success path strategies are decided upon.

In principle, database development as outlined above should be independent of the user interfaces since there may be multiple interfaces for a database system.

In the sequel, assume a small portion of a relational database developed within the above framework. The database models the world of donations to New York University by alumni and other donors. The class of donor is distinguished by a "source code."

DONORS (name, address, source_code, . . .)

DONATIONS (name, amount, fiscalyear, . . .)

3.2. Linguistic Development

The process begins with the determination of the application-dependent words that users may employ. In addition, for each such word its "meaning" and expected usage have to be determined. Extensive interviewing is necessary for this task.

A general methodology for an analysis of the vocabulary is to study word associations by sentence frames. These are canonical graphs that show the expected relationships linked to concepts of a given type. For instance, in the following frame for the verb "to donate," the concepts [PERSON], [MONEY], and [YEAR] play the roles of agent (AGNT), object (OBJECT), and point-in-time (PTIM), respectively, in a sentence. Note also the correspondence with the columns NAME, AMOUNT, and FISCALYEAR of the table DONATIONS:

[DONATE] -
(AGNT) -> [PERSON]

$$(OBJ) \rightarrow [MONEY]$$
$$(PTIM) \rightarrow [YEAR]$$

The processes of analyzing the vocabulary and making the connection with the database system are now exemplified for the prototype NLQS described in Section 2.

3.2.1. Work on the NLQS: Lexical Entries. As already noted, most application-specific words are defined upon installation of the natural language system. Basically, the words that must be defined for the user in the application vocabulary are *nouns, verbs,* and *adjectives.* Values that appear in the database do not have to be defined; they are automatically recognized. In some cases, though, e.g., conflict with a noun, such database values must also be defined in the application vocabulary. Other application-specific words may be added after implementation, as the users' manners of expression demonstrate the need for additional words. In either case, the syntactic information that the NLQS requires for these words must be presented in the form of lexical rules, as in the following example:

```
<NOMEN:+NTR,+SG,LAB=28,+LP:FPE-NOMEN('ALUM')> < = 'ALUMNUS';
<NOMEN:+NTR,+PL,LAB=28,+LP:FPE-NOMEN('ALUM')> < = 'ALUMNI';
<VERB:TYP=NA,+SUF,+LP,+TP:FPE-VERB('DONATE')> < = 'DONAT';
```

As can be seen above, these rules consist of a character string to the right of the final equals sign and syntactic information to the left of the sign. A rule is invoked if a character string in the query matches the character string in the rule. Regular morphological variations are handled automatically, as can be illustrated with the lexical entry, 'DONAT.' The third rule will be invoked by both queries, 'Who donat*ed* more than 1000000?' and, 'Did Schwartz donat*e* more than 1000000?'. Irregular morphological forms have to be defined explicitly with separate entries, as demonstrated by the entries 'alumnus' and 'alumni' for the singular and plural forms.

The left-hand side of the equals sign begins by specifying a syntactic category (in these examples, 'NOMEN' and 'VERB'). Much of the other information refers to the various features that may apply. *Features* are sequences of code that serve to further refine the lexical and grammar rules. They enable the NLQS grammar to handle the irregularities and exceptions of the English language, as well as efficiently diminishing the number of rules inappropriately invoked. Features can be either syntactic or semantic in nature, and contain information on such things as number and gender of nouns, and transitive or intransitive use of verbs.

The NLQS employs three types of features: *logical, interger,* and *case* features. Logical features take either a plus or a minus value. For example, the "+LP" seen in all of the above rules allows a dimension of place, permitting phrases such as "alumnus *in New York.*" The "+TP" is another logical feature which allows the user to ask queries involving a time dimension, such as "Did Jennis donate *in 1981?*" In addition, there are integer features, such as "LAB = 28," which take natural numbers as their values. Since "LAB" is used to indicate the prepositions that can be used with the defined word, and 28 is the code value for the 'from', "LAB = 28" means that the preposition "from" may be used with the defined word. Likewise, "LAB = 22" indicates that the preposition "for" may accompany the defined word. "TYP = NA" is an example of a case feature; it specifies that the verb is transitive. As will be discribed later, an interactive program helps the developer to formulate these definitions.

Looking back at the rules, the constructs with the prefix "FPE" are names of interpretation routines. The words which appear as arguments in these routines (in these examples, "ALUM" and "DONATE") refer to the relational views to which the character strings apply. The views are utilized during semantic analysis, and will therefore be discussed in more detail subsequently.

3.2.2. Work on the DBMS: Semantic Views. Just as words have lexical entries for the syntactic analysis to be performed, the NLQS requires that the words have entries with semantic information so that the semantic analysis may be performed. In the NLQS, a word's "meaning" is defined relative to the database. Rather than taking shape from a collection of properties (hard, soft, animate, inanimate), a word's meaning is formulated with respect to the database tables and fields.

The semantic information is represented in the form of database views which are derived from base tables. Views are virtual tables (their extensions are not stored) and are built up from the base tables of the database. Since words are semantically defined by relating them to the database, views become an effective method for defining words. For example, the word "alumnus" could be defined as the NAME value of those entries in the base relation named "DONORS" where the SOURCE_CODE is "AL." When the word "alumnus" appears in a query, this view definition would be called from the database, and would indicate that the query is referring to all NAME values that are in the same row (tuple) of the base table as an SOURCE_CODE value of "AL."

When defining a view, each column (field) in the view must be assigned a data type (also called domain) and a role name. Data types include character string, number, code, date, and quantity. Role names include nominative case, dative case, point in time, etc. The data type and role name abbreviation form a prefix to the column name during view definition.

Continuing with our definition of alumnus, the view named "ALUM" would appear as follows:

Define view ALUM (WNOM_NAME) as
    Select NAME
    From DONORS
    Where SOURCE_CODE = 'AL';

This specifies that the view ALUM will consist of one column named NAME whose entries are character strings ("W") from the base table named DONORS and are used in the nominative case ("NOM").

Likewise, the view "DONATE" could be defined:

Define view DONATE (WNOM_NAME, ZACC_AMOUNT,DTP_FISCALYEAR) as
    Select NAME, AMOUNT, FISCALYEAR
    From DONATIONS

The values in AMOUNT column are numbers used in the accusative case. The NAMEs are character strings which are used in the queries in the nominative case. FISCALYEAR is a column whose entries are of the data type "date," and indicates a point in time. The above definition asserts "Donors *donate* money in a year." Recalling the lexical entry "DONAT," this definition enables queries such as "Did Jones donate more than 3000 in 1980?," or, "How much did Jones donate?."

It is no coincidence that many of the role names are identical to the features specified in the lexical rules. These establish a connection between the syntactical constructs and the columns of the views. Another connection between the lexical information and the semantic analysis is the view name. As mentioned previously, each lexical entry contains the name of the corresponding view. The lexical rules for "alumnus" and "alumni" both referred to the view named "ALUM," while the verb "donate" referred to the view "DONATE."

4. Issues for Application Development

Our experience in evaluating a NLQS[10,21,22] revealed the following research issues concerning application development.

4.1. Sharing the Database: Multiple Natural Language Users

A number of hypotheses were formulated early in our project with natural language at New York University. The first was that natural language users may differ in their conceptions of a word's meaning and usage. The second was that the number of different words employed by the users may be very large, thus making the process of application development tedious and introducing difficulties in verifying the internal consistency of the natural language lexicon. Moreover, the large number of required terms might cause system performance degradation.

Our study had mixed results concerning these hypotheses. In the field test,[10,21] there were several terms used by different users with different meaning (e.g., "donors *give* donations" and "schools *give* degrees"). Such situations caused some problems in application development since multiple views for a verb were not allowed by the system; however, the application domain was not complex and rich enough to cause any serious conflicts that could not be handled.

The laboratory study[22] demonstrated a high commonality of word usage among 26 users. Specifically, on the average, 55% of the most commonly used words were shared by all users. Furthermore, the top five words were used by 93% of the users. Words that were used infrequently could have been dropped without serious loss of overall performance. For instance, 44% of the words occurred less than three times, accounting for only 6% of all word occurrences (more details and the study methodology are given in Table 1).

These positive results concerning word usage are mainly attributed to the high degree of focus (specific database quering) for the laboratory experiment and the format of the users' task. General requests were presented to the laboratory subjects. Each request was three sentences long and presented a situation and an implicit need for the generation of database queries to carry out a task. For example, one request was: "The dean of the Graduate School of Business is planning a trip to Japan. He has expressed interest in meeting with alumni currently living in that country. Where can he find them?" From this request, subjects generated queries such as: "What are the addresses of all alumni in Japan?", or, "Where do the Japanese alumni live?"*

Naturally, the users tended to use terms from those occurring in the request, thus contributing to the high degree of word usage commonality. The situation may be different in unrestricted database querying, as the field test partially demonstrated, and depends on the number of natural language users and the domain's conceptual complexity.

A possible solution to the problem of word usage is to allow each user

---

*This query demonstrates a possible ambiguity. In the lexicon, "Japanese" was defined as an adjective for alumni whose country of residence is Japan." On the other hand, it can also be defined as an adjective for alumni whose nationality is Japanese.

TABLE 1
Methodology and Language Usage Results[a,b]

Categories	Unique words	Occurrences
Application-dependent		
Verbs (nonimperative)	45	440
Nouns/adjectives	101	1592
SUBTOTAL	146 (56%)	2032 (44%)
Application-independent		
Verbs (imperative)	8	195
Pronouns	11	247
Operators	7	86
Comparatives	4	221
Connectives (conjuctives)	6	216
Articles	4	120
Prepositions	12	748
Modifiers	10	131
SUBTOTAL	62 (24%)	1964 (45%)
Database values		
Constant values (Nos.)	21	304
Constant values (strings)	30	178
SUBTOTAL	51 (12%)	482 (11%)
TOTALS	259 (100%)	4478 (100%)

[a]The words used in natural language queries were grouped in three major word types: Application-dependent words (must be defined for each new application, e.g., verbs, adjectives); Application-independent words (are predefined in the core lexicon of the system, e.g., operators, prepositions); and Database-values (stored in the database, e.g., numbers and proper names).
[b]Methodology to assess word usage commonality: Miller's method[14] was used (Application-dependent words only). A list of the top 25 words in frequency of use by all subjects was created. The list accounted for 6% of unique and for 49% of all words occurrences. For each subject, an individual list was made for their 25 most frequent words. The commonality was assessed by contrasting all lists.

to employ his or her own terminology. Considering the scarcity of application development tools (see Section 4.3) this may not be easy.

An alternative solution requires the ability to question the system's capabilities, and in particular, allow for questions about the database (tell me about donations) or about the vocabulary (what is an alumnus?)

Knowledge of a word's definitions could prove helpful to users during error detection and correction, as well as during query formulation. For example, an infrequent user might appreciate knowing that "donor" has been defined as an alumnus, organization, or other entity that has contributed money since 1968. The user could use this information to correct a query referring to donors in 1965. A user might also appreciate access to this definition before formulating the query in order to verify the word's appropriateness within the contemplated query (if, for example, the user wanted assurance that "donor" referred to organizations as well as to alumni). In both of the above cases, the user would benefit from the ability to simply ask: How do you define "donor?"

We have discussed the problems of sharing the database among natural language users. The situation is more serious when other interfaces are also used on the same database.

4.2. Sharing the Database: Multiple User Interfaces

The use of a NLQS for databases may follow one of two scenarios. In the simplest scenario, NLQS provides the only interface to the database, therefore all the flexibility for a custom-made database design exists. In a more complex situation, yet one that is more likely, natural language is only one of several user interfaces. For instance, database users may have the option to use formal database languages (e.g., SQL), programming languages, graphics-based browsers, or menus.

Theoretically, NLQSs are intended to serve the database environment, regardless of whether or not the database was desinged with NLQS in mind. However, since the system's semantics have their roots in the database, the database design must facilitate the semantics of the application. An example is in order.

The NYU alumni database originally contained a DONATIONS relation in which were listed the alumni names and their donations during the years 1982, 1981, 1980, 1979, 1978, 1977, and a consolidation of years 1968–1976 (Fig. 3). This organization was chosen because it maps directly to some needed reports (using a programming language interface). Note that the relation is in third normal form despite its awkwardness. Unfortunately, this original relation does not map easily into English.

Queries such as "When did 139386074 donate $5,000,000?" could not have been answered directly since the time information is not represented with database values. The attribute 1982-AMT captures both year and amount information.

The relation DONATIONS(name, amount, fiscalyear)—see also Sec-

NAME	1982 AMT	1981 AMT	1980 AMT	1979 AMT	1978 AMT	1977 AMT	BEFORE1977 AMT
Jones	0	0	0	0	0	375.00	5000.00
Smith	100.00	0	0	0	0	0	0
Black	0	0	0	0	0	1785.00	0

FIGURE 3. Original DONATIONS relation.

tion 3.2—is structurally more flexible. It more easily maps to a frame for the concept type DONATE and hence to English sentences containing the verb *donate*. The argument that the new form of the DONATIONS relation can be given as a relational view is not valid for current database systems and languages. No database view defining mechanism is powerful enough to express directly this new form as a query on the orignal DONATIONS relation.

Like the NLQS prototype, INTELLECT[2] roots its semantics on database fields and tables. It therefore has the same potential problem of having to redesign an existing database to fit the semantics of an application. In addition, INTELLECT requires the creation of database indices (data inversion) from the base tables. These indices facilitate INTELLECT's handling of database values, but they may degrade the performance of database modifications through a separate user interface (e.g., data manipulation language).

An important research topic is the identification of database design characterstics for natural language databases (a natural language normal form?). Our conjecture is that although a database could potentially require major restructuring if it were not designed with the natural language query system in mind, most well-designed databases will in fact also be well suited for NLQS semantics, and will therefore require only minor restructuring. Furthermore, the development of new, more expressive database languages will help overcome some design problems. In particular, returning to our running example, using the logic-based language PROLOG the new DONATIONS relation can be expressed as a view (query) on the original relation. This is because PROLOG allows for delayed binding of variables in view definitions.[23]

In general, the database sensitivity issue for natural language is rooted to the fact that natural language semantics correspond to the semantics of a "conceptual" schema, yet natural language must be mapped to the "internal" schema for processing. Logical data independence is only partially available in current database systems.

4.3. Evolving Application: Meeting the User's Requirements

Since application development is an evolving process, several enhanced versions of the NLQS will be developed. If the user does not see a fairly competent system early in the development, he/she may become disenchanted with the system and refrain from using it.

A related issue is the inadvertent influence on the structure of users' queries. While users are learning to query the database they may try many inappropriate language constructs. They tend to seize upon the first one that works and use it exclusively even though enhanced versions of the system will handle many others.[10]

The goal then is to present a very competent system to the users of natural language. The scarcity of good methodologies for linguistic analysis of the application and of effective application development tools is a big obstacle.

To the authors' knowledge, no comprehensive methodology that allows for an early linguistic development has been described. At a minimum, such a methodology might include one or more of the following steps:

1. Use a controlled study to determine the way future natural language users ask questions on a hypothetical database. Of importance here is not only the words used but the context and meaning of their use.
2. In case there is a conversion from another database language to natural language, ask direct database users to rephrase their queries in English. For database users working through intermediaries, examine the language usage at the request level.
3. Develop and test a prototype implementation as soon as possible. Note that this should be done with care since prolonged use of the prototype by prospective users may establish undesirable usage patterns as discussed earlier in this section.

The subject of application development tools is much maligned but still very important for anyone who attempts to set up a natural language application. Unfortunately, NL systems developers put little emphasis on providing facilities for this process. These facilities should range from tools for the "database linguist" (administrator) to tools for the end-user.

The NLQS prototype has an interactive program to help the developer or user formulate the lexical entires for application-specific vocabulary (Fig. 4.) The program interrogates the developer about the morphological forms of the word to be entered, other words (such as prepositions and particles) that often appear with the new entry, and roles that the new entry can take on. Although it can be helpful in holding a naive user's hand through the definition process, it may become (as do most such application development tools) too verbose and tedious for expert users. Currently, it provides no assistance to the application developer in defining the semantic information associated with words (the views). This is a substantially more difficult problem. Developers must therefore have knowledge of databases and be proficient in SQL, since this is the language in which views are defined. In addition, basic linguistic knowledge could prove helpful. Although this does not violate the NLQS's goal of freeing users from

The example application development tool interrogates the developer about the morphological forms of the word to be entered, other words (such as prepositions and particles) that often appear with the new entry, and roles that the new entry can take on.

An illustration of how the tool aids in defining the verb 'to start' is given. Although the developer would have to know that 'start' was a verb, the remaining technical and linguistic knowledge is extracted through questions about the use of the word within queries.

Illustration

WHAT TYPE OF WORD DO YOU WANT TO DEFINE?
*NOUN
*VERB
*ADJECTIVE
*NAME
IF YOU DO NOT WANT TO DEFINE ANY MORE WORDS, WRITE 'END'.
  verb
(Editing note: here the system provides a paragraph explaining
                how one should enter the verb.)
  start
NOW YOU SHALL SEE DIFFERENT FORMS OF YOUR VERB. IF THERE ARE ERRORS, PLEASE ENTER THE NUMBERS OF THE INCORRECT FORMS. YOU WILL THEN BE PROMPTED FOR THE CORRECT FORM...
1 START
2 STARTS
3 STARTING
4 (HE) STARTED   5 (HE HAS) STARTED
IS YOUR VERB A PARTICLE VERB, LIKE TAKE OFF, OR GIVE UP?
ENTER YOUR PARTICLE OR TYPE NO.
  no
IS THE NAME OF YOUR RELATION START? IF SO, ENTER YES. OR.,
ENTER THE NAME OF THE RELATION...
  yes
DO YOU USE YOUR WORD WITH TIME SPECIFICATIONS?
  yes
WHAT KIND OF TIME SPECIFICATIONS DO YOU USE?
1 WHEN...?
2 FROM WHEN...?
3 UNTIL WHEN....?
  1
(editing note: acceptable prepositions are now determined)
HOW WILL YOU USE THE VERB; PLEASE ENTER THE TYPE:
TYPE NI: WHO STARTS...?
TYPE NA: WHO STARTS WHAT...?
TYPE NOA: WHO STARTS WHAT TO WHOM...?
  na
...

FIGURE 4. Sample word definition illustration.

learning a formal query language and having database and linguistic skills, it does prohibit users from defining their own words.

Finally, one of the technical issues that face designers of natural language systems is the representation and independent maintenance of dictionary knowledge in two places; the database system and the natural language system. Most NLQSs keep database values duplicated in their lexicon. This allows for system efficiency, in particular when spellers are employed. The problem arises with very volatile databases, where frequent updates of database values which are not propagated in the NLQS's lexicon result in representation and usage inconsistencies. Furthermore, integration of metadata descriptions may be needed so that a major advantage of modern database systems (i.e., store metadata as you store data) is not lost.

5. Concluding Remarks

The process of application development for natural language query systems was examined and several issues that may have a major influence on the practical use of natural language as a database interface were identified. The paper provides a fairly general framework for the initial development of the linguistic component of a NLQS and some suggestions for facilitating the evolution of a system during use.

In the authors' opinion these operational issues have as much importance as the more often raised issues of natural language usability (e.g., ambiguity, precision, user misconceptions, etc.)

Acknowledgment

Part of this work was carried out in a joint study with the IBM Corporation.

References

1. A. Albano, L. Cardelli, and R. Orsini, Galileo: A strongly-typed interactive conceptual language, Technical Report, 83-11271-2, Bell Laboratories, Murray Hill, New Jersey, 1983.
2. Artificial Intelligence Corp., Intellect query system, Reference Manual, 1982.
3. M. M. Astrahan, and D. D. Chamberlin, Implementation of a structured English query language, *Commun. ACM* 18(10), 580–588 (1975).
4. G. J. Barber, Supporting organizational problem solving with a workstation, *ACM Trans. Office Inf. Systm.* 1(1), 45–67, (1983).

5. M. BATES and R. J. BOBROW, A transportable natural language interface for information retrieval, *Proceedings of the 6th International ACM SIGIR Conference*, ACM special interest group on information retrieval, Washington, June, 1983.

6. O. P. BERTRAND, J. J. DAUDENNARDE, and B. DU CASTEL, USAGE: A user application generator, META user's guide, IBM France Scientific Center, 1978.

7. S. GREENSPAN, Requirements modelling: A knowledge representation approach to software engingeering, Ph.D. thesis, Department of Computer Science, University of Toronto, March, 1984.

8. B. J. GROSZ, TEAM, A transportable natural language interface system, in *Proceedings of the Conference in Applied Natural Language Processing*, ACL and NRL, Santa Monica, California, February (1983).

9. L. R. HARRIS, User oriented database query with the ROBOT natural language query system, *Int. J. Man-Mach. Stud.* **9,** (1979).

10. M. JARKE, J. TURNER, E. STOHR, Y. VASSILIOU, N. WHITE, and K. MICHIELSEN, A field evaluation of natural language for data retrieval, *IEEE Trans. Software Eng.* **SE-11,** (January 1985).

11. M. JARKE and Y. VASSILIOU, Choosing a database query language, NYU Working Paper Series, CRIS #67, New York, April (1984).

12. H. LEHMANN, Interpretation of natural language in an information system, *IBM J. Res. Dev.* **22**(5), September, (1978).

13. V. LUM *et al.*, 1978 New Orleans database design workshop report, IBM Technical Report RJ2554, 1978.

14. L. A. MILLER, Natural language programming: Styles, strategies, and contrasts, *IBM Sys. J.* **20,** 2 (1981).

15. R. C. MOORE, Practical natural language processing by computer, Technical Report Technical Note 251, SRI International, October (1981).

16. S. R. PETRICK, On natural language based computer system, *IBM J. Res. Dev.* **20,** 4 (1976).

17. B. SHNEIDERMAN, *Software Psychology*, Winthrop, Cambridge, Massachusetts (1980).

18. S. P. SHWARTZ, Natural language processing in the commercial world, in *Artificial Intelligence Appliations for Business*, W. REITMAN, Ed., Ablex, Norwood, New Jersey, 1984, pp. 235–248.

19. E. A. STOHR, J. A. TURNER, Y. VASSILIOU, and N. H. WHITE, Research in natural language retrieval systems, *15th Annual Hawaii International Conference on System Sciences*, Hawaii, 1982.

20. H. R. TENNANT, Evaluation of natural language processors, Ph.D. dissertation, University of Illinois, December, 1980.

21. J. TURNER, M. JARKE, E. STOHR, Y. VASSILIOU, and N. WHITE, Using restricted natural language for data retrieval—A plan for field evaluation, in *Human Factors and Interactive Computer Systems*, Y. VASSILIOU, Ed., Ablex, Norwood, New Jersey, 1984.

22. Y. VASSILIOU, M. JARKE, E. A. STOHR, J. TURNER, and N. WHITE, Natural language for database queries: A laboratory study, *MIS Q.* **7**(4), 47–61, (1983).

23. Y. VASSILIOU, J. CLIFFORD, and M. JARKE, Access to specific declarative knowledge in expert systems: The impact of logic programming, *Decisions Support Syst.* **1**(2), Spring (1985).

24. D. L. WALTZ, An English language question answering system for a large relational database, *Commun. ACM* **21**, (1978).

25. W. A. WOODS, R. M. KAPLAN, and B. NASH-WEBBER, The lunar sciences natural language information system, Bolt Beranek and Newman, Cambridge, Massachusetts, June, 1972.

7

SEMANTIC REPRESENTATIONS FOR NATURAL LANGUAGE QUERY PROCESSING

Stephen D. Burd, Shuh-Shen Pan, and Andrew B. Whinston

1. Introduction

One of the recommendations of the CODASYL committee[7] was the development of a nonprocedural language for accessing a database system. The goal of this recommendation was to allow a larger number of users to access a database. Ideally, a query language should allow a user to access the database without procedurally specifying exactly how this access must be performed. In addition, the language should be simple to use and should require a minimal amount of training and prerequisite knowledge on the part of the user.

Since this recommendation was made, many such languages have been developed which have met the ideal standards with varying degrees of success. Examples of such languages include SQUARE,[2] SEQUEL,[3] SQL, and Query by Example.[23] Most of these languages are based on context-free languages with fairly rigid syntactic and semantic requirements. Query-by-Example and SQUARE differ significantly from the others in that queries are expressed via data entries into tables rather than via a language per se.

Although many of these query languages are quite powerful, they are not entirely free of procedural specification and require that users be trained in their use. In a study evaluating the learnability of SQUARE and SEQUEL, Reisner, Boyce, and Chamberlin[18] showed that reasonable profi-

STEPHEN D. BURD • Anderson School of Management, University of New Mexico, Albuquerque, New Mexico 87131   SHUH-SHEN PAN • Bell Communications Research, Holmdel, New Jersey 07733   ANDREW B. WHINSTON • Krannert Graduate School of Management, Purdue University, Lafayette, Indiana 47907

ciency in either language required 12–14 hours of formal instruction. In addition, considerable difficulty was experienced by some subjects in learning and retaining some of the more complex language features. Thomas and Gould[20] studied Query-by-Example and showed that the system could be learned more quickly than SQUARE and SEQUEL, but the number of query formulation errors encountered was still quite high.

Other researchers sought to bypass these and other problems associated with artificial languages by designing natural language query processing systems. Winograd[21] developed a natural language processing system to interface to "blocksworld." The system was capable of recognizing both commands and queries within this limited domain. Harris[11] developed ROBOT, which employed a parser based on augmented transition networks to recognize database queries. This system was the basis for a commercial product. Kaplan[12] developed the CO-OP system, which provided sophisticated error-handling capabilities for natural language queries. McCord[15] developed a Prolog formulation partially based on his slot grammar[16] and used it to process natural language database queries.

This chapter will examine the role of semantic data modeling techniques in the development of a natural language query system. We will critically examine previous natural language query systems in terms of the tools used to build them and the semantic knowledge and representations employed. We will largely ignore issues of syntax in order to concentrate on semantic problems. This is not to imply that an adequate syntactic model is unimportant for natural language understanding. We simply wish to insulate ourselves from the present competition between alternative syntactic models.

We will consider a query to be a single interrogative statement which can be answered by retrieving data fields from the database. By restricting our attention to single-sentence queries, we can utilize a simpler recognition model than would be required if we were concerned with an entire discourse. We will require that the query be syntactically correct and that the data necessary to answer the query exist within the database.

2. Semantic Issues in Parsing

Most of the more successful attempts at natural language understanding have integrated syntactic and semantic analyses to parse the input string and generate an internal semantic representation thereof. McCord[15] built a system employing this approach using the definite clause grammars of Pereira and Warren.[17] Definite clause grammars and McCord's approach will be discussed in greater depth later in this chapter.

The need for semantic analysis in parsing is not a new idea. Although most early attempts at automated natural language processing were based on purely syntactic models, the need for semantics had been debated in the linguistics literature for many years. Perhaps the most extreme argument at the semantics end of this spectrum was given by Bloomfield.[1] He argued that in order to know the meaning of any nontechnical term, one must have complete information about everything in the universe of discourse.* Although Bloomfield was not directly addressing the issue of parsing, the logical conclusion of his argument is that no semantic representation of a sentence can be generated even if the sentence can be recognized on a syntactic level.

Much of the early work in natural language processing was concerned with natural language translation. Many of the attempts at automating this process were based on some variation of Chomsky's transformational grammar.[5] Most of the work in programming languages used simplified models of phrase structure grammar [e.g., precedence grammars, LR(k) grammars, LL(k) grammars, etc.] and treated this problem separately from the problem of generating a semantic representation (i.e., code generation). It is perhaps this history that led so many later researchers in natural language processing to maintain this dichotomy. It is our contention that the problems of syntactic recognition and generation of a semantic representation of a natural language sentence must be addressed simultaneously.

Fillmore's case grammar[10] was one of the early attempts to integrate a syntactic and semantic model. We follow his line of reasoning in arguing that we must make use of semantic information in order to adequately recognize and interpret natural language sentences. Thus, we seek to define a natural language recognition model in which semantics helps to reduce the complexity of syntactic recognition and syntax helps to overcome the complexity of the semantic recognition problem.

Given that we choose such an approach, we encounter design problems concerning the type of semantic knowledge to be utilized and the method of utilization. The complexity of the first problem is reduced somewhat by the scope of our problem. Since that is limited to natural language recognition as concerns queries to a specific database, we can use the contents of that database and other relevant information about it to define the semantic knowledge needed.

To define a method of utilizing this semantic knowledge, we will build

---

*Thus, the meaning of a term such as "asynchronous" exists independently of any context while the meaning of a nontechnical term is always dependent on contextual factors (e.g., adjacent terms, speaker intentions, etc.).

on the work of other researchers in the areas of logic programming, definite clause grammars, and semantic analysis. While we will adopt a specific recognition model for purposes of discussion and presenting examples, we emphasize that the semantic issues discussed and the database techniques presented are general and can be applied in the context of recognition systems with similar design philosophies.

The remainder of the chapter will be organized as follows. In the next section, we will review three implementations of a natural language query processor. We will also review the tools with which these systems were built, namely, augmented transition networks and definite clause grammars. We choose these three particular systems because they represent diverse techniques of semantic analysis and because they have each been implemented to at least a limited extent.

Section 4 will compare and contrast the semantic analysis techniques employed in these systems. We will examine relevant research in linguistics in order to determine the suitability of the techniques employed and possible extensions thereto. Section 5 will examine the role of semantic modeling techniques in natural language query processing and discuss appropriate methods of representing such models. The final section will summarize the research and discuss its limitations and possible extensions.

3. Prior Research in Natural Language Processing Systems

Woods[22] proposed a parsing strategy for natural language that he termed augmented transition networks (ATNs). The model consists of a network of nodes interconnected by directed arcs. The nodes correspond to states in a finite state machine and the arcs represent transitions from state to state. Associated with each arc is a set of conditions and a set of structure-building actions. The conditions specify restrictions which must be met if the arc is to be followed. These conditions are usually concerned with determining the presence or absence of syntactic constructs (e.g., noun phrase, verb phrase) or lexical items. The structure-building actions are used to construct the internal representation of the input sentence (i.e., the parse tree).*

As formulated by Woods, the model was used to generate parse trees for input sentences. Through the use of recursion, he was able to capture much of the linguistic complexity found in a transformational grammar. The model can represent higher-order transformations involving inser-

---

*A simple example of natural language recognition using an ATN is given in the appendix to this chapter.

tion, deletion, replacement, and copying of syntactic items. Woods demonstrated that the model is an efficient parsing mechanism that is roughly equivalent in power to transformational grammar.

The specific formulation presented by Woods was solely concerned with syntactic recognition of sentences. The model can, however, be extended to generate a semantic representation as well as (or in place of) a syntactic representation by modifying the structure-building actions in the arcs. Thus, it is possible to use the ATN formalism to implement semantic recognition models such as Fillmore's[10] case grammar.

Harris[11] developed a natural language query system called ROBOT which used an ATN for parsing. The ATN for this application is fully recursive and has an automatic backup facility. Utilizing these two features, the system manifests both lexical and structural ambiguity by generating as many parses (interpretations) of the sentence as possible. The rationale given by Harris (Ref. 11, p. 20) for this strategy is as follows:

> . . . we employ the philosophy of letting the parser run, providing a fast syntactic analysis, and ask questions about semantics later. In this way we avoid asking questions needed by interpretations that will eventually fail syntactically anyway.

In order to manifest lexical ambiguity, the ATN iteratively tries all known meanings of a lexical item. The only restriction on these meanings is that they be consistent with the syntactic use of the word in the query (e.g., is "experiment" used as a verb or a noun?). The dictionary contains entries for basic lexical items such as "who," "what," "is," etc., and the names of files and fields in the database. Items stored in the database are not stored in the dictionary, but are accessed directly by utilizing the inverted structure of the database.

Semantic analysis is performed by two separate modules within the system. The first module isolates semantically valid interpretations produced by the parser. This is accomplished by interacting with the database to determine which interpretations are semantically correct with respect to the database. The second module takes the set of semantically valid interpretations from the first module and selects the interpretation intended by the user. This is accomplished by determining the likelihood of a given interpretation and, if necessary, querying the user of his intended meaning.

Kaplan[12] developed a natural language query system that provided extensive error-handling capabilities. The parsing mechanism consisted of a lexical scanner and an ATN which produced a semantic representation called the meta query language (MQL). The syntactic model employed was rather restricted as queries were required to be Wh-questions.

Unlike the ROBOT system, Kaplan's CO-OP system utilized the database as little as possible in order to make the system as portable as possible. Thus, the lexical contents of the dictionary are similar to those of ROBOT, but the system does not make use of the database to look up terms which exist only in the database. The CO-OP system attempts to determine the appropriate database referent (attribute, relation, etc.) from the context in which the lexical item appears. For those words defined in the lexicon, a data structure is maintained which provides information regarding possible referents of the item in the database. Thus, all of the domain-specific knowledge required to answer a query is present in the lexicon.

The error-handling capabilities of the system are designed to provide cooperative responses to users when errors are encountered in processing the query. The system accomplishes this task by determining the beliefs and presuppositions implicit in the query through syntactic analysis. For example, the query

How many students failed English 433 last semester?

would produce the response:

English 433 wasn't offered last semester.

rather than the response "none" or "nil" if the course had not been offered in the previous semester. In this example, the system would recognize that the query implies that English 433 was offered last semester and would reject it on those grounds rather than providing a misleading answer or halting entirely.

Pereira and Warren[17] describe a natural language recognition formalism based on earlier work by Colmerauer.[9] They describe the implementation of definite clause grammars using Prolog and compare this strategy with ATNs implemented using LISP. The motivation for this strategy is described by Pereira and Warren (Ref. 17, p. 232) as follows:

> The problem of recognizing, or parsing, a string of a language is then transformed into the problem of proving that a certain theorem follows from the definite clause axioms which describe the language.

The formulation of a language as a definite clause grammar (DCG) constitutes both a representation for the language and a program for recognizing it. When a DCG is executed by Prolog, the Prolog inference mechanism behaves as an efficient top-down parser for the language. Comparative performance data between this system and LUNAR (an ATN-based system) shows that the Prolog implementation can be at least twice as fast as LUNAR. The definite clause grammar is also more textually compact than the programmed version of its ATN counterpart.

Another significant advantage of definite clause grammars, as compared to ATNs, exists with regard to building syntactic and semantic representations (i.e., structure building). Structure building in an ATN-based system is restricted to closely follow the execution path of the algorithm. This is not too much of a problem for generating syntax trees but can cause considerable difficulty in generating semantic representations.

McCord[15] developed a DCG for recognizing database queries that took great advantage of this structure-building power. McCord's system generated a representation of the semantic meaning of a query which did not necessarily have to follow the syntactic derivation of the query. By not binding the semantic representation to the syntactic representation, McCord demonstrated that he was able to adequately deal with the problem of modifier scoping.

The McCord system is implemented as a three-pass system. The first pass takes a sentence as input and produces a special type of syntax tree as output. Each node of the syntax tree consists of a common data structure called a syn. In addition to containing the syntactic category of an item, a syn also contains information about the semantic interpretation of the item and about the way the item can be used to modify other items. The latter information is contained in a data item called the determiner.

The second pass of the system takes the syntax tree produced by the first pass as input and reshapes this tree such that the scoping of modifiers is semantically correct. In order to perform this reshaping, McCord posits a hierarchy of determiners. The reshaping algorithm proceeds by recursively examining each syn in the tree and raising/lowering it or moving it left/right according to the precedence relations between the determiner of the syn and the determiners of its daughter and sister syns. Thus, if the determiner of a syn has a lower precedence than the determiner of one of its daughter syns, the algorithm alternates the positions of the two syns in the tree. Similarly, daughter syns are moved such that their left-to-right order in the tree corresponds to the decreasing precedence of their determiners.

The third pass takes the reshaped tree produced by the second pass and produces a formal representation of the query. This is accomplished through translation rules which examine the determiner and predication of each syn and translate these into a format suitable for processing the actual query. The operation of this pass is dependent on the target query language being used, and the translation rules are constructed such that they can be easily modified to generate different query formats. Theoretically, it would be possible to use this system to map a natural language query to any formal query language (e.g., SQL, MDBS III, etc.).

In addition to enforcing precedence restrictions between syns, the system is also able to enforce semantic restrictions. In an earlier paper, McCord[16] developed a formalism called slot grammars which allowed the lexical entry for an item to specify a list of possible syntactic modifiers for that item. The parsing algorithm for these grammars utilized these lists (called slot lists) to aid in the parsing. The system described in McCord[15] also utilized these slot lists and extended the concept to include semantic restrictions on the slots. This technique allows the DCG to properly determine which arguments belong to which predicates, where the argument should be placed within the predicates, and if the arguments meet semantic typing restrictions.

4. Comparison of Implementation Strategies

The three systems discussed above (McCord,[15] Harris,[11]) and Kaplan[12] cover a continuum ranging from the total separation of syntactic and semantic analyses to their total integration. The ROBOT system lies at the separation end of this spectrum, the McCord system at the integration end, and the Kaplan system near the middle.

The ROBOT approach corresponds roughly to the theoretical approach of Katz and Fodor[13] in which a sentence is disambiguated by generating every possible reading of the sentence (via projection rules) and eliminating those readings which contain violations of selectional restrictions between lexical items. Katz and Fodor considered examples such as verbs which required animate subjects, etc. Harris' implementation of this strategy considered selectional restrictions pertaining to the particular database being queried. The semantic projection rules of Katz and Fodor are implicitly represented within the ROBOT ATN via the backtracking mechanism which forces the generation of a separate parse for each possible meaning of a lexical item.

Although the use of selectional restrictions in this manner is an intuitively appealing approach, it is not always sufficient in and of itself as recognized by Katz and Fodor (Ref. 13, p. 191):

> We may observe that the occurrence of a (sentence) without readings is a necessary but not sufficient condition for a sentence to be semantically anomalous.

McCawley (Ref. 14, p. 130) examined the Katz and Fodor approach and concluded

> ... the violation of selectional restrictions is only one of many grounds on which one could reject a reading as not being what the speaker had intended and that it moreover does not hold any privileged position among the various criteria for deciding what someone meant.

One of the primary reasons given by Harris for separating syntactic and semantic analyses was processing efficiency. He stated that integrating the two analyses would force the system to semantically check interpretations which might eventually fail on a syntactic basis anyway. However, this inefficiency works in both directions. Generating parses for a query that is semantically invalid is also a type of processing inefficiency. Consider the query.

> What employees are being depreciated over five
> years using the double declining balance method?

If this query were analyzed on a purely syntactic level, the entire sentence would be processed and the analysis would succeed. If syntactic and semantic analyses were combined, the system would reject the sentence after the fifth word, thereby saving the processing time necessary to analyze the rest of the sentence.

Harris argues further that syntactic analysis is relatively easy as compared to semantic analysis as the latter requires interface to the database. This argument is, however, more a function of his entire system design (specifically his ATN and lexicon) than of semantic analysis in general. Thus, while the separation of syntactic and semantic analyses may be an optimal design choice in ROBOT, it is not necessarily optimal in the context of other designs.

By choosing to use an ATN without a semantic component, Harris loses the opportunity to use semantic knowledge to guide the parsing process. Kaplan (Ref. 12, p. 102) also used an ATN, but chose to add a semantic component and gave his reasons as follows:

> Simply stated, the syntax–semantics interaction is a mutual one: semantics must guide the parse while the parse must guide the semantics. Consequently, the selection of word sense and the selection of a parse in CO-OP occur simultaneously, each process suitably constraining the other at critical points.

Although both Kaplan and Harris used an ATN for parsing, the structures of their lexicons were fundamentally different. Harris viewed the database as an extension of the dictionary and, thus, did not store any lexical items from the database in the dictionary. Kaplan sought to avoid accesses to the database until the actual data retrieval process was to begin. His dictionary contained not only all of the terms in the database, but also a representation of their meaning. Thus, he did not have to access the database in order to utilize semantic information.

McCord also followed this strategy for designing his lexicon. The lexical entry for any lexical item in his system contains both syntactic and semantic data about the term. The database in McCord's system was a simple relational database consisting of unit clauses. Most of the lexical entries were also unit clauses and both the lexicon and the database were accessed in the same manner by the resolution mechanism of PROLOG.

On a functional basis, the designs of McCord's and Kaplan's systems are somewhat similar. Both systems use a parser which analyzes the input sentence both in semantic and syntactic terms. Both systems translate a sentence into a formal query language and employ this language for database retrieval. Although the McCord system was interfaced to a relational database and the CO-OP system to a CODASYL database, the techniques used were general enough to allow either system to interface to either type of database.

By utilizing Prolog, McCord was able to achieve economy in processing that was not achieved in either ROBOT or CO-OP. Both of these systems required separate mechanisms for parsing and semantic interpretation. McCord expressed both of these processes as Horn clauses and used the Prolog inference mechanism as the common control strategy.

This design has the advantage of maintaining a degree of independence between the different parts of the system. Changes to either the grammar description or the translation mechanism require changes to only a subset of the clauses in the system. Since the only domain-oriented knowledge used by the system is contained in the lexicon, the grammar and translation mechanisms are also independent of the database.

For the purposes of this chapter, we will follow the basic approach of McCord. That is, we will assume a logic-based recognition system which combines both syntactic and semantic analysis and a lexicon which contains both syntactic and semantic information. This will allow us to realize the aforementioned advantages of processing efficiency/economy and functional independence, and will allow us to extend the system to utilize additional semantic information which will be covered later in this chapter. The database is assumed to be relational, and can be viewed as a collection of logical assertions for our purposes.[8]

5. Semantic Representations and Database Extensions

One of the primary problems involved in natural language processing is the problem of multiple meanings. The disambiguation of lexical items within a sentence often requires the use of knowledge not contained in an ordinary dictionary (i.e., encyclopedic knowledge). Consider the following sentence:

(a) The picture will be shown for the remainder of the week.

In this example, it is not possible to determine whether the word "picture" refers to a movie being shown in some theatre or a photograph or painting being displayed in an exhibit. If the sentence were one of the following:

(b) The picture will be playing for the remainder of the week.
(c) The picture will be on exhibit for the remainder of the week.

then the proper meaning of the term "picture" could be implied from the predicate (verb) of the sentence.

If we were trying to develop a system that could understand and interpret the whole of natural language, then the ambiguity of (a) would be an unsolvable problem at the sentential level. However, when we restrict our attention to natural language queries to a specific database, occurences of this type of problem tend to be greatly reduced or eliminated entirely. For example, if we are developing a query system for the database of a theatre chain, then there will be only one possible meaning for the word "picture" (i.e., film or movie). Thus, in general, we need only define as many meanings for a lexical item as are relevant to a particular database (the domain of discourse).

If only one meaning of "picture" is defined in the database, then no inference mechanism is required to disambiguate the term "picture" in (b) and (c). If there is more than one meaning for picture, then we must have some formal way of representing the subject restrictions implicit in the verbs "play" and "exhibit" and we must utilize some inference mechanism to derive the proper meaning (database referent) of "picture." This is precisely the problem addressed by McCord[15] in his PROLOG formulation of slot grammars.

Given a relational database which represents the meaning of lexical items as expressions in predicate logic, a query processor must be able to map from the predicates in the lexicon to the relations (predicates) in the database. The contents of the database and the domain of discourse provide us with the information necessary to construct such a mapping. In the domain of theatres and films, an actual database might contain a relation representing the showing of a picture in a theatre as

Playdates(Picture, Theatre, Date)

If we consider the terms "play" and "show" to be references to this relation then we can utilize this predicate in the lexical entries for these terms.

McCord utilized database predicates to represent the meaning of terms in the lexicon. He also stored information about the arguments of such predicates. Consider the following examples:

1. What picture is playing in the Bijou on 4/25/83?
2. What theatres will *Star Wars* be shown in on 4/25/83?
3. When will *Star Wars* start playing in the Bijou?

In each of these queries, the picture acts as the subject, although this fact is somewhat obscured by the passive transformation in (2). Also note that the theatre (or absence thereof) is marked by the preposition "in" in all three queries. In (1) and (2) a particular playdate is marked by the preposition "on."

These regularities can be attributed to the specific meaning of the verb "play" in this database. A slot grammar formulation of the lexical entry for "play" would recognize these regularities and store them along with the more obvious fact that "play" is a verb. A simplified version of McCord's format for verb entries is

verb(lexical-item, predication, logical-subject, [slot-list])

Using this notation, a lexical entry for the verb "play" embodying the above restrictions is

verb(play, Playdates(Picture, Theatre, Date), Picture,
[pobj (in): Theatre,pobj(on):Date])

which states that "play" is a verb which can be interpreted in terms of the predicate 'Playdates' in which the argument 'Picture' is the logical subject and the arguments 'Theatre' and 'Date' are marked as objects of their corresponding prepositions.

McCord's grammar rules utilize such entries to derive a relational query. For example, in examining query (1) the grammar rules will examine the set of lexical entries and find the entry above for the verb "play." The rules will recognize the correspondence between the terms "What picture" and the logical subject of the predicate "Playdates." The Wh determiner will cause this argument to be left undefined and it will be marked as the object of the query. The grammar will then utilize the slots list to attempt to bind the other arguments of "Playdates" to values in the sentence. Thus, "Theatre" will be assigned the value "Bijou" and "Date" will be assigned the value "4/25/83." Conceptually, the output of this process can be represented as

$$\text{Playdates(Picture?, 'Bijou', '4/25/83')}$$

Which may then be processed against the database in order to find matching occurrences of the predicate. Thus, the answer to the query may be determined as the value of the argument "Picture" for all occurrences of the relation "Playdates" with Theatre = 'Bijou' and Date = '4/25/83'.

This approach to the design of the lexicon provides the natural language understanding system with the ability to enforce selectional restrictions. However, it goes beyond the simple restrictions of Katz and Fodor. By tying subject, object, and other semantic restrictions to an actual database predicate and combining these with semantic typing restrictions, the system obtains the ability to disambiguate word sense with regard to the semantic restrictions implied by a specific database.

Such an approach works well in situations where the meaning of a lexical item corresponds to a specific predicate in the database. Problems arise, however, when such a correspondence does not exist. Such predicates are sometimes referred to as derived relations. In the theatre domain, contracts are written for the showing of a picture in a theatre. Thus, a database in this domain may contain a relation such as

$$\text{Contract(Number, Terms, Start-date, Stop-date, Booker)}$$

where "Number" is a unique integer identifier, "Terms" is a field containing the contract price (expressed as a percentage of theatre admissions), and "Booker" is the employee responsible for the contract. Since a contract represents an $n{:}m$ relation between picture(s) and theatre(s) (e.g., double features and multiple theatre showings) the database may also contain the following relations:

$$\text{Picture-playdates(Contract.Number, Picture.Name)}$$
$$\text{Theatre-playdates(Contract.Number, Theatre.Name)}$$

Now consider the lexical entry for the verb "play." If the database contains the three relations just described instead of the relation "Playdates," then the predication for "play" is no longer directly represented in the database. However, the information necessary to fill the arguments in this predicate does exist within the database in the three relations just defined. McCord ignored this problem entirely by restricting the class of lexical items represented to those which corresponded directly to relations in the database (or arguments of those relations). By using the database itself as the lexicon, it seems that Harris forces such derived relations to be explicitly represented in the database (the cited reference is unclear as to the method used to deal with this problem). We choose to deal with this problem by utilizing virtual relations as an extension to the lexicon.

A virtual relation can be defined as a relation without explicitly stored tuples. It can, however, be defined in terms of relations that are explicitly stored. The predication for "play" is an example of such a relation. A definition for "play" in terms of the database relations "Contract," "Picture-playdates," and "Theatre-playdates" is*

> Play(Picture.Name, Theatre.Name, Date) :-
>   Contract(Number, Terms, Start-date, Stop-date, Booker),
>   Picture-playdates(Number, Picture.Name),
>   Theatre-playdates(Number, Theatre.Name),
>   Inclusive(Date, Start-date, Stop-date)

where "Contract," "Picture-playdates," and "Theatre-playdates" are database relations as defined above and "Inclusive" is a Boolean function that determines if "Date" lies in the time interval spanned by "Start-date" and "Stop-date."

In a natural language query system, the existence of multiple database views is manifested in the lexicon. Different users may view the same entity at different levels of abstraction or may be concerned with different attributes of the same entity. Thus, the problem of multiple database views can be addressed as a problem in linguistic (lexical) diversity.

Since the database will, in general, maintain only one common representation of data, some of these user vocabularies may not correspond directly to relations in the database (e.g., the lexical items "play" and "show" in the previous example). Where such a correspondence does not exist, semantic models (and their representation via virtual relations) serve to bridge the gap between the user's vocabulary and the actual representation of data within the database.

For example, consider a database containing the following relations:

> Salesman(SS#, Name, Address, Salary, [Customer-list])
> Manager(SS#, Name, Address, Salary, Education, District)
> Janitor(SS#, Name, Address, Salary)

If the predication of the lexical item "employee" is defined as

> Employee(SS#, Name, Address, Salary)

---

*We will utilize Horn clause notation to represent virtual relations.

then a mapping is required to retrieve the arguments of "Employee" from the actual database relations. Such a mapping is accomplished via the virtual relation

Employee(Name, Address, SS#, Salary):-
    Salesman(SS#, Name, Address, Salary, [Customer-list]),
    Manager(SS#, Name, Address, Salary, Education, District),
    Janitor(SS#, Name, Address, Salary)

This mapping would allow the processing of queries such as

What employees earn more than $20,000?

and other queries which reference the generalized entity "Employee."

Various researchers have proposed extensions to traditional database models to allow the implicit and explicit storage of additional semantic information. Examples of such research include Smith and Smith's[19] aggregation/generalization model, Chen's entity-relationship model,[4] and various approaches to modeling temporal information and restrictions (e.g., Clifford and Warren[6]). While such research has provided a rich variety of semantic modeling techniques, the motivation for using such techniques and the actual procedures for their utilization have often been neglected.

Codd[8] considered several of these models and developed an extension to the relational database model which incorporated much of this research. He summarized the usefulness of such models as follows:

> . . . a meaning-oriented data model stored in a computer should enable it to respond to queries and other transactions in a more intelligent manner. Such a model could also be a more effective mediator between the multiple external views employed by application programs and end users on the one hand and the multiple internally stored representations on the other. (Ref. 8, p. 398)

Natural language query processing provides both a use and a justification for data modeling techniques such as the ER model and aggregation/generalization. The user's natural language implicitly embodies concepts such as generalization as they apply to a given domain of discourse. The concepts of entity and relationship are explicitly represented in English and other natural languages as nouns and predicates. Temporal information is represented through the use of tense.

Thus, a natural language interface to a database must either access representations of these concepts in the database or provide explicit representations of these concepts in order to map to actual database relations.

This assertion is supported by Smith and Smith (Ref. 19, pp. 107–108), who state " . . . explicitly representing generalizations . . . will allow naming conventions in the model to correspond with natural language and enable the users to employ established thought patterns in their interactions with the database."

The virtual relations defined previously are examples of such representations. The determination of whether or not they must be present in the natural language understanding system depends on whether or not the semantic models they are intended to represent are present in the database. Codd developed a system of operators which allowed the manipulation of a database in which semantic models were explicitly stored. In the case where such a database exists, the semantic models necessary to process natural language queries are already present in the database and operators defined thereon may be utilized for retrieval. In the case of a database in which no semantic models are stored, the models would have to be provided by the natural language understanding system.

6. Conclusion

We have examined Harris,[11] Kaplan,[12] and McCord[15] and described the methods of semantic analysis and semantic representations utilized therein. A review of research in linguistic theory has shown that the semantic techniques utilized by McCord are preferable to the other techniques reviewed as they employ a wide variety of semantic information in order to disambiguate and understand natural language queries.

An analysis of semantic modeling techniques has shown that representations of semantic knowledge such as generalization are also necessary for natural language query recognition. The choice of representing such knowledge within the database or as an extension of the lexicon is dependent on the availability of semantic modeling tools and operators with the database system. In the situation where such models are explicitly represented and opeators thereon are provided, the task of natural language query system design is simpler. Where such models and/or operators are not provided in the database system, they must be provided in the natural language understanding system.

This chapter has concentrated on queries which require the retrieval of database contents (or derivations thereof) for answering. Further research in this area is needed to develop methods for deriving answers to queries which implicitly or explicitly require transformation of data in the database. For example, the query:

Predict sales for next month based on prior month's sales.

implicitly requires a transformation of past sales data consistent with the term "predict." Several research questions can be posed as regards this type of query:

1. What should the lexical entry for "predict" be?
2. Should "predict" be tied to a procedure that performs the act of "predicting"?
3. What about multiple definitions of "predict"? (e.g., regression, time series analysis, etc.)
4. How can we handle queries which require the execution of more than one procedure?

These and other questions push the frontier of natural language query processing research beyond the realm of database and linguistics and toward the areas of generalized knowledge representation and decision support systems.

Appendix: Augmented Transition Network Example

Consider the following simplified English grammar:

$$
\begin{array}{l}
S \;\; \text{---}> NP\ Aux\ VP \\
NP \text{---}> Det\ N \\
NP \text{---}> N \\
VP \text{---}> V\ PP \\
PP \text{---}> Prep\ NP \\
VP \text{---}> V\ NP
\end{array}
$$

where S is a sentence, NP is a noun phrase, Aux is an auxiliary verb (e.g., is, was, has, etc.), VP is a verb phrase, Det is a determiner (e.g., the, a, etc.), N is a noun, V is a verb, Prep is a preposition (e.g., under, in, through, etc.), and PP is a prepositional phrase. The symbols S, NP, VP, and PP are defined by at least one production rule. The remaining symbols would also be defined by one or more production rules. These symbols, however, would always be defined in terms of actual English words. For example:

$$
\begin{array}{l}
Det \text{---}> the \\
Det \text{---}> a \\
Aux \text{---}> is \\
Aux \text{---}> was
\end{array}
$$

Now consider the diagram of an ATN for this grammar presented in Figure 1. Note that each of the symbols representing words is written in lowercase. Each of the remaining symbols is noted in uppercase and corresponds to a portion of the ATN.

Recognition of an input sentence will always start at state S. Execution proceeds by traveling along any arc from the current state by meeting the conditions specified in the arc. In the case of an arc such as the one between state 1 and state 2, the condition(s) consist of reading an auxiliary verb from the input stream and building an internal representation thereof. In practice this would be implemented as a search through the dictionary for the current word to determine whether or not it is an auxiliary verb. If the current word is an auxiliary verb then the condition succeeds and an internal representation of the word is constructed.

In the case of an arc such as the arc from state S to state 1, the condition consists of successfully executing the portion of the ATN starting

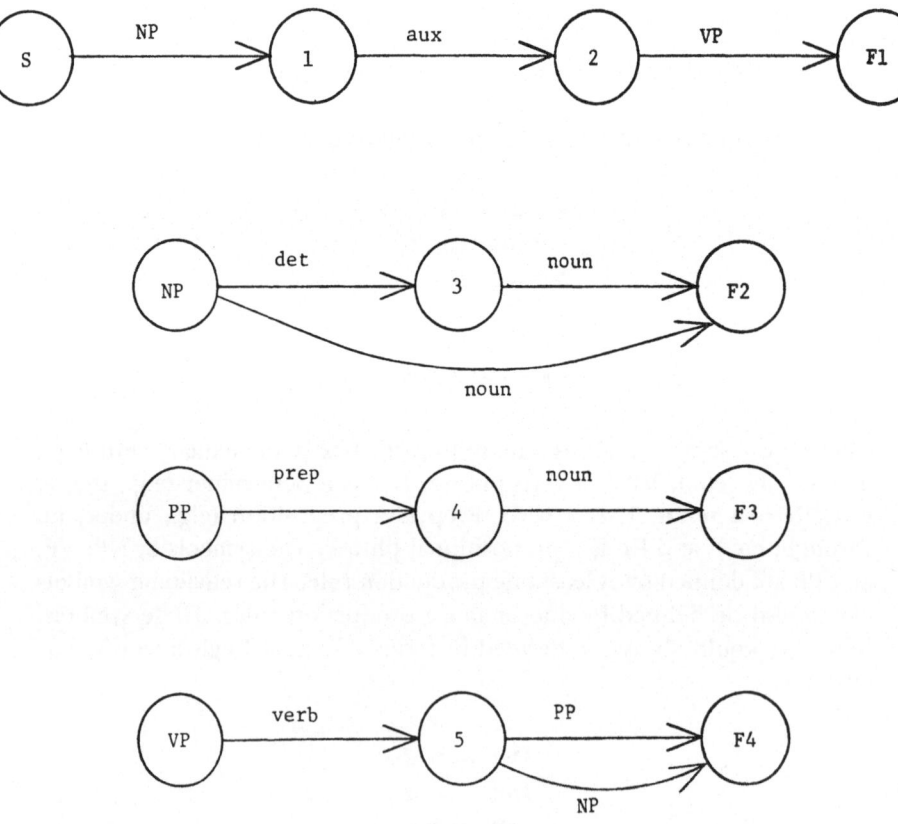

with state NP. If this execution is successful, then control would return to and proceed from state 1. Any structures built during the execution of the NP portion of the ATN would be passed to state 1.

Consider the sentence:

Who is enrolled in CS442?

The recognition of this sentence by the ATN would start at state S. Since the only arc that proceeds from state S is the NP arc, control would transfer to state NP. At this point there is a choice of arcs. If the det arc is tried, the condition will fail since the current word (Who) is not a determiner. The noun arc is then tried and succeeds. The structure noun (Who) is created and passed to state F2. Since this is the last state for this portion of the ATN, control passes back to arc NP (proceeding from state S). This arc succeeds and passes the structure noun (who) to state 1, which uses it to construct

s (noun(Who))

As execution continues through the ATN additional structures will be created and added to the s(. . .) structure until an entire representation of the input sentence has been constructed. The execution of the ATN is successful if the state F1 is reached. The entire execution path for this sentence is given below:

```
S --> 1:
    NP --> 3: fail
    NP --> F2: noun(who)
S --> 1: s (np (noun(who)))
1 --   2: s (np (noun(who),aux(is)))
2 --> F1:
    VP --> 5: verb(enrolled)
    5   --> F4:
        PP --> 4: prep(in)
        4 --> F3:
            NP --> 3: fail
            NP --> F2: noun(CS442)
        4 --> F3: np (noun(CS442))
    5   --> F4: pp(prep(in),np(noun(CS442)))
2 --> F1: s(np(noun(who)),aux(is),vp(verb(enrolled),
            pp(prep(in),np(noun(CS442)))))
```

References

1. L. BLOOMFIELD, Linguistic aspects of science, in *The International Encyclopedia of Unified Science*, O. Neurath, R. Carnap, and C. Morris, Eds., University of Chicago Press, Chicago, 1955, Vol. 1, pp. 1–5.
2. R. F. BOYCE, D. D. CHAMBERLIN, W. F. KING, and M. M. HAMMER, Specifying queries as relational expressions: SQUARE, IBM Research Report No. RJ-1291, IBM Research Laboratory, San Jose, California, October, 1973.
3. D. D. CHAMBERLIN, and R. F. BOYCE, SEQUEL: A structured English query language, *Proceedings of ACM SIG-FIDET Workshop*, Ann Arbor, Michigan, May, 1974.
4. P. P.-S. CHEN, The entity-relationship model—Toward a unified view of data, *ACM Trans. Database Syst.* March, 9–36 (1976).
5. N. CHOMSKY, *Aspects of the Theory of Syntax*, MIT Press, Cambridge, Massachusetts, 1965.
6. J. CLIFFORD, and D. S. WARREN, Formal semantics for time in databases, *ACM Trans. Database Syst.* **8**(2) 214–254 (1983).
7. CODASYL COMMITTEE, Data base task group report, Association for Computing Machinery, 1971.
8. E. F. CODD, Extending the database relational model to capture more meaning, *ACM Trans. Database Syst.* **4**(4), 397–434 (1979).
9. A. COLMERAUER, Metamorphosis grammars, in *Natural Language Communication with Computers*, L. Bolc, Ed., Springer-Verlag, Berlin, 1978.
10. C. J. FILLMORE, The case for case, in *Universals in Linguistic Theory*, Emmon Bach and Robert T. Harms, Eds., Holt, Rinehart and Winston, New York, 1968.
11. L. R. HARRIS, ROBOT: A high performance natural language interface for data base query, Dartmouth College Technical Report No. TR77-1, February, 1977.
12. S. J. KAPLAN, Cooperative responses from a portable natural language data base query system, Ph.D. thesis, Department of Electrical Engineering, University of Pennsylvania, 1979.
13. J. J. KATZ, and J. A. FODOR, The structure of a semantic theory, *Language* **39**, 170–210 (1963).
14. J. D. McCAWLEY, The role of semantics in a grammar, in *Universals in Linguistic Theory*, E. Bach and Robert T. Harms, Eds., Holt, Rinehart, and Winston, 1968.
15. M. C. McCORD, Using slots and modifiers in logic grammars for natural language, *Artif. Intell.* **18**, 327–367 (1982).
16. M. C. McCORD, Slot grammars, *Am. J. Comput. Linguistics* **6**(1), 31–43 (1980).
17. F. C. N. PEREIRA, and D. H. D. WARREN, Definite clause grammars for language analysis—A survey of the formalism and a comparison with augmented transition networks, *Artif. Intell.* **13**, 231–278 (1980).
18. P. R. REISNER, R. F. BOYCE, and D. D. CHAMBERLIN, Human factors evaluation of two data base query languages—SQUARE and SEQUEL, *AFIPS National Computer Conference Proceedings*, 1975, pp. 447–452.
19. J. M. SMITH, and D. C. P. SMITH, Database abstractions: Aggregation and generalization, *ACM Trans. Database Syst.* **2**(2), 105–133 (1977).
20. J. C. THOMAS, and J. D. GOULD, A psychological study of query by example, *AFIPS National Computer Conference Proceedings*, 1975, pp. 439–445.
21. T. WINOGRAD, *Understanding Natural Language*, Academic Press, New York, 1972.
22. W. A. WOODS, Transition network grammars for natural language analysis, *Commun. ACM* **13**(10), 591–606 (1970).
23. M. M. ZLOOF, Query by example, *AFIPS National Computer Conference Proceedings*, 1975, pp. 431–438.

8

DESIGN OF AN EASY-TO-USE QUERY LANGUAGE FOR OFFICE AUTOMATION

KAZUO SUGIHARA, TOHRU KIKUNO, AND NORIYOSHI YOSHIDA

1. Introduction

The increasing extent to which applications of computer systems have been applied has made the design of a user-friendly interface a more critical problem in the development of computer systems. In particular, a database system for office automation must provide a query language that allows users without any skill to access data.

The relational query languages[8] such as SQL, QUEL, and QBE achieve a user-friendly interface in the sense that they support the physical data independence. This independence removes users from the requirement of knowing physical storage structures. However, none of these relational query languages supports logical data independence which would make queries and programs independent of a conceptual database structure.

Several researchers have attacked the logical data independence.[2-4] However, they have concentrated only on the elimination of relation names from a query description. Furthermore, most of them rely on a so-called universal relation assumption.[1,8] Although universal relation systems such as System/U[8] provide good user interfaces, there still exist the following two problems:

1. It fails to take account of all meaningful connections among attributes in a query.

KAZUO SUGIHARA, TOHRU KIKUNO, and NORIYOSHI YOSHIDA • Electronic Circuits and Systems Laboratory, Faculty of Engineering, Hiroshima University, Shitami, Saijo-cho, Higashi-Hiroshima, 724, Japan

2. There is no way for a user to know which answer to his or her query is the one he or she expected.

This chapter attempts to design an easy-to-use query language for office automation. We first propose a semantic data model that is suitable for user interface. It consists of domains, user views, and interdomain connections. The domains are sets of entities. The user views are intuitive relationships among entities. The interdomain connections represent the semantics of a database. Then, we introduce the concept of domain calculus, which enables a database system to provide user views in the semantic data model. Based on the domain calculus, we design a query language that has the following features:

(1) A user can specify a query without knowledge of a conceptual database. Thus, the query language supports the logical data independence that makes queries independent of a conceptual database structure.

(2) The semantics of a query in the query language is given by the interpretation scheme which maps the query to a query in a relational query language. The scheme can provide a distance measure for answers to ambiguous queries, so that a user can choose the answer he expects.

(3) A user can directly retrieve relationships among entities such that the relationships are not in a database, but can be derived from it.

2. Preliminaries

A relation scheme is a finite set of attributes, and a database scheme is a finite collection of relation schemes. Associated with each attribute A is a domain, denoted $dom(A)$, which is a set of entities. Note that different attributes may share the same domain.

A relation r_i for a relation scheme $R_i = \{A_1, A_2, \ldots, A_n\}$ is a finite set of mappings that assign an entity in $dom(A_j)$ to each attribute A_j in R_i. A mapping in the relation r_i is called a tuple of r_i. A database d for a database scheme $D = \{R_1, R_2, \ldots, R_n\}$ is a collection of relations $r_1, r_2, \ldots, r_n$, where r_i is a relation for R_i $(1 \leq i \leq n)$.

Let X and Y be subsets of a relation scheme R_i. If t is a tuple of a relation r_i for R_i, then the X component $t[X]$ of the tuple t is a mapping t' defined on X such that t' assigns $t(A_j)$ for each attribute $A_j \in X$. A functional dependency $X \rightarrow Y$ is the assertion that for all tuples t and s in r_i for R_i, if $t[X] = s[X]$ then $t[Y] = s[Y]$. X is said to be a key of R_i if $X \rightarrow R_i$ holds.

For the sake of notational simplicity, we also regard a tuple t of r_i for $R_i = \{A_1, A_2, \ldots, A_n\}$ as an n-tuple $(e_1, e_2, \ldots, e_n)$, where $e_i = t(A_i)$ $(1 \leq i \leq n)$ unless the order of attributes is significant. Then r_i can be regarded as a finite subset of $dom(A_1) \times dom(A_2) \times \cdots \times dom(A_n)$.

In this chapter, we consider relational expressions in which three operators,[8] selection "σ_F," projection "π_X," and Cartesian product "$\times$," are allowed, and the operands are relation schemes. Let $d = \{r_1, r_2, \ldots, r_n\}$ be a database for a database scheme $D = \{R_1, R_2, \ldots, R_n\}$ and E be a relational expression on D. An evaluation $E(d)$ of E on d is the relation that is obtained by assigning r_i to each R_i ($1 \le i \le n$).

In what follows, a string of uppercase letters stands for an attribute name and a string of lowercase letters surrounded with quotation marks stands for a domain name. An attribute A in a relation scheme R is denoted by $R \cdot A$ when we distinguish the attribute A in R from an attribute A in another relation.

3. User Views in a Semantic Data Model

3.1. What Is a User View?

When a user composes a query, the user usually considers intuitive relationships among entities and wishes to describe the query on his or her own semantic structure of a database. Such a query is what we mean by a "user view." That is, a user view is a query that is described in terms of domains instead of relations and attributes.

A user view provides the logical data independence for users and queries (see Figure 1). Therefore, it is no longer necessary for the user to have explicit knowledge of a conceptual database. Conversely, a database designer can modify the conceptual database without regard to user views.

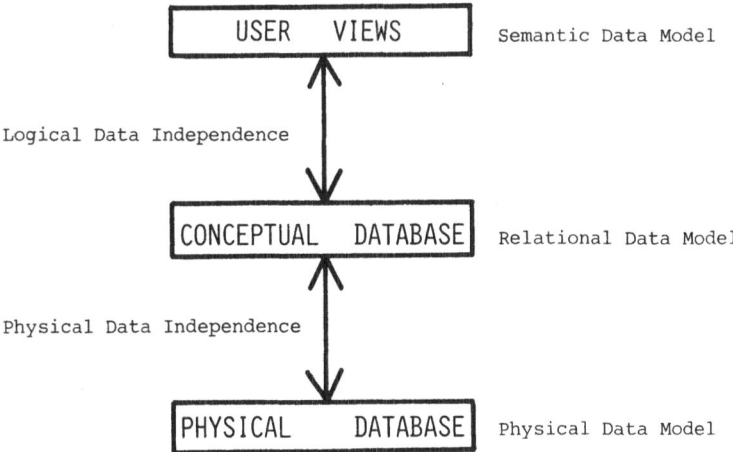

FIGURE 1. Levels of abstraction of a database.

Most previous approaches to enhancement of the logical data independence remove relation names from a query description. However, they require the user to describe queries by attribute names which are specified in a conceptual database. That is, these approaches are still based on the relational data model. Therefore, they support the logical data independence in a limited way.

In this chapter, we develop a semantic data model that is suitable for user interfaces (see details in Section 3.2). Based on the semantic data model, a notion of domain calculus is introduced to allow a user to describe a user view without knowledge of a conceptual database (see details in Section 4.1).

3.2. The Semantic Data Model

A user view is described in the semantic data model which consists of domains, user views, and interdomain connections.

(1) Domains. An object is a thing of interest to us in an application environment. A domain is a collection of objects that can be identified in the domain. Each object in a domain has the semantics which is provided by the domain. Note that a domain D_i is distinguished from a domain D_j which is the same collection of objects as D_i if the semantics of D_i is different from that of D_j.

(2) User Views. An association among domains $D_1, D_2, \ldots, D_n$ is a set $\{ (D_1, d_1), (D_2, d_2), \ldots, (D_n, d_n) \}$, where d_i is an object in the domain D_i $(1 \leq i \leq n)$. The association represents a semantical connection among the objects $d_1, d_2, \ldots, d_n$ whose roles in the association are implicitly defined by the domains $D_1, D_2, \ldots, D_n$. We say the association covers a set $\mathcal{S}$ of domains if $\mathcal{S}$ is a subset of $\{D_1, D_2, \ldots, D_n\}$. A user view over domains $D_1, D_2, \ldots, D_n$ is a set of associations that cover $\{D_1, D_2, \ldots, D_n\}$.

(3) Interdomain Connections. The domains may be semantically related by means of interdomain connections that are derived from such data abstractions[7] as generalization, aggregation, and classification. In this chapter, we treat interdomain connections derived from generalization and joinability. We say a domain D_i is a generalization of a domain D_j, or D_j is a specialization of D_i, if each object in D_j is regarded as an object in D_i at an appropriate level of abstraction. In other words, the semantics of D_i includes the semantics of D_j. Two domains D_i and D_j are said to be joinable if there exists an object that belongs to D_i from one point of view and D_j from another point of view. Note that the interdomain connections should be hidden from users.

Example 1. Let us describe a bank database in the semantic data model. Assume that there are eight domains: "bank," "account,"

"deposit," "loan," "customer," "depositor," "borrower," and "address."
In general, a user can define a user view over an arbitrary set of domains.
For example, user views v_1 over "bank" and "account," v_2 over "loan,"
"borrower" and "address," and v_3 over "bank" and "customer" (see Figure 2).

In Figure 2, circles represent domains and diamonds represent user
views. Arcs between two domains represent interdomain connections
derived from generalization (i.e., a generalization hierarchy). For example,
the domain "customer" is a generalization of both the domains "depositor" and "borrower," and the domain "account" is a generalization of
both the domains "deposit" and "loan." Dotted lines between two
domains represent interdomain connections derived from joinability. For
example, "depositor" and "borrower" are joinable.

3.3. Ambiguous User Views

A user view may cause ambiguity on its semantics for the underlying
conceptual database, since it is described without an explicit correspondence of structures between the user view and the conceptual database.

To show an example of ambiguous user views, we consider the bank

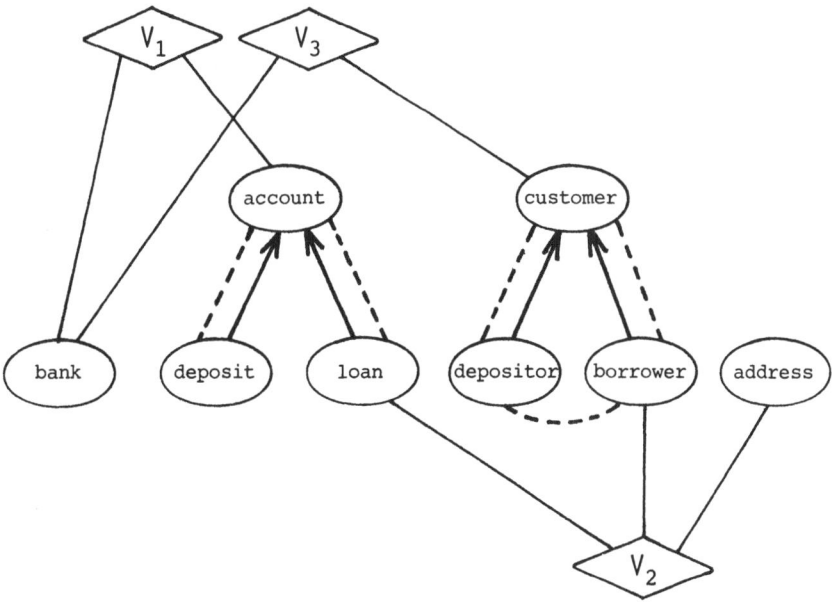

FIGURE 2. A bank database.

database in Example 1. Assume that the underlying conceptual database consists of the following relations: R1(BANK, DEPT), R2(DEPT, DEPR), R3(BANK, LOAN), R4(LOAN, BORR), and R5(CUST, ADDR).

The relation R1 is a list of deposits at each bank and the relation R2 specifies the depositor for each deposit. The relation R3 is a list of loans at each bank and the relation R4 specifies the borrower for each loan. The relation R5 specifies the address for each customer. The domains associated with the attributes BANK, DEPT, LOAN, DEPR, BORR, CUST, and ADDR are "bank," "deposit," "loan," "depositor," "borrower," "customer," and "address," respectively. We assume that there exist interdomain connections as shown in Figure 2. Figure 3 shows a hypergraph representation[8] of structure of the conceptual database.

Example 2. Let us consider the following query Q1 on the user view v_3 over the domains "bank" and "customer" (see Figure 2):

Q1: List all customers at the bank "*B*."

The query Q1 is ambiguous, since there are two connections in Figure 3 between two domains "bank" and "customer": one through the relations R1, R2, and R5, and the other through the relations R3, R4, and R5. Thus, we can consider four kinds of semantics for the query Q1:

"List all customers who have an account in the bank *B*."
"List all customers who have a loan in the bank *B*."
"List all customers who have both an account and a loan in the bank *B*."
"List all customers who have either an account or a loan in the bank *B*."

Most researchers have resolved the ambiguity by making a so-called universal relation assumption.[1,8] However, the assumption does not always hold.[1] Therefore, we will explore an alternative approach for resolving the ambiguity, which does not rely on the universal relation assumption.

A straightforward approach for resolving the ambiguity is to take all possible interpretations for the semantics into account. However, there

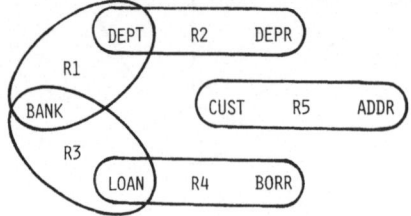

FIGURE 3. Structure of a conceptual database.

are two major disadvantages in the approach. First, it may include an obviously meaningless interpretation. Second, there is no way for a user to know which answer is the one he or she expected.

Osborn has defined meaningful interpretations by means of a lossless join criterion.[4] Maier and Ullman have incorporated multiple interpretations into the universal relation assumption by using maximal objects.[3] While these approaches resolve the first disadvantage mentioned above, the second one still remains. Furthermore, they can give a user no answer or intuitively incorrect answers.[1,3]

To cope with problems in the previous approaches, we introduce a distance between a view and its interpretation (see details in Section 4.2).

4. An Easy-to-Use Query Language

To design an easy-to-use language, we first introduce the concept of domain calculus. Next, the semantics of domain calculus is defined by an interpretation scheme which is a mapping from views to relational expressions. Finally, we present the syntax of an easy-to-use query language that is based on the domain calculus.

4.1. Domain Calculus

Let a formula $\mathcal{F}$ be a conjunction of clauses $D_i\theta d_i$ or $D_i\theta D_j$, where D_i and D_j are domains, d_i is an object in the domain D_i, and $\theta \in \{ =, \neq, <, \leq, >, \geq \}$. Let $\mathcal{X}$ be a set of domains. A view selection, denoted $\Sigma_{\mathcal{F}}$, is the set of all associations such that the association $\{(D_1,d_1), (D_2,d_2), \ldots, (D_n,d_n) \}$ is included in a database and satisfies the formula $\mathcal{F}$. A view projection, denoted $\Pi_{\mathcal{X}}(\Sigma_{\mathcal{F}})$, is the set of all associations such that the association $\{ (D_i,d_i) \mid D_i\in\mathcal{X}\}$ is the subset of an association in the view selection $\Sigma_{\mathcal{F}}$. For notational convenience, an association $\{(D_1,d_1), (D_2,d_2), \ldots, (D_m,d_m)\}$ in a view projection is denoted by $(d_1,d_2,\ldots,d_m)$. We also refer to a view selection or a view projection as a view. A domain D_i is said to be a source domain in $\Sigma_{\mathcal{F}}$ or $\Pi_{\mathcal{X}}(\Sigma_{\mathcal{F}})$ if D_i has an occurrence in $\mathcal{F}$. A domain D_i is said to be a target domain in $\Pi_{\mathcal{X}}(\Sigma_{\mathcal{F}})$ if $D_i\in\mathcal{X}$.

An atom is of either type I or type II.

Type I: $x_i\theta x_j$, where $\theta\in\{=,\neq,<,\leq,\geq,>\}$, and x_i and x_j are domain variables for domains D_i and D_j, respectively.

Type II: $(x_{i1}, x_{i2}, \ldots, x_{im})\in\Pi_{\mathcal{X}}(\Sigma_{\mathcal{F}})$, where $x_{i1},x_{i2},\ldots,x_{im}$ are domain variables for domains $D_{i1}, D_{i2}, \ldots, D_{im} \in\mathcal{X}$, respectively.

A *D*-formula is recursively defined as follows:

1. Every atom is a *D*-formula.
2. If ψ_1 and ψ_2 are *D*-formulas, then $(\psi_1) \wedge (\psi_2)$, $(\psi_1) \vee (\psi_2)$, $\neg (\psi_1)$ are *D*-formulas.
3. If ψ is a *D*-formula, then $^\forall x[\psi]$ and $^\exists x[\psi]$ are *D*-formulas.
4. Nothing else is a *D*-formula.

A domain calculus expression is an expression of the form either $\Sigma_{\mathcal{F}}$ or $\{(x_1, x_2, \ldots, x_n) \mid \psi(x_1, x_2, \ldots, x_n)\}$, where ψ is a *D*-formula in which every domain variable appears in an atom of type II and $x_1, x_2, \ldots, x_n$ are the only free domain variables in ψ.

Example 3. Recall the database in Example 1. The query Q1 is expressed by the following expression V_1 in the domain calculus:

$$V_1 = \{(x_1) \mid (x_1) \in \Pi_{\text{customer}} (\Sigma_{\text{bank}=\text{``}B\text{''}})\}$$

The following expression V_2 represents the query Q2 "List all customers and their addresses such that the customer is one of the bank *B*":

$$V_2 = \{(x_1, x_2) \mid ((x_1) \in \Pi_{\text{customer}} (\Sigma_{\text{bank}=\text{``}B\text{''}})) \wedge ((x_1, x_2) \in \Pi_{\text{customer, address}} (\Sigma))\}$$

The following expression V_3 represents the query Q3 "What relationships are there between the bank *B* and the customer Smith?":

$$V_3 = \Sigma_{\text{bank}=\text{``}B\text{''} \wedge \text{customer}=\text{``Smith''}}$$

4.2. Semantics of the Domain Calculus

4.2.1. Interpretation of Domain Calculus Expressions.
Let *S* be a set of relation schemes $R_1, R_2, \ldots, R_n$. A connection between R_i and R_j in a relational expression $\sigma_F(R_1 \times R_2 \times \cdots \times R_n)$ is the conjunction of all clauses of the form $A = B$ in *F* such that $A \in R_i$ and $B \in R_j$. A relational expression $\sigma_F(R_1 \times R_2 \times \cdots \times R_n)$ is said to be an implicit join of *S* if it satisfies the following conditions 1–3:

1. *F* is a conjunction of clauses of the form $A = B$ such that $A \in R_i$, $B \in R_j$ $(1 \leq i, j \leq n, \ i \neq j)$, and the domains dom(*A*) and dom(*B*) are joinable.
2. For any nonempty subset S' of *S*, there exists a clause $A = B$ in *F* such that $A \in \bigcup_{R_i \in S'} R_i$ and $B \in \bigcup_{R_j \in S - S'} R_j$.
3. For any R_i and R_j in *S* $(i \neq j)$, the connection *C* between R_i and R_j in $\sigma_F(R_1 \times R_2 \times \cdots \times R_n)$ is a maximal conjunction of clauses of the

form $A = B$ such that dom(A) and dom (B) are joinable and every attribute has at most one occurrence in C.

Note that there can exist more than one implicit join of S.

Let $\mathcal{V}$ be the set of all views and $\mathcal{E}$ be the set of all relational expressions. Let v be a view and $\mathcal{D}$ be a set of all source domains and target domains in v. An interpretation scheme I is a mapping from $\mathcal{V}$ to $2^{\mathcal{E}}$. If $v = \Sigma_{\mathcal{J}}$ or $v = \Pi_{\mathcal{X}}(\Sigma_{\mathcal{J}})$, then $I(v)$ is a set of relational expressions of the form $\sigma_F(E)$ or $\pi_X(\sigma_F(E))$, respectively, which satisfy the following conditions C1–C3:

C1: There exists a mapping $\alpha : \mathcal{D} \to Z$ such that each domain $D_i \in \mathcal{D}$ is a generalization of the domain dom($\alpha(D_i)$), where Z is the set of all attributes in a database scheme.

C2: F and X are obtained by substituting $\alpha(D_i)$ for each domain $D_i \in \mathcal{D}$ in $\mathcal{J}$ and $\mathcal{X}$, respectively.

C3: E is an implicit join of S, where S is a minimal set of relation schemes such that $X \subseteq \bigcup_{R_i \in S} R_i$.

We refer to a relational expression $\sigma_F(E)$ or $\pi_X(\sigma_F(E))$ in $I(v)$ as an interpretation of v.

By conditions C1 and C2, interdomain connections derived from generalization are incorporated into the domain calculus. By condition C3 and the concept of implicit joins, interdomain connections derived from joinability are incorporated into the domain calculus.

An interpretation scheme I is naturally extended from views to domain calculus expressions. Let V be a domain calculus expression that includes views $v_1, v_2, \ldots, v_m$. An interpretation Δ of V with respect to I is defined as the expression that is obtained by substituting an interpretation $E_i \in I(v_i)$ for each v_i $(1 \leq i \leq m)$. The evaluation $\Delta(d)$ of Δ on a database d is the relation that is obtained by substituting $E_i(d)$ for v_i $(1 \leq i \leq m)$. An evaluation $I(V)$ (d) of V on d with respect to I is defined as follows:

$$I(V)(d) = \bigcup_{\Delta \in I(V)} \Delta(d)$$

This implies that we intend to take the semantics of a domain calculus expression in the broadest sense. However, the "broadest" sense does not imply to take account of all possible interpretations of the domain calculus expression, since these interpretations may include obviously meaningless ones. The meaningfulness will be defined by the distance between a view and its interpretation (see details in Section 4.2.2).

Example 4. Consider interpretations of the domain calculus expression V_1 for the query Q1 in Example 3. The following relational expressions E_1 and E_2 satisfy C1–C3 for the view $v = \Pi_{customer} (\Sigma_{bank} = {}^{\cdot \cdot}B'')$.

For the mapping α_1 such that $\alpha_1(\text{bank}) = \text{BANK}$ and $\alpha_1(\text{customer}) = \text{DEPT}$,

$$E_1 = \pi_{R2 \cdot DEPR}(\sigma_{R1 \cdot BANK = {}^{\cdot \cdot}B''}(\sigma_{R1 \cdot DEPT = R2 \cdot DEPT}(R1 \times R2))).$$

For the mapping α_2 such that $\alpha_2(\text{bank}) = \text{BANK}$ and $\alpha_2(\text{customer}) = \text{BORR}$,

$$E_2 = \pi_{R4 \ BORR}(\sigma_{R3 \ BANK} = {}^{\cdot \cdot}B''(\sigma_{R3 \ LOAN = R4 \ LOAN}(R3 \times R4))).$$

Let I be an interpretation scheme such that $I(v) = \{E_1, E_2\}$. Then the evaluation $I(V_1)(d)$ on the database d shown in Figure 4 is the following:

$$
\begin{aligned}
I(v)(d) &= E_1(d) \cup E_2(d) \\
&= \{\text{Smith, Peterson}\} \cup \{\text{Smith, Johnson}\} \\
&= \{\text{Smith, Peterson, Johnson}\}
\end{aligned}
$$

Note that it corresponds to the fourth semantics for Q1 mentioned in Example 2.

4.2.2. Distance Between a View and Its Interpretation. An attribute A in a relation scheme R_i is said to be a pseudosource under an interpretation $\pi_X(\sigma_F(E))$ or $\sigma_F(E)$ if at least one of the following conditions C4–C6 holds:

C4: F includes a clause $A = c$ for some constant c.

C5: F includes a clause $A = B$ where $B \in R_j (i \neq j)$ and B is a pseudosource under the interpretation.

C6: There is a set Y of attributes such that all attributes in Y are pseudosources under the interpretation and a functional dependency $Y \rightarrow A$ holds.

A connection C between R_i and R_j is said to be lossless if any pair of relations r_i for R_i and r_j for R_j satisfies $r_i = \pi_{R_i} (\sigma_C (r_i \times r_j))$ and $r_j = \pi_{R_j} (\sigma_C(r_i \times r_j))$. A connection C between R_i and R_j in an interpretation is said to be strong if at least one of the following conditions C7 and C8 holds, otherwise weak.

C7: Every attribute in C is a pseudosource under the interpretation, and the set of all attributes $A \in R_k$ in C is a key of R_k for either $k = i$ or $k = j$.

C8: C is lossless and there is no relation scheme R_k in a database scheme such that $R_i \cup R_j \subseteq R_k$.

Let I be an interpretation scheme. Let v be any view and $E \in I(v)$ be its interpretation. The distance dist(v, E) between v and E is defined as the

r_1

BANK	DEPT
A	123
B	234
B	456
C	789

r_2

DEPT	DEPR
123	Jones
234	Smith
456	Peterson
789	Jones

r_3

BANK	LOAN
B	987
B	654
C	321

r_4

LOAN	BORR
987	Smith
654	Johnson
321	Hall

r_5

CUST	ADDR
Hall	Chicago
Johnson	New York
Jones	Los Angeles
Peterson	New York
Smith	Hiroshima

FIGURE 4. Database in Example 4.

number of weak connections in E. For two interpretations E_1 and E_2 in $I(v)$, an evaluation of E_1 may bring out more "meaningful" and "desirable" answers than that of E_2 if $dist(v,E_1) < dist(v,E_2)$. In other words, the distance $dist(v,E)$ can be regarded as a measure on meaningfulness of the interpretation E for the view v.

Example 5. Recall the database in Example 1 and the conceptual database shown in Fig. 3. Assume that the connection R1.DEPT = R2.DEPT between R1 and R2 and the connection R3.LOAN = R4.LOAN between R3 and R4 are lossless, and the following functional dependencies hold: DEPT→BANK, DEPT→DEPR, LOAN→BANK, LOAN→BORR and CUST→ADDR.

Let us consider the view $v = \Pi_{\text{customer, address}} (\Sigma_{\text{loan}=\text{"987"}})$. There are five interpretations of v as follows: For the mapping α_1 such that α_1 (customer) = CUST, α_1(address) = ADDR and α_1(loan) = LOAN,

$$E_1 = \pi_{\text{R5 CUST, R5 ADDR}} \left(\sigma_{\text{R4.LOAN} = \text{"987"}} \left(\sigma_{\text{R4 BORR} = \text{R5 CUST}}(R4 \times R5) \right) \right)$$

and

$$E_2 = \pi_{\text{R5 CUST, R5.ADDR}} \left(\sigma_{\text{R3 LOAN} = \text{"987"}} \left(\sigma_{\text{R1 BANK} = \text{R3 BANK} \wedge \text{R1.DEPT} = \text{R2.DEPT} \wedge \text{R2.DEPR} = \text{R5 CUST}} (R1 \times R2 \times R3 \times R5) \right) \right).$$

For the mapping α_2 such that α_2(customer) = DEPR, α_2(address) = ADDR and α_2(loan) = LOAN,

$$E_3 = \pi_{\text{R2.DEPR, R5 ADDR}} \left(\sigma_{\text{R4 LOAN} = \text{"987"}} \left(\sigma_{\text{R2.DEPR} = \text{R4 BORR} \wedge \text{R4.BORR} = \text{R5 CUST}}(R2 \times R4 \times R5) \right) \right)$$

and

$$E_4 = \pi_{\text{R2.DEPR, R5.ADDR}} \left(\sigma_{\text{R3 LOAN} = \text{"987"}} \left(\sigma_{\text{R1.BANK} = \text{R3.BANK} \wedge \text{R1 DEPT} = \text{R2.DEPT} \wedge \text{R2.DEPR} = \text{R5.CUST}} (R1 \times R2 \times R3 \times R5) \right) \right).$$

For the mapping α_3 such that α_3(customer) = BORR, α_3(address) = ADDR and α_3(loan) = LOAN,

$$E_5 = \pi_{\text{R4 BORR, R5.ADDR}} \left(\sigma_{\text{R4 LOAN} = \text{"987"}} \left(\sigma_{\text{R4 BORR} = \text{R5.CUST}} (R4 \times R5) \right) \right).$$

In the interpretation E_1, there is only the connection R4.BORR = R5.CUST between R4 and R5. By condition C4, R4.LOAN is a pseudo-source under E_1. By condition C6 and the functional dependency LOAN$\rightarrow$BORR, R4.BORR is a pseudosource under E_1. Then, by condition C5, R5.CUST is a pseudosource under E_1. Thus, by condition C7, the connection R4.BORR = R5.CUST is strong in E_1. Therefore, dist(v, E_1) = 0.

In the interpretation E_2, there are three connections: R1.BANK = R3.BANK between R1 and R3, R1.DEPT = R2.DEPT between R1 and R2, and R2.DEPR = R5.CUST between R2 and R5. The second connection is strong in E_2, since it is lossless. Other connections are weak in E_2. Therefore, $\mathrm{dist}(v,E_2) = 2$.

In the interpretation E_3, there are two connections: R2.DEPR = R4.BORR between R2 and R4, and R4.BORR = R5.CUST between R4 and R5. The connection between R4 and R5 is strong in E_3, since R4.BORR and R5.CUST are pseudosources under E_3 and R5.CUST is a key of R5. The connection between R2 and R4 is weak in E_3. Therefore, $\mathrm{dist}(v,E_3) = 1$.

In the interpretation E_4, there are three connections. Only the connection R1.DEPT = R2.DEPT between R1 and R2 is strong in E_4. Therefore, $\mathrm{dist}(v,E_4) = 2$.

In the interpretation E_5, there is only the connection R4.BORR = R5.CUST between R4 and R5. It is strong in E_5, since R4.BORR and R5.CUST are pseudosources under E_5 and R5.CUST is a key of R5. Therefore, $\mathrm{dist}(v,E_5) = 0$.

4.2.3. Interpretation Scheme. To obtain meaningful interpretations of a view, we use the distance between the view and its interpretation in two ways. The first one is to remove meaningless interpretations by applying a threshold to their distances. The second one is to define an ordering of interpretations, which provides a user with a measure on the meaningfulness of interpretations.

Let v be a view and $\mathcal{E}(v)$ be a set of relational expressions that satisfy C1–C3 in Section 4.2.1. Assume that an integer δ is a threshold on the distance $\mathrm{dist}(v,E)$. It is natural to consider the interpretation scheme I^* such that $I^*(v) = \{E_i \in \mathcal{E}(v) \mid \mathrm{dist}(v,E_i) \leq \min\{\mathrm{dist}(v,E_j) \mid E_j \in \mathcal{E}(v)\} + \delta\}$. The integer δ represents a threshold on the meaningfulness of interpretations.

Example 6. Recall the database in Example 5 and let a threshold $\delta = 1$. Let us consider the domain calculus expression $V = \{(x_1, x_2) \in \Pi_{\text{customer, address}}(\Sigma_{\text{loan}=``987"})\}$. As mentioned in Example 5, there are five possible interpretations $E_1$–$E_5$ of the view $v = \Pi_{\text{customer, address}}(\Sigma_{\text{loan}=``987"})$. The minimum distance between v and these interpretations is equal to $\mathrm{dist}(v,E_1) = 0$ (see Example 5). Thus,

$$I^*(V) = \{E_i \in \mathcal{E}(v) \mid \mathrm{dist}(v,E_i) \leq 1\} = \{E_1, E_3, E_5\}$$

Since it is obvious that $E_1(d) = E_5(d)$ and $E_3(d) \subseteq E_1(d)$ for any database, the evaluation $I^*(V)(d)$ of V is the following:

$$I^*(V)(d) = E_1(d)$$

Note that this evaluation is intuitively correct.

Next, we define an ordering of interpretations. Assume that a domain calculus expression V includes m views $v_1, v_2, \ldots, v_m$. Let Δ be an interpretation of V with respect to I^* such that Δ is obtained by substituting $E_i \in I^*(v_i)$ for v_i ($1 \leq i \leq m$). We will define an integer $\mu(\Delta)$ as follows:

$$\mu(\Delta) = \sum_{i=1}^{m} \text{dist}(v_i, E_i)$$

By using $\mu(\Delta)$, we can define a measure of the meaningfulness of interpretations. That is, we will consider that an evaluation of Δ is more meaningful than that of Δ' if $\mu(\Delta) < \mu(\Delta')$. Therefore, a user can know which evaluation includes the more desirable answer.

4.3. Syntax of a Domain Calculus Language

We will present the syntax of a query language based on the domain calculus by showing examples. We assume that the reader is familiar with QBE,[9] since the syntax of our query language is similar to that of QBE. Figure 5 shows a table skeleton. Unlike QBE, domain names will be filled in the first row of the table skeleton. Each row except the first one is used to express a view.

Figure 6 shows query descriptions for Q1, Q2, and Q3 in Examples 2 and 3. The descriptions in Figures 6a, 6b, and 6c correspond to the domain calculus expressions V_1, V_2, and V_3 in Example 3, respectively. In Figure 6a and 6b, the entry "P." indicates to output objects that satisfy the conditions specified in rows except the first one. In Figure 6c, the entry "P.X" in the leftmost column on the first row indicates to output associations that satisfy the specified conditions.

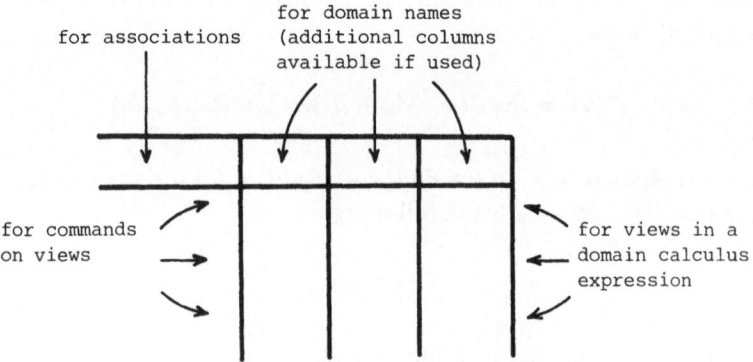

FIGURE 5. Table skeleton.

	bank	customer
	B	P. X̱

(a) Q1

	bank	customer	address
	B	X̱	
		P. X̱	P. Y̱

(b) Q2

P. X̱	bank	customer
	B	Smith

FIGURE 6. Query descriptions for Q1,   (c) Q3
Q2, and Q3.

4.4. Enrichment of Interdomain Connections

The concept of a default specialization is useful to provide a sophisticated interpretation scheme. For example, the domain "account" may imply "deposit" rather than "loan." In such a case, the domain "deposit" is said to be a default specialization of "account." Then, it is more meaningful to consider a mapping α_1 as α in the condition C1 than a mapping α_2, where dom $(\alpha_1(\text{account})) = $ deposit and dom$(\alpha_2(\text{account})) = $ loan.

The incorporation of aggregation into the domain calculus enhances ease of use of the domain calculus language. To do so, we need to modify the definition of a mapping $\alpha: \mathcal{D} \rightarrow Z$ in the condition C1 (see Section 4.2.1) as follows:

C1: There exists a mapping $\alpha: \mathcal{D} \rightarrow 2^Z$ such that for each domain $D_i \in \mathcal{D}$ {dom(A) | $A \in \alpha(D_i)$} is a minimal set that covers D_i, where a domain D_i is said to be covered by a set $\mathcal{S}$ of domains if one of the following conditions 1–4 holds:
1. $D_i \in \mathcal{S}$.
2. D_i is a generalization of a domain covered by $\mathcal{S}$.
3. D_i is an aggregation of domains covered by $\mathcal{S}$.
4. An aggregation of domains which include D_i is covered by $\mathcal{S}$.

In addition, we need to define a transformation from instances for D_i to instances for $\mathcal{S}$ that covers D_i.[6]

5. Conclusion

In this chapter we have proposed a semantic data model that is suitable for user interfaces, and we have introduced the concept of domain calculus which enables a database system to provide user views in the semantic data model. Based on the domain calculus, we have designed an easy-to-use query language that allows a user to specify a query without knowledge of a conceptual database. The semantics of the query language is given by using interdomain connections in the data model which represent the semantic structure of a database. Furthermore, we have presented a distance measure for answers to an ambiguous query so that a user can choose the answer he expects. We believe that this navigates the user to intuitively correct answers for most of queries while there may be an anomalous query.

We are implementing the query translator which translates a query in the domain calculus into a query in QBE,[9] according to the interpretation scheme in Section 4.2.3. The query translated into QBE is processed by the relational database system PRIM[5] which has been developed as a part of LIMS (Laboratory Information Management System).

References

1. P. ATZENI and D. S. PARKER, JR., Assumptions in relational database theory, *Proceedings of the ACM Symposium on Principles of Database Systems*, 1982, pp. 1–9.
2. D. MAIER and D. S. WARREN, Specifying connections for a universal relation scheme database, *Proceedings of the ACM SIGMOD International Conference on Management of Data*, 1982, pp. 1–7.
3. D. MAIER and J. D. ULLMAN, Maximal objects and the semantics of universal relation databases, *ACM Trans. Database Syst.* **8**(1), 1–14 (1983).
4. S. L. OSBORN, Towards a universal relation interface, *Proceedings of the 5th VLDB*, 1979, pp. 52–60.
5. K. SUGIHARA, J. MIYAO, T. KIKUNO, and N. YOSHIDA, A semantic approach to usability in relational database systems, *Proceedings of the International Conference on Data Engineering*, 1984, pp. 203–210.
6. K. SUGIHARA, J. MIYAO, T. KIKUNO, and N. YOSHIDA, Formal specification of data formats for user-friendly interfaces, *Proceedings of the 1984 IEEE Workshop on Languages for Automation*, 1984, pp. 209–214.
7. D. C. TSICHRITZIS and F. H. LOCHOVSKY, *Data Models*, Prentice-Hall, Englewood Cliffs, New Jersey, 1982.
8. J. D. ULLMAN, *Principles of Database Systems*, Second Ed., Computer Science Press, Rockville, Maryland, 1982.
9. M. M. ZLOOF, Query-by-example: A database language, *IBM Syst. J.* **16**(4), 324–343 (1977).

DATA MANAGEMENT

THE STRUCTURE OF ABSTRACT DOCUMENT OBJECTS

Gary D. Kimura and Alan C. Shaw

1. Introduction

A current research problem of great intellectual and commerical interest is the construction of interactive document processing systems that can easily handle the large variety of objects that typically appear in electronic and paper documents. These objects can be divided roughly into four classes: textual, tabular, mathematical, and pictorial. It has proven difficult to gracefully integrate the treatment of these different classes so that dissimilar objects may be used together, for example, mathematics within pictures, pictures within text, or text within mathematics inside a table. Ideally, there should be a common underlying structural model, similar editing and specification languages, and a uniform way to store, traverse, and display all objects.

Systems do exist that provide some partial forms of integration.[1] Perhaps the most comprehensive of these is the UNIX† package,[2] which has preprocessors and macros for all four classes of objects; however, the system is batch-oriented, has different languages for each class, and handles structure only through macros. Modern interactive systems such as the Xerox STAR,[3-5] permit the creation, editing, and display of a limited number of objects from all classes, but have no facilities for grouping objects into larger structures or for creating new object types easily. A major problem is that none of these systems is based on any comprehensive model describing the structure and relations among their constituent doc-

---

†UNIX is a Trademark of Bell Laboratories.

GARY D. KIMURA and ALAN C. SHAW • Department of Computer Science, University of Washington, Seattle, Washington 98195

ument objects. The purpose of this chapter is to present and justify our structuring methods and model.

Some of our initial ideas were reported earlier.[6,3] The emphasis is on the logical or abstract components of documents as distinct from their concrete (e.g., two-dimensional) realizations. (TEX's model of nested boxes connected by stretchable glue[7] applies to concrete document objects.) The principal concepts are the notion of abstract objects, the hierarchical composition of objects, sharing of components, reference links connecting related elements, and ordered and unordered objects (i.e., sequences and sets, respectively). The simplest parts of this model have been used for many years in such areas as document preparation systems, file systems, and database management systems. For example, XS-1[8] and PEN[9] are both based on a pure tree structure, NLS[10] uses tree-structured composition and reference links, and HES[11] allows a general graph of links; a recent paper describes a system with a tree representation for both illustrations and text.[12] Our addition of shared objects for document systems and our combination of ordered and unordered sets† seem to be new and useful ideas.

In the next section, we define abstract objects within the context of a general document processing model that is interesting in its own right. Section 3 describes our basic hierarchical composition methods and the use of ordered and unordered elements. The model is extended with shared objects in Section 4 and with reference links in Section 5. The final section describes some of our current work and lists a number of interesting research problems. Throughout, many examples are presented, taken primarily from the content and structure of this chapter itself.

2. A Document Processing Model

The model is an elaboration and extension of the one introduced in previous papers.[6,3] A document is defined as a hierarchy of *objects*, each of which is an *instance* of some *class* that specifies the possible constituents and other attributes of the instances. For example, this chapter could include instances of the following classes: techreport, header, section, bibliography, paragraph, string, itemized list, figure, circle, graph, table, and formula.

Abstract objects are the logical entities comprising a document, such as those listed above. An abstract object can be identified by a [classname,

---

†The benefits of using both ordered and unordered objects have also been recognized recently in a draft of a proposed ANSI standard for document structure descriptions.[13]

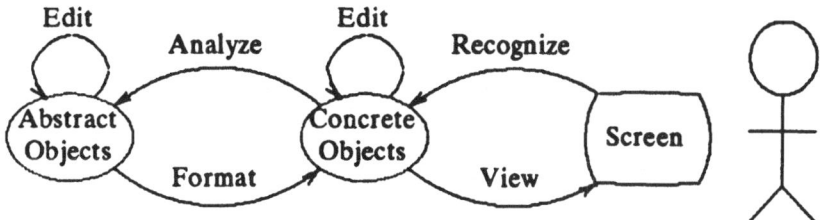

FIGURE 1. Document processing activities.

instancename] pair and may also have some data attributes and structure associated with it. This section could be denoted, for example, as [section, S2] or perhaps [section, DocModel]. The external representations are called *concrete objects*. These are two-dimensional formatted images which may be specified mathematically as picture functions from coordinates to intensities or as arrays of pixels.† Corresponding to each abstract object, there may be one or more concrete objects. A string of text, for example, could have several corresponding concrete strings with different fonts and other concrete attributes, such as line positions, spacing, and hyphenation.

Using the notions of abstract and concrete objects, we define document processing as the activities illustrated in Figure 1. *Editing* operations are mappings from either abstract to abstract objects or concrete to concrete objects. The first case, where a user (system or human) specifies commands for, say, creating, inserting, deleting, and copying abstract objects, is the most common. For editing concrete objects, one deals directly with two-dimensional images and their layout. *Formatting* is the process of transforming abstract objects to concrete ones; thus one speaks of formatting a paragraph, table, figure, or equation when the concrete form of the component is desired. The final *viewing* transformation is concerned with the output of concrete objects. This is a machine-dependent mapping which isolates the properties of particular output devices, such as video displays, typesetters, and printers. Our emphasis here, as shown in the figure, is on video.

For completeness we have also defined the inverse operations, recognition and analysis. *Recognition* is the determination or classification of a concrete object from its screen image; often, this is a low-level process in a terminal, such as recognizing a particular character at the cursor or mouse position. A concrete object is *analyzed* to produce its corresponding abstract object; the nontrivial case of this occurs after direct editing of a concrete object, such as erasing some part of the image.

---

†We will assume two-dimensional images but other possibilities for concrete objects include three-dimensional images, animated pictures, and sound.

This model of processing has proven useful for classifying and discussing various aspects of document preparation systems.[3] It has also been used as a basis for organizing algorithms and data structures in a general abstract object editor.[14] The model can handle any of the object types that occur in documents and may be more generally applicable to other kinds of objects that appear in interactive interfaces. This paper emphasizes abstract objects but within the framework of our processing model.

3. Basic Structures

3.1. Hierarchies of Objects

An abstract document object has a *name* (*N*), some *data* (*D*), and a *composition* (*C*). *N* is the [class, instance] identification mentioned in the last section. *D* contains nonstructural information describing the semantics of the object and is written as a quoted string, figure, or mathematical entity. The composition specifies the composite objects of *N*; if *N* is an atomic object, than *C* is empty. *N* is called the *parent* of any of the elements of *C*. We will use the notation

$$N\,(D)\,=\,C$$

to denote our objects.

Examples 1.

1. The top-level structure of a technical report for this chapter could be given by the description

[TechReport, TR#83-09-02] ("The Structure . . . Objects") =
    [Header, H] [Section, S1] [Section, S2] [Section, S3] [Section, S4]
    [Section, S5] [Section, S6] [Acknowledgment, A]
    [Bibliography, B]

The document object has the name [TechReport, TR#83-09-02], data equal to "The Structure of Abstract Document Objects," and composition consisting of a header (H), six sections (S1, . . . , S6), an acknowledgment (A), and a bibliography (B).

2. One of the authors, an atomic object, could have N and D components:

$$[\text{Authorname,K}](\text{``Gary D. Kimura''})$$

3. Each ellipse in Figure 1 could have the obvious image as its data:

$$[\text{Ellipse, E1}](\text{``}\bigcirc\text{''})$$

4. The stick figure at the right in Fig. 1 could be decomposed into three parts: a head (circle), body and arms (cross), and legs (caret):

$$[\text{Graphic, Person}](\ ) = [\text{Circle, C}], [\text{Lines, Cross}], [\text{Lines, Caret}]$$

Where there is no ambiguity or chance of confusion, we will occasionally use class and/or instance names only instead of complete pairs for simplicity. The first example above can then be abbreviated

$$\text{TechReport} = \text{H S1 S2 S3 S4 S5 S6 A B}$$

3.2. Sequences and Sets

Conventional paper documents are decomposed naturally into structures of *ordered* objects (i.e., *sequences*). Thus the body of an article could consist of a sequence of sections, each of which contains a sequence of paragraphs which, in turn, are each comprised of sequences of atomic text blocks (sentences, enumerated items, strings, etc.). The expression

$$N(D) = N_1N_2 \cdot \cdot \cdot N_n,$$

where the N_i are object names, will mean that the C component of N is the sequence of objects $N_1N_2 \ldots N_n$. In Example 1.1 above, the TechReport object consists of a header followed by six section objects followed by an acknowledgment and a bibliography.

With the ordering relation, objects can be easily and implicitly selected by their absolute or relative position using designators such as first, last, predecessor, successor, and nth. Similarly, one can use sequence constructor, selector, and inquiry operations to concatenate sequences, insert and delete elements and subsequences, and ask questions about their contents.

In order to handle objects that either are unrelated (except by common parenthood) or have more complex relations than simple sequencing, we also permit C to be *unordered* objects (i.e., a *set*). The notation:

$$N(D) = N_1, N_2, \ldots, N_n$$

indicates that the C component of N consists of the (unordered) set $\{N_1, N_2, \ldots, N_n\}$.

Examples 2.

1. We might use

$$[\text{Authors, A}]( ) = [\text{Authorname, K}](\text{``Gary D. Kimura''}),$$
$$[\text{Authorname, S}](\text{``Alan C. Shaw''})$$

to show the equally shared authorship of this chapter.

2. In the mathematical domain, the fraction el/e2, where el and e2 denote expressions, could be given as

$$[\text{Fraction, F}]( ) = [\text{Numerator, el}], [\text{Denominator, e2}]$$

3. Example 1.4 above decomposes the person of Figure 1 into three unordered elements: a circle, a cross, and a caret.

Obvious manipulations on these sets are: the standard union, intersection, and complement operations; insertion and deletion of elements; enumeration of members; and membership and subset queries.

Often, document parts are composed of *mixtures* of ordered and unordered objects. Such mixtures are specified using parentheses for grouping in a straightforward way; for example:

$$N(D) = N_1(N_2, N_3, N_4) N_5$$

and

$$N(D) = N_1, (N_2 N_3 N_4), N_5$$

represent a set within a sequence and a sequence within a set, respectively. Arbitrarily deep nestings are also permitted. Another way of representing this document (see Example 1.1 above), which reflects the nonsequential nature of the header, acknowledgment, and bibliography relative to the rest of the paper, is

$$\text{TechReport} = \text{H, (S1 S2 S3 S4 S5 S6), A, B}$$

It is often convenient and clearer to represent object structures pictorially in a graphlike form resembling a tree. Sets are drawn as unadorned composition trees while sequences have an additional directed arc across the edges (Figures 2a, 2b). Dummy nodes (*d*) are inserted when sequences and sets are mixed (Figures 2c,2d).

Examples 3.

1. The header part of the technical report for this document has the structure (Figure 3)

$$[Header, H]( ) = [Title, T], [Authors, A], [Techrep, TR], [Date, D],$$
$$[Abstract, AB]$$

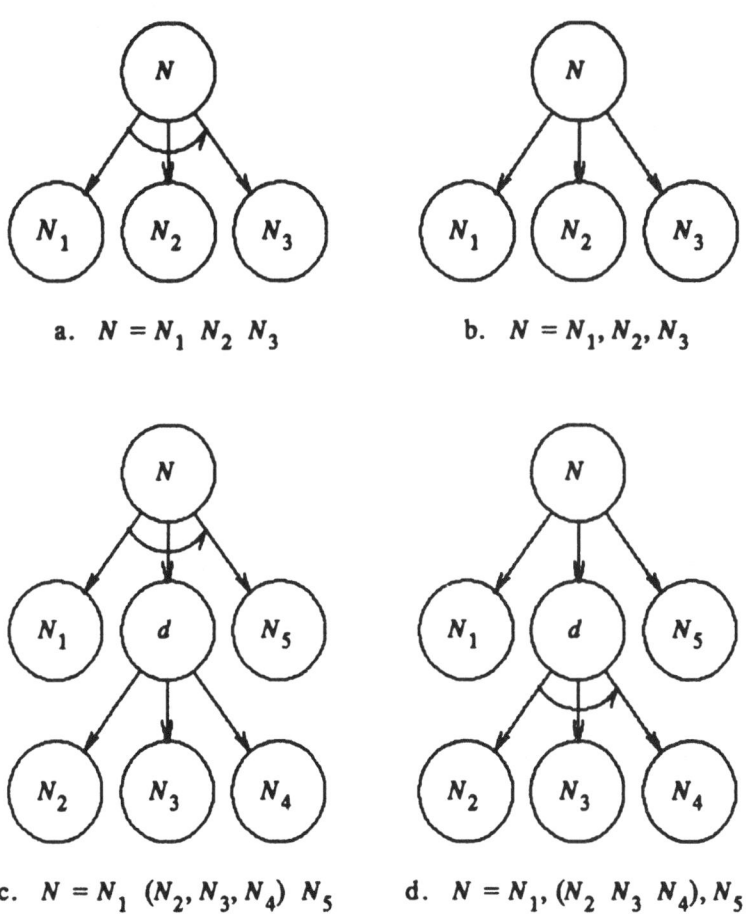

a. $N = N_1 \, N_2 \, N_3$ b. $N = N_1, N_2, N_3$

c. $N = N_1 \, (N_2, N_3, N_4) \, N_5$ d. $N = N_1, (N_2 \, N_3 \, N_4), N_5$

FIGURE 2. Graph representations of basic structures.

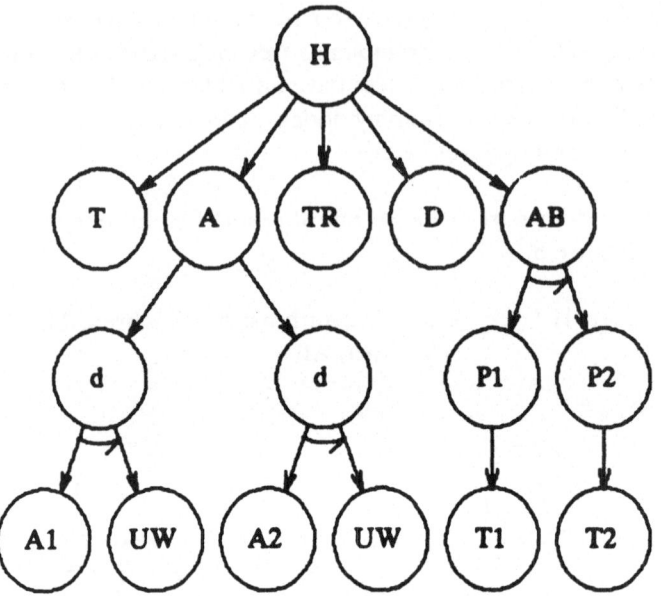

FIGURE 3. Header structure of this chapter.

[Title, T]("The Structure · · · Objects")
[Authors, A]( ) = ([Authorname, K] [Affiliation, UW]),
                        ([Authorname, S] [Affiliation, UW])
[Techrep, TR]("Technical Report No. 83-09-02")
[Date, D]("September 1983")
[Abstract, AB]("ABSTRACT") = [Paragraph, P1] [Paragraph, P2]
[Paragraph, P1]( ) = [Textblock, T1]
[Paragraph, P2]( ) = [Textblock, T2]
[Textblock, T1]("Underlying every · · · structure.")
[Textblock, T2]("The principal concepts · · · system.")

2. One of the authors wrote a simple program in Modula-2[15] to print
a text file on hard copy. It has the structure

PrintFileProgram = MainModule, ImportedModules
ImportedModules = Terminal, FileNames, Line, FileSystem
MainModule = ImportList VariableDeclarations StatementBody

The four ImportedModules are defined as unordered since they are independent and, consequently, have no sequencing constraints. The objects
of the MainModule are shown as a sequence since, for example, a variable
may be declared to be a new type that appears in the ImportList.

3.3. Discussion

The principal application of our structural description scheme is to specify documents; the specifications are meant for human-to-human communications, human-to-machine descriptions, and as a guide for computer system design. We have tested the human communications aspect with many examples and have also used the model as a description and design tool for a prototype document processing system.[14]

The composition of objects into sets and sequences leads to natural methods for storing, formatting and displaying, traveling through and searching, and editing documents. Table 1 illustrates these ideas. Object sequences can be stored and accessed much like a sequential file while unordered sets can be stored, accessed, and manipulated both associatively and with set operations. There is usually a standard geometric ordering of concrete objects corresponding to a sequence of abstract objects; for example, in the simple case of linear text, we have a concrete ordering based on position within a line (x within y) within a page. Sequential objects can then usually be formatted using the schema

$$\text{format\_sequence}(N_1\ N_2\ \cdots\ N_n)$$
$$= \text{format } (N_1); \text{format } (N_2); \ldots ; \text{format } (N_n)$$

where format (N_i) is a function that formats the object N_i (i.e., the objects are formatted in sequence). Unordered objects potentially require more general interactions for formatting, since the parts may appear in separate windows or may involve intricate geometric relations. We can represent this case by the schema

$$\text{format\_set } (N_1, N_2, \ldots , N_n)$$
$$= \text{format } (N_1) \parallel \text{format } (N_2) \parallel \ldots \parallel \text{format } (N_n)$$

where the symbol $\parallel$ indicates parallelism or interleaving with potential communications between objects.

TABLE 1
Applications of Basic Description Method

	Sequence	Unordered set
Storage	sequential	associative
Formatting and display	sequential	parallel or interleaved
Traveling and searching	sequence operations	set operations
Editing	sequence editing	set editing

There are two extremes of granularity for objects. At the macro end of the spectrum, there are the "large" document objects for which our structural methods are most clearly applicable; these include sections, figures, equations, and tables. At the other extreme, the micro level, are "small" objects such as parts of a figure (e.g., line segments), words or characters of text, entries in tables, and elements of formulas. Current research is concerned with determining the best granularity and choices for primitive objects.

The basic structuring scheme treats a document as a tree of ordered and unordered objects. This tree is certainly sufficient for many applications, but it does have some limitations that are especially apparent when describing electronic documents. In particular, the scheme fails to handle explicitly the case where the same object is used at several places and it has no mechanism for treating "cross references" between objects. The next two sections describe two straightforward but interesting extensions, shared objects and links, that remove these limitations without adding undue complexity.

4. Shared Objects

The child of an object is interpreted as being part of its parent. A *shared object* is one that has more than one parent and is part of each of its parents. In the corresponding graph, a shared object has an incoming edge for each of its parents, making the graph resemble a directed acyclic graph (DAG). We have discovered two major applications of shared objects.

The first and simplest occurs when a single instance of an abstract object appears in several places within a document. For example, it is common for mathematical formulas or equations to appear in different places in the same document. Without sharing, a new but identical instance of the equation's abstract object must be inserted at each use. Sharing, in principle, uses only one copy, which helps maintain consistency and can potentially save space. Having only one abstract instance of an object guarantees that regardless of the number of times it is used and modified within a document each occurrence is consistent with the others.

A second and more complex application of sharing is for the description of document parts that have a tabular or arraylike organization. For example, the internal structure of a table is easier to specify by sharing entries among rows and columns. Without sharing, a table is usually organized in either row major or column major order. With sharing, it need not rely on either row or column major order (see Example 4.3).

An object is shared if its name appears within the composition speci-

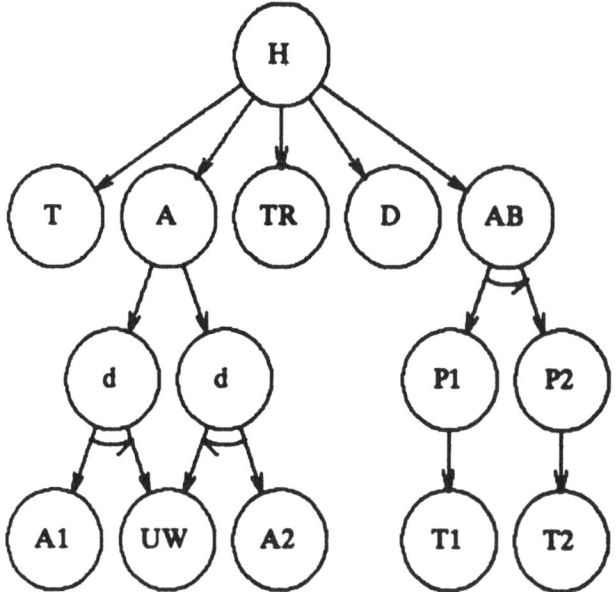

FIGURE 4. Header structure of this chapter with sharing.

fications more than once.† In Example 3.1, the [Affiliation, UW] object appears twice in the composition of [Authors, A]. In the graph, the [Affiliation, UW] object can be represented by a single node (cf. Figs. 3 and 4).

Examples 4.

1. Example 3.1 of this document is used by Section 3 and Section 4. The example appears only in Section 3, but is also referenced here and in the preceding paragraph. This is standard practice for paper documents. Logically the example is shared. In an electronic document, the text of the example could appear several times. (Each occurrence could have a different concrete representation.)

2. Some of the formulas used in Section 3 appear in the text and in Figure 2. For example, "$N(D = N_1(N_2, N_3, N_4)N_5$" appears twice:

[Section, S3]( ) = · · · [Formula, FORM1] · · · [Figure, FIG2]
[Figure, FIG2]( ) = · · · [Formula, FORM1] · · ·
[Formula, FORM1]("$N(D) = N_1(N_2, N_3, N_4)N_5$")

---

†Note that this extension changes our sets to multisets or bags, since formally sets do not permit repeated elements.

Figure 5 shows this shared structure.

3. Table 1 of this document has the following possible structure, where object names are abbreviated with R for row, C for column, and E for entry (Figure 6):

[Table, T1]("Applications · · · Method") = (C1 C2), (R1 R2 R3 R4)
C1("Sequence") = E1, E1, E5, E7
C1("Sequence") = E1, E1, E5, E7
C2("Unordered Set") = E2, E4, E6, E8
R1("Storage") = E1, E2
R2("Formatting and Display") = E1, E4
R3("Traveling and Searching") = E5, E6
R4("Editing") = E7, E8
E1("sequential")
E2("associative")
E4("parallel or interleaved")
E5("sequence operations")
E6("set operations")
E7("sequence editing")
E8("set editing")

Note that entry E1 is shared by rows R1 and R2, and appears twice in the concrete representation of the table. Also note that an entry corresponding to R_i and C_j is found by taking the set intersection of the components of R_i and C_j.

4. Many programs have parts that are logically shared: each shared part is used by more than one program component; having more than one

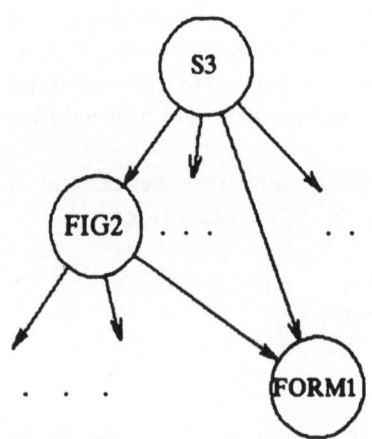

FIGURE 5. Shared formula.

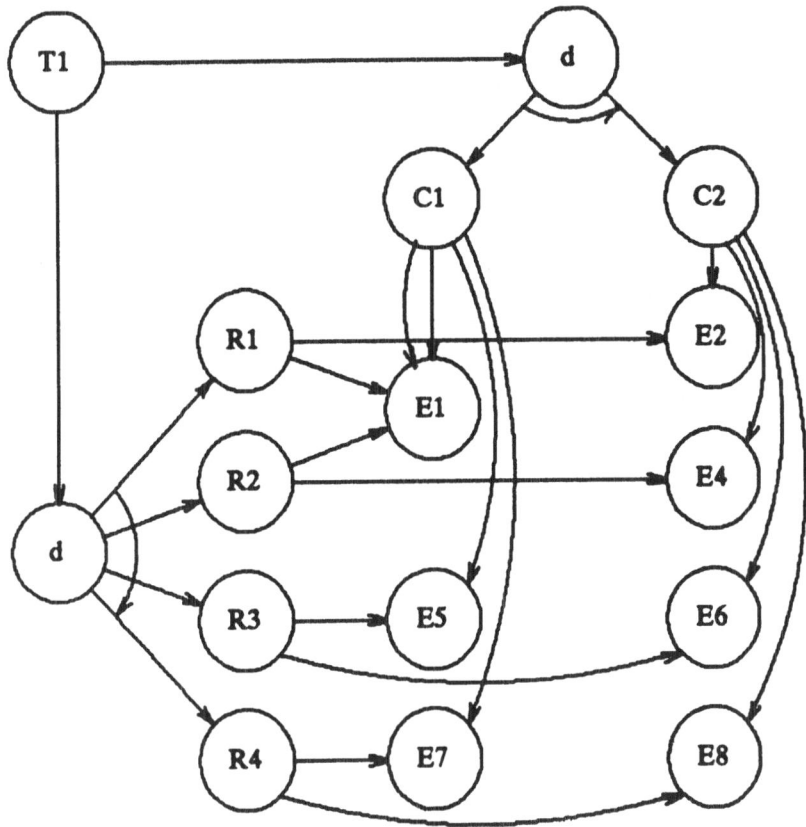

FIGURE 6. Structure of Table 1.

copy of a shared part is redundant and can easily lead to errors (inconsistencies). The most obvious examples of potentially shared parts are procedures and data. A procedure is shared by all of the other procedures that call it. In FORTRAN, a common block of data is shared. Sharing of modules also occurs naturally; for example, in the program mentioned in Example 3.2, objects of the Terminal module are shared by both the MainModule and the FileNames module.

5. Links

Some common document objects are not conveniently describable using only hierarchic composition with sharing, in particular, references to citations and cross references between objects. A reference between two parts of a document does not imply that one part is composed of the other. They are merely related, most likely via their data. For example, the second paragraph of this chapter cites Ref. 3 in the Bibliography (Reference section). The reference in the second paragraph does not contain the data found in the Bibliography (Reference section); it merely points to an object within the section that the reader can refer to for enlightenment. Even more general traversals of the document are implied by cross references. We define *links* to handle these kinds of relations.

A link is a pointer from one object to another object and is independent of the composition structure. We use the notation

$$N(D) = C : L$$

to denote an object with links, where L can be of the same form as C, i.e., ordered, unordered, and mixed. (Most of our examples use only unordered links and it is not yet clear whether this full generality is needed.) Links are drawn as dashed arcs in our graph representation scheme.

Examples 5.

1. The second paragraph of this chapter has the structure (Figure 7)

```
[Paragraph, P2] = TB1 TB2 TB3 TB4 TB5 TB6
TB1("Systems do exist · · · of integration") = : Niev82
TB2(". Perhaps the most · · · is the")
TB3("UNIX") = [Footnote, FN1]
TB4("package") = : Kern82
TB5("which has · · · Xerox STAR") = : Furu82, Smit82, Meyr82
TB6(", permit the creation · · ·")
[Footnote, FN1]("UNIX is a trademark · · · ")
[Bibliography, B1] = Niev82, Kern82, Furu82, Smit82, Meyr82, · · ·
Niev82( )
Kern82( ) = · · · : Niev82
Furu82( )
Smit82( )
Meyr82( )
```

A reference such as Furu82 is a short form for [Citation, Furu82] and TB1 is short for [Textblock, TB1]. As illustrated by this example footnotes and references may be naturally associated with sentences, phrases, or individual words (uniformly called textblocks here).

2. The first sentence of the preceding example (Example 5.1) cross

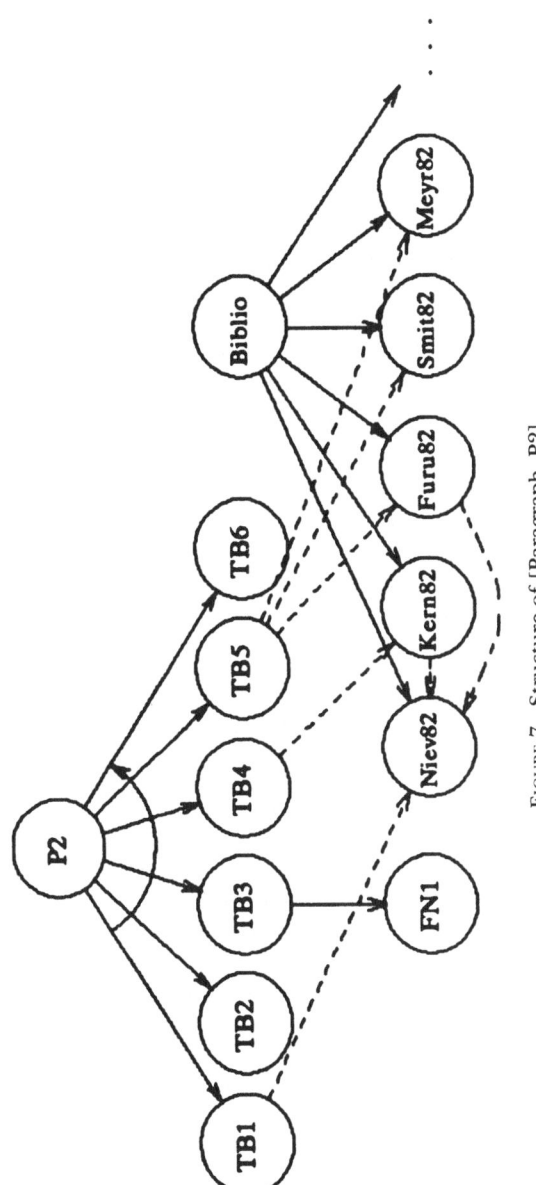

FIGURE 7. Structure of [Paragraph, P2].

references the second paragraph of this chapter; and this example refers to Example 5.1. The structure of these links can be written as

$$[Example, Ex5\text{-}1]\ (\ ) = \cdot\ \cdot\ \cdot : [Paragraph, P2]$$
$$[Example, Ex5\text{-}2]\ (\ ) = \cdot\ \cdot\ \cdot : [Example, Ex5\text{-}1]$$

Links and shared objects can often be used for the same purpose, and it is not yet evident which choice of the two is most desirable. For example, the structure of Table 1 as given by Example 4.3 and Figure 6 employs shared entries. However, the same table could be composed of rows, columns, and entries, with the rows and columns containing links to the entries.

In *paper* documents, links but not shared objects are an obvious and natural concept. For example, in this chapter "see Example 4.3" and "(3)" are two naturally occurring references for which a link representation is appropriate. A link to Example 3.1 is also implied by the reference in Section 4 (see Example 4.1); a shared object is more natural in this case. In electronic documents, both links and shared objects are obvious and natural concepts.

6. Summary and Current Work

We have presented a method for structuring documents using the ideas of abstract and concrete objects, hierarchic composition, ordered and unordered objects, sharing, and links. It has been demonstrated that a large variety of document objects can be described with this specification scheme.

A prototype abstract object editor, based on the model of this chapter, has been running since the fall of 1983.[14] The system has been used to write various technical reports, including this one, and can handle textual, tabular, mathematical, and graphical data. It uses the model as a basis for storing, formatting, traveling through, searching, and editing documents. Object classes are specified with grammars that generate instances of our structures. Such object generations are syntax directed; they are controlled by concrete templates associated with the grammar rules. Granularity is chosen by the author.

A window is associated with each rule of the grammar and object. Data are textual with some exceptions such as graphical boxes and circles. For example, math is similar to a structural version of eqn-like structures.[2] The document structure also may be executed in quasi-real-time to produce formatted two-dimensional concrete images and viewed, as per Fig. 1.

Current work is in the following related areas:

1. Object Attributes. In addition to the data attribute used here (the *D* component of the document description), more general attributes are desirable, for example, to define the abstract to concrete object mappings for formatting. (These attributes are similar to ones used in other systems, e.g., Refs. 9 and 12.)

2. Graph Representations. Objects can be represented by several superimposed graphs. The hierarchic composition of an object is mapped into a DAG in a straightforward way. The objects at each level of a hierarchy are related by yet another DAG when the order/unordered relations (a partial order) are included. Finally, there is also the graph defined by the links. Further study of these graphs may yield efficient computer storage methods and processing algorithms.

3. Applications Development. In addition to further testing of the method on diverse types of conventional document objects, we are also studying applications to other kinds of objects, such as programs, spread sheets, forms,[16] and interactive interfaces.

Acknowledgments

We are grateful to R. Furuta, G. Swart, and J. Zahorjan for their helpful comments on this paper.

This paper was originally published in the Proceedings of the Second ACM-SIGOA Conference on Office Information Systems, Toronto, 1984. Copyright 1984, Association for Computing Machinery, Inc.

References

1. J. NIEVERGELT, G. CORAY, J.-D. NICOUD, and A. C. SHAW, (Eds.). *Document Preparation Systems,* North-Holland, New York, 1982.
2. B. W. KERNIGHAN and M. E. LESK, UNIX document preparation, Ref. 1, pp. 1–20.
3. R. FURUTA, J. SCOFIELD, and A. SHAW, Document formatting systems: survey, concepts, and issues, *ACM Comp. Surv.* 14(3), 417–472 (1982); also contained in Ref. 1, pp. 133–220.
4. N. MEYROWITZ and A. VAN DAM, Interactive editing systems: Parts I and II. *ACM Comput. Surv.* 14(3), 321–415, (1982).
5. D. C. SMITH, C. IRBY, R. KIMBALL, and B. VERPLANK, Designing the Star user interface. *Byte* 7(4), 242–282 (1982).
6. A. C. SHAW, A model for document preparation systems. Technical Report 80-04-02, Dept. of Computer Science, Univ. of Washington, Seattle, April 1980.
7. D. E. KNUTH, *TEX and Metafont: New Directions in Typesetting,* Digital Press and the American Mathematical Society, Bedford, Massachusetts, and Providence, Rhode Island, 1979.
8. H. BURKHART and J. NIEVERGELT, Structure-oriented editors. Berichte des Instituts fuer Informatik 38, Eidgenoessische Technische Hochschule Zuerich, Zurich, Switzerland, May, 1980.

9. T. ALLEN, R. NIX, and A. PERLIS, PEN: A hierarchical document editor. Proc. ACM SIG-PLAN SIGOA Symp. Text Manipulation, *SIGPLAN Not. (ACM)* **16**(6), 74–81 (1981).
10. D. C. ENGELBART, R. W. WATSON, and J. C. NORTON, The augmented knowledge work-shop. ARC Journal Accession Number 14724, Stanford Research Center, Menlo Park, Calif., March 1973. Paper presented at the National Computer Conference, June 1973.
11. S. CARMODY, W. GROSS, T. E. NELSON, D. RICE, and A. VAN DAM, A hypertext editing system for the /360, Center for Computer and Information Sciences, Brown Univ., Prov-idence, Rhode Island, March 1969. Also contained in *Pertinent Concepts in Computer Graphics,* M. Faiman and J. Nievergelt, Eds., Univ. of Illinois, Urbana, 1969, pp. 291–330.
12. R. BEACH and M. STONE, Graphical style towards high-quality illustrations, Proc. SIG-GRAPH 83, *Comp. Graphics (ACM)* **17**(3), 127–135 (1983).
13. C. F. GOLDFARB, (Ed.). Information processing systems processing languages text inter-change and processing part six: Document markup metalanguage, Fifth Working Draft, International Standard, ISO TC97/SC5/EG CLPT N173-6, May 1983.
14. G. D. KIMURA, A structure editor and model for abstract document objects. Ph.D. dis-sertation, Computer Science Dept., Univ. of Washington, Technical Report 84-07-02, July 1984.
15. N. WIRTH, *Programming in Modula-2,* Springer-Verlag, New York, 1982.
16. D. TSICHRITZIS, Form management. *Communications of the ACM* **25**(7), 453–478 (1982).

10

AN OBJECT MANAGEMENT SYSTEM FOR OFFICE APPLICATIONS

STANLEY B. ZDONIK

1. Introduction

The field of data processing has traditionally been concerned with *data intensive applications*. That is, the complexity of the data in these applications dominates the complexity of the processes. A program that prints a report listing the outstanding accounts receivable is relatively simple compared to the structure of the general accounting data of the firm.

Database management systems have been successful in providing an environment for controlling the complexity of data processing applications. Programming time is reduced in an environment in which there exists a uniform view of data expressed in a language that presents a high-level semantic interface.

Office information systems (OISs) often present the user with a small collection of generic applications. These applications often include text editors, graphics editors, electronic mail systems, and spreadsheets, applications that address the "common denominator" of office functions. The new generation of office system (e.g., LISA, Macintosh, 1-2-3, Alis) has made pioneering steps toward an integrated environment in the sense that data produced by one application is available to other applications.

We believe that the greatest productivity gains can be achieved by focusing on the individual needs of a particular office. Offices are not all alike. They exist to perform some critical function within the overall context of an organization.[11] An understanding of this function will often lead

STANLEY B. ZDONIK • Department of Computer Science, Brown University, Providence, Rhode Island 02912 to the definition of specialized application programs that address procedures that are specific to the individual office. We will call these programs *custom applications*. A well-designed set of custom applications can lead to improved OIS performance because they reflect the information handling patterns of the individual office and relieve the users from the burden of remembering these details. They can embody some level of knowledge about the business of the office and, thereby, augment the decision processes that occur in that environment. These custom applications must, however, display the same level of integration as the generic applications.

Of course, the construction of custom applications involves programming, and programming is an expensive activity. If we are to see the wide spread emergence of custom applications as a part of OISs, new techniques must be developed for making this task easier. Software engineering, in general, and better programming environments, in particular, all have something to contribute to this goal. We believe, however, that one of the most promising areas is the development of a programming environment that includes better data management tools, since office applications are data intensive in the same way as data processing applications.

An object management system should support the development of custom office applications in much the same ways that conventional database systems have supported applications development. An object management system should allow users to store data of many different types in a single logical location. Objects of arbitrary type can, then, be retrieved using a single uniform language interface. This approach avoids the common practice of keeping one's record-oriented data (e.g., employee and department relations) in a database managed by a DBMS and keeping one's documents and graphics in files that are managed by a file system. The files that represent the documents and graphics typically do not present any common interface.

This chapter will describe an approach to modeling office data that forms the basis for a kind of office database system that we will call an *Object Management System* (OMS). The particular OMS that we have built based on these principles is called ENCORE. The basic features and philosophy of the model will be described.

2. Object Management Systems

We feel that many office applications have characteristics that distinguish them from conventional data processing applications. These distinctions serve to delineate more clearly the areas in which new systems techniques can make an impact. In this section, we will try to understand the peculiarities of office applications by analyzing them in distinction to the applications that have been addressed by data processing. Some of these observations have to do with the character of the applications themselves, and some have to do with the characteristics of the data on which these applications operate.

2.1. Characteristics of DP Applications

Data processing applications and their data often display the following characteristics.

1. Use of a Single Modeling Type. Data processing applications have developed in a world in which the record construct has driven the way in which objects are modeled. Everything (e.g., employees, departments) is seen as a record, thereby inheriting the semantics of this ubiquitous structure.

2. Flat View of Objects. The record is a flat structure that aggregates a number of fields, each of which is used to assert something about an object of interest. The information that is required by most DP applications is easily expressible as a set of short coded assertions about objects [e.g., salary (Smith) = $20K] that are external to the machine (e.g., employees, departments).

3. More Static Environment. The data changes to reflect a change in the state of the external world of the application. It is usually assumed that when such a change is noticed, the data should be updated as soon as possible with little need to track the previous values. The structures that capture the basic data change very infrequently. When the schemas change significantly, a major disruption in the system is endured in order to convert to the new view.

4. Rigid Object Usage. The data in a DP application is designed carefully to reflect the ways in which it will be used. A great deal of planning is required to predict future requirements, because once built, the database objects are usually used in only these ways.

5. One-Time Development. The database is usually designed, constructed, loaded with data, and brought on-line once. Further changes to this design are cumbersome and costly.

2.2. Characteristics of OA Applications

In contrast to conventional data processing applications, many office applications and their associated data have a somewhat different set of characteristics. These differences will influence the design of a successful OMS and will distinguish an OMS from the current state of the art in database management. Some of these distinguishing characteristics are listed below.

1. Integration of Many Types. All of the many information types that will be stored in the database of an office workstation must be usable together effectively. The interactions of these types must be made as smooth as possible. It must be possible to create connections between them dynamically as well as to treat them differently in different contexts.

2. Highly Structured Objects. Many of the objects that occur in an office environment are built out of other objects. This kind of structure can be seen in reports that contain chapters and the chapters that further contain paragraphs. Conventional data models do not provide any means for building *molecular* objects out of more primitive pieces like records. Although it is possible to connect records, it is not possible to have the system treat these connections as a single unit that has identity as a whole. These more complex objects tend to be objects that exist and are manipulated within the boundaries of the machine (e.g., reports, graphs, programs).

3. Change. Office workers are surrounded by many complex objects that support the effective execution of their duties. They change these objects frequently to reflect new understandings or conditions. Not only does the content of these objects change, but often the basic structure will change as well. Consider the many revisions that a typical report will undergo.

4. Flexibility. The objects in an office are often used in many different ways. The behavior of an object can depend on how it is being used. A chapter in a final report might be printed differently than the same chapter in a working paper.

5. Incremental Development. The development of new object types never stops. An office environment constantly produces exceptions to the previous way of expressing things. New definitions of what constitutes a report are always being discovered.

An office workstation environment should contain data management tools that are sensitive to the differences cited above. We will explore a model of data and an approach to data management that reflects some of these differences.

2.3. Object Management System Goals

Traditional file systems provide facilities for the storage and retrieval of objects of arbitrary type that are created in user programs, but the semantics of these objects is not known to the file system. Database management systems provide a means of capturing the semantics of objects using a single basic paradigm, the record. This model is inadequate for describing the richer semantics of office objects. An object management system combines the advantages of both a file system and a database management system in that it can store instances of arbitrary user-defined object types and, at the same time, maintain a high-level description of their meaning.

2.4. Other Uses for an Object Management System

ENCORE has been designed to address a large class of office applications. As was noted above, this class of applications can best be characterized by a need to manage many large composite objects whose content undergoes a lengthy history of change. Furthermore, we have assumed that the definitions of the object types will evolve as the needs of the office application change.

It is interesting to note that many of these same characteristics pertain to other application domains as well. We have come to believe that an OMS is more generally suited to design applications. In the design process, one is interested in constructing new artifacts from other older artifacts and perturbing these structures until an object that meets all of the stated objectives is achieved. It happens that office applications in which we build reports, graphics or other such artifacts are really special cases of design applications.

For this reason, we have begun to investigate the suitability of ENCORE to other areas of design such as CAD and software enginnering. In the CAD area, for example, the activity of designing a new integrated circuit can be viewed as assembling already existing cells or modules. In software engineering, we view an OMS as forming the basis for an intelligent program library. The process of constructing a large system is one of selecting the proper program modules from this library and specifying the interconnections

3. Overview of the Data Model

The basic capabilities of our OMS can best be understood by looking at the data model which it embodies. This data model provides a set of modeling constructs that we feel address the basic issues of office information management. The use of the model is based on our assumptions about how the information needs of an office can best be served.

In this chapter, we will present some of the more central concepts of our OMS. Space does not permit a complete treatment. Some aspects of the model resemble other entity-based data models that have been proposed recently.[12] We have adopted an approach that starts with this style of data model and extends it in a direction that more closely matches office applications. The most important aspects that we will be describing here are composite objects, version histories, and the ways in which these two mechanisms interact.

It is assumed that there is some programming environment that accompanies the OMS. Ideally, the programming language would also be object-oriented in that it supports data abstraction and type hierarchies.[6,10]

3.1. Objects and Types

The OMS, much like a conventional DBMS, is responsible for managing a large space of objects. An object corresponds to any entity in the application domain that has some significance to the application designer.

Our model of objects is similar to that of object-oriented programming languages such as SMALLTALK.[9,10] Objects are instances of one or more types. A *type* is a template that defines the behavior of its instances by specifying a set of operations that can be applied to any instance of that type.

One type can be defined to be related to another type. We say that a type T is a *subtype* of some other type S if all instances of T are also instances of S. S is said to be a *supertype* of T. In this way, all the operations that are defined on S must also be defined on T. We say that T *inherits* all operations from its supertypes (e.g., S).

A type can have more than one immediate supertype. For example, the type *Letter* might have two immediate supertypes *Formattable-Object* and *Communication* (See Figure 1). L1 is a particular instance of the type Letter which might inherit from the type Formattable-Object the attribute *Justi-*

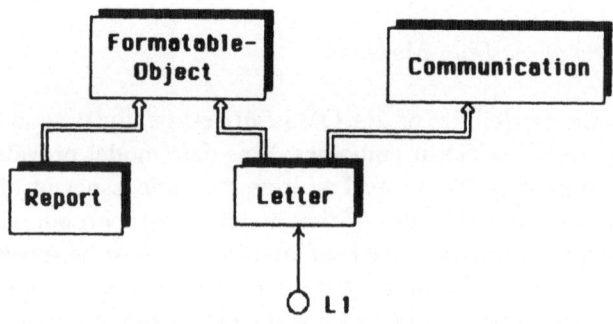

FIGURE 1. Example of multiple inheritance.

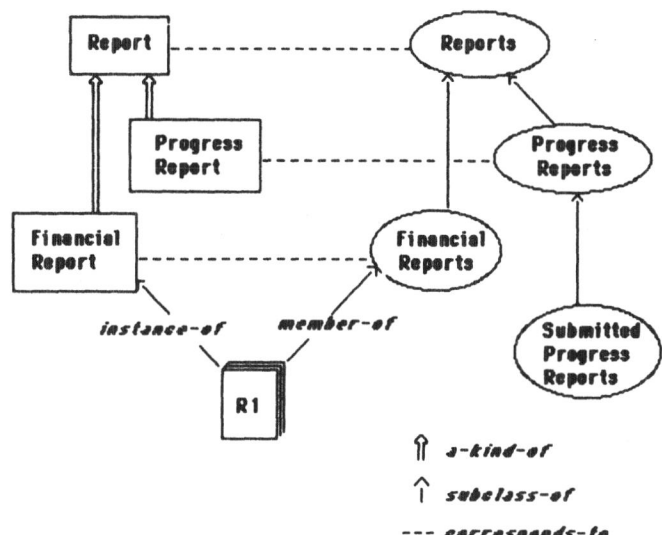

FIGURE 2. Example of type/class distinction.

fied?, which has a value of *true* if the right margin is even, and which might inherit from the type Communication the attribute *Sender-Name*.

In the database world, it is useful to have a mechanism for talking about interesting sets of objects whose membership is persistent. Objects are defined as members of *classes*. A class is a collection of objects that is maintained by the OMS. By being a member of a class, an object will be remembered from session to session. It will not disappear when the process that created it disappears. Our notion of a class is very similar to that used in the SDM.[12]

Since classes are sets, derived classes can be defined using any set-forming operations. New classes can be created from other OMS classes by predicate selection, manual selection, set union, set intersection, and set difference.[12] We will call the expression that defines a derived class in terms of its parent classes its *derivation expression*. From this, we can see that a given object can reside in an arbitrary number of classes.

What is the relationship between types and classes? A type is a description, while a class is a set. It is not possible to iterate over all members of a type, while iteration is one of the most common operations on a class. In our view, each type that is known to the OMS has an associated class which collects all its instances. There can also be classes that do not correspond directly to a type. This can best be seen by an example (see Figure 2). If the user defines a type called *Report,* there would be a corresponding class called *Reports* that contains all current instances of the type *Report*. Each time a new instance of *Report* is created it is automatically inserted into the corresponding class. In the figure, there are also two other types, *Financial-Reports* and *Progress-Reports,* defined as subtypes of the type Report. The classes *Financial-Reports* and *Progress-Reports* correspond, respectively, to these two subtypes. There might also be a class called *Submitted-Progress-Reports* that contains those progress reports that have been submitted to management for review. The members of this class add no new operations to the type *Progress-Report.* They are simply a distinguished collection of entities that have special meaning to the application. There is, therefore, no type that directly corresponds to this class. We will call classes like Reports *type-defined classes* and classes like Submitted-Progress-Reports *derived classes.*

A type-defined class has a new member added to it whenever the create operation for the corresponding type is invoked. A derived class is defined in terms of some previously defined class. In order for an object to be a member of a derived class, it must first be a member of some other class. Invoking the *create* operation for a type T may or may not add a new member x to class S, where S is a subclass of class R and R is the class that is defined by T. Object x will be a member of S if and only if x satisfies the derivaton specification for S.

The most general type is type *Entity,* since all other types are subtypes of this type. It provides the root of the type hierarchy. The class that corresponds to this type is the class *Entities.* This class corresponds to the whole database since it contains all objects that are known to the system.

All things that the user can refer to are objects and, therefore, are instances of some type. This includes the modeling objects that the system supplies (i.e., objects, attributes, and operations). These types are related to each other as shown in Figure 3. The type *Entity* is partitioned into three fundamental subtypes. The type *Program* corresponds to the active elements in the database and is subdivided into two additional subtypes. An *Operation* is a program that can be invoked on the instances of a particular type. It is an operation that defines the behavior of a type. The type *Transaction* corresponds to programs that are atomic (i.e., are consistent and recoverable). Transactions are built from operations. The type *Attribute* defines the objects that are used to relate other entities in the database. An attribute is an entity and, therefore, can have attributes of its own. The type *Object* is available for everything else (e.g., employees, reports). A type is considered to be an object and, therefore, enjoys all of the properties of an object or an entity. It should also be noted that all of the boxes in Figure 3 represent types and, therefore, are instances of the type *Type.*

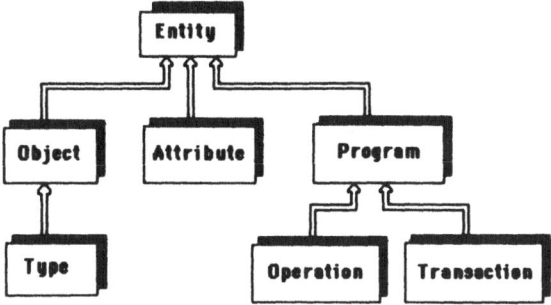

FIGURE 3. Basic system types.

3.2. Attributes

A type also defines a named set of attributes that are associated with each instance of that type. The name must be unique within the set of attributes for a given object type. Each attribute definition specifies some other OMS class from which its value is drawn. This class is called the *value class* for the attribute. A particular attribute for some object has a single value or a set of values that are drawn from that attribute's value class. Users define these attributes in a textual specification called an *object schema*. The object schema describes the attributes for each object type and their associated value classes. Each attribute that is defined in the object schema induces a pair of operations on the definition of the corresponding type. These two operations will be used to get and set the value of the attribute. For example, if the attribute *salary* is defined in the object schema for the type *Employee,* there must exist two operations, *get-salary-value* and *set-salary-value.*

An attribute is used to make an assertion about some object. For example, the salary attribute of an employee asserts that the salary of some employee is some amount (e.g., Salary [Smith] = $20K). In traditional DBMSs, the object that the assertion describes exists in the application environment, external to the machine, and a surrogate record is created in the database to represent that object.

Attributes are entities and as such can be accessed and referred to by other objects. An attribute can also have attributes of its own. For example, an instance of the type Report might have an attribute named *Reviewer* that relates a given report to the person who was assigned to review it. Reviewer might have an attribute named *Date* that expresses the date on which this association was made. The *Date* attribute really describes the relationship and not either the report or the reviewer. It is, therefore, more meaningful to place an attribute such as this on the entity that it actually modifies.

An attribute relates two database entities. All attributes have two special attributes called *subject* and *direct-object*. The subject attribute has as a value the object to which this attribute is attached, and the direct-object has as a value the object or objects that are the value of the attribute. If an instance of the attribute type *employs* relates General Motors and John, then General Motors is the subject and John is the direct object of the attribute.

3.3. Versions

Much of the work that happens in an office is characterized by the evolution of objects through intermediate states, each reflecting one's thoughts at some point in time. For example, a research proposal is not created out of whole cloth; it is modified many times by many people until it reaches a state of completion. We, therefore, believe that the data model should include a notion of versions as a primitive concept. The version scheme should make it easy to refer to current and previous versions of an object. It should also allow us to maintain several competing versions at the same point in time.

The OMS collects versions of objects in *version sets*. A version set is an ordered collection of objects that each represent a snapshot of the state of that object at some point in time. Each of the members or versions of a version set is created by some transaction which is remembered by the version set. In this way, one can ask how a given version was created. The order represents the ordering in time in which these snapshots are produced. If V_2 succeeds V_1, then V_2 was created later in time than V_1. An object schema specifies whether or not objects of that type participate in version histories.

3.3.1. Linear Version Sets. The most common way of collecting versions is to produce a linear sequence in which each version, except the oldest, has a unique predecessor, and each version, except the newest, has at most one successor. A new version V is always added at the "end" of this sequence such that V becomes the successor of the previous latest version. There is always a unique "latest version" in any linear version set. A pictorial representation of a linear version set is shown at the top of Figure 4.

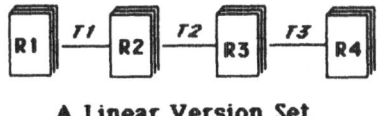

A Linear Version Set

A Branching Version Set

FIGURE 4. Basic version set types.

3.3.2. Alternatives. The version mechanism also allows users to build a more complicated kind of version set that includes the branching of *alternatives.* (See the bottom of Figure 4.) An alternative is a version whose predecessor has more than one successor (i.e., it has sibling versions). An alternative will typically produce two or more chains of versions that each emanate from the same starting point. These chains represent the changes that each of the alternatives experience as they evolve independently of their respective associated alternatives. It is also possible for an alternative or one of its successors to fork another set of alternative paths.

A branch in a version set represents the case in which several versions are being considered simultaneously as legitimate variations of an object. It provides a way to isolate simultaneous changes from each other. Perhaps, two office workers have each decided to work on a report at the same time, thereby producing their own version sequences that both originated from the same starting point.

It is possible to merge two paths that have been generated as alternatives. This merging, however, must occur manually by means of negotiation among the interested parties. This corresponds to the situation in which a given version was actually derived from pieces of several alternatives. If one asked for the predecessor of this new version, the set of versions that were the last entries in the alternatives' version chain would be returned. The custodians of each alternative would decide which pieces of the competing versions will be retained as the one "official" latest version.

3.3.3. Version Set Operations. In what follows, operations will be specified as the operation name followed by the types of its arguments in parentheses. This is followed by a right-pointing arrow (i.e., "→") and the type of the object that is returned. An argument that is enclosed in square brackets (i.e., "[]") is optional. A type name enclosed in curly brackets (i.e., "{}") indicates a set whose members are of the indicated type. The operations that are allowed on a version set are as follows:

1. Create-version-set ( ) → Version-set
2. Add-new-version (Version-set, Object, [Object]) → Version-set
3. Delete-version (Version-set, Object) → Version-set
4. Latest-version (Version-set) → {Object}
5. Derived-by (Version-set, Object) → Transaction
6. Antecedent-for (Version-set, Object) → Transaction

The *Add-new-version* operation makes the second argument the latest version of the first argument. This is straightforward in the case of a linear version set. A branching version set that has more than one latest version requires that the third argument be specified to indicate to which of the several latest version objects this new version is to be connected.

The *Delete-version* operation removes the given object from the given version set returning the version set with the object removed. This operation does not reuse the "slot" in the version set to store other versions. In other words, things that referred to the deleted object should not all of a sudden refer unexpectedly to some other object as a side effect of the deletion. An appropriate error condition should be raised that indicates that the old version set member no longer exists. This involves conceptually keeping "tombstones" in the version set.

Since version sets are also instances of the type *Set,* they have available to them all the operations that can be performed on sets. This includes operations such as *Iteration, Predicate-Selection, Union, Intersection,* and *Set-Difference.* Using these operations on version sets will return an object of type Set (not a version set) since the ordering information is lost. The set operation, *Add-new-member,* has been refined in this context, however, to allow addition to occur to the set only in very specialized ways (i.e., at the "fringe").

The *Latest-version* operation is one of the most useful, since it returns the most up-to-date versions of an object. They are the version set members that are the current fringe of the "tree". For alternatives, this is different from the single version that was produced latest in time. The additional operation, *Newest-version (Version-set)* will accomplish that. In most default cases, the object or set of objects that is returned by Latest-version is exactly what the user is interested in. It should be noticed that this operation is somewhat redundant. It can be constructed out of the *Set-selection* operation and a predicate that expresses the notion of latest element. Since it is so important in our system, though, we have chosen to list it separately.

The operation *Derived-by* is used to determine the transaction that created this version. Whenever an Add-new-version is executed, a reference to the transaction that invoked it is recorded in the version set. This is possible since transactions are entities in the database and can, therefore, be referenced by other objects. In Figure 4, this corresponds to the labels on the arcs that connect the versions. For example, in the branching version set case, derived by $(R_4) = T_5$, which indicates that transaction T_5 issued the Add-new-version operation which added R_4 to the version set. Notice that the two arcs that lead to R_7 from R_4 and R_6 are labeled with T_6 since a merging of alternatives must be done by a single transaction (i.e., the one that adds R_7).

Information about changes that are made by transactions is often useful for understanding the current state of an object. Some transactions can assert things about their result. For example, a spelling corrector program processes a document and produces a new version that (in theory) has no spelling errors. Therefore, if we know that a document has been processed by the spelling corrector and no other changes have been made since then, we can assert that the document is spelling-error free. As soon as any changes that add new words are made to the document, this assertion cannot be made any longer. Assume that there is another transaction called *Publish-Document* that submits a document to the publication process. This transaction might have a precondition that states that the spelling must be correct before the transaction can proceed. The validity of this condition can be tested by appealing to the transaction history of the version set.

3.3.4. Conceptual Objects. Not all objects accumulate version histories. As a consequence, there are two additional subtypes that partition the type, Entity. These are the types *Primitive-Object* and *Conceptual-Object*. Conceptual objects have a version set that accumulates all the versions of that object associated with them. The conceptual object stands for the idea or concept that has evolved through all of the intermediate versions contained in its version set. Primitive objects do not have associated version sets; however, they can be members of version sets if a conceptual object of the right type has been defined. This can be specified in the object schema for the primitive object type. Thus, the object schema for the type *Letter* might begin as follows:

Define Type Letter **to be a-kind-of** History-Bearing-Object

This automatically declares the additional type *Conceptual-Letter*. Attributes for this new type can be optionally defined within the schema as well.

Conceptual objects can have attributes of their own that make statements about the entire collection of versions as opposed to about any individual one. For example, the conceptual object type, *Conceptual-Report*, might have an attribute, *Associated-project*, which makes a statement about the project for which a given conceptual report was created. It is not a statement about the individual primitive report objects, but rather a statement about the evolving object as a whole.

3.3.5. Type Changes. We have already expressed the idea that office environments are characterized by change. The changes that we have discussed so far all pertain to the instance level. But changes can also occur to the structure of the system itself. Our conceptualization of the basic types that the system manages can also change. For example, new attributes or operations can be added or deleted, or the semantics of an operation can change.

The OMS also makes use of the version mechanism to keep track of changes to the definition of a type as well as changes to the structure of the type hierarchy. This is accomplished by treating types as conceptual objects (i.e., type-defining objects), thereby making all of the version mechanism available to record changes to the type definitions. Since each OMS object always has access to the type-defining object under which it was created, it is always possible to find the correct version of a type in order to get the code for its operations. For example, there could be several versions of the type *Report*, each with a different set of operations. Maybe one has appendices and the other does not. Any report instance can access its corresponding type-defining object; therefore, each instance preserves its proper behavior. All the variations of the *Report* type object are related by the fact that they belong to the same version set. It is possible to find all reports as all those objects that have been created as an instance of some version of the *Report* type.

There also exists a mechanism that can be applied in some situations for making objects that were created under one version of a type appear as if they were created under a different version of that type (see Figure 5). In this scheme, the programmer supplies a set of *filter programs* (the box labeled *f* in Figure 5). A filter program makes use of the operations that have been defined on one version of a type to manifest the behavior of some operation on some other version of that type. The filter is defined in the context of some operation and links it to some other version of that type. Typically this is the previous version (as in the figure); however, it need not be. A filter that converts an operation of a newer version of a type from an older version of the type is called a *backward filter*.

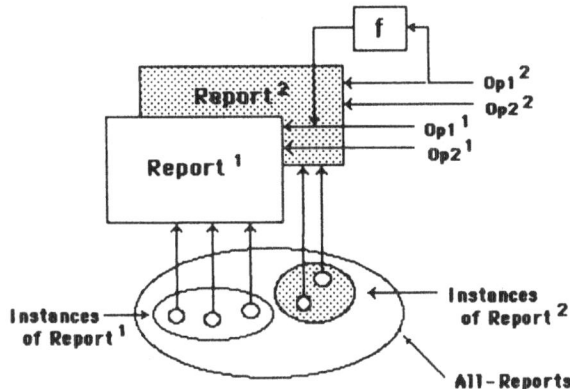

FIGURE 5. Tracking type changes.

It is also possible to create filters that extend from an operation on an older version of a type to some newer version of that type. We will call this a *forward filter*. The forward filter is useful for old programs that were written in terms of operations on older versions of a type and that need to work in the context of objects that are instances of newer versions of that type.

It is not always possible to create these filter programs. In those cases for which the appropriate filter is not defined, the system cannot iterate over the set of all instances of the conceptual type. The application can detect this and inform the user.

3.4. Content

Objects can contain other objects as components. It is common in an office environment to build new objects out of old ones, which leads to a hierarchical view of objects based on the containment relationship. The set of objects that make up the entire set of direct components of an object are called its *components*. The components of a report, for example, might be its title page, its table of contents, its chapters, and its appendices. All of these objects are the components of the report, but it is convenient to be able to group these components into sets and give these sets names. Therefore, we would have the *Chapters* component, and the *Appendices* component. In this case, the Chapters component consists of a set of objects of type *Chapter*. This is illustrated in Figure 6.

The OMS also allows for the sharing of objects by two or more other objects. For example, two chapter objects can incorporate the same section object as a component of each. This can also be seen in Fig. 6. In this way, changes to the shared section object can be reflected in the content of both chapter objects that are sharing it.

What distinguishes the content of an object from any of its other relationships? The content of an object is that part of the object that defines it. In conventional database systems, one constructs records which make assertions about objects or concepts in the external world of the application. Thus, an Employee record contains important information about a particular employee, but the employee is not captured in the machine. As the employee changes, users of the system must be responsible for keeping the shadow (i.e., the record) up to date. In distinction to this, many office objects are captured in the machine. The essence of a report is contained within the office workstation. The paper copies that are printed are artifacts of the real report that is in the computer. It is this computer object that changes. In fact, "editing" programs are used to change these objects. The thing which one edits is the object content; therefore, we consider the content to be the object itself.

Changing the content of an object is seen as creating a new object since it is the content that determines its identity. If the content is different, the object is different. Yet, there is still something that relates an object to the result of editing it. We use the version mechanism as the way of expressing a changing content. In other words, when a transaction mod-

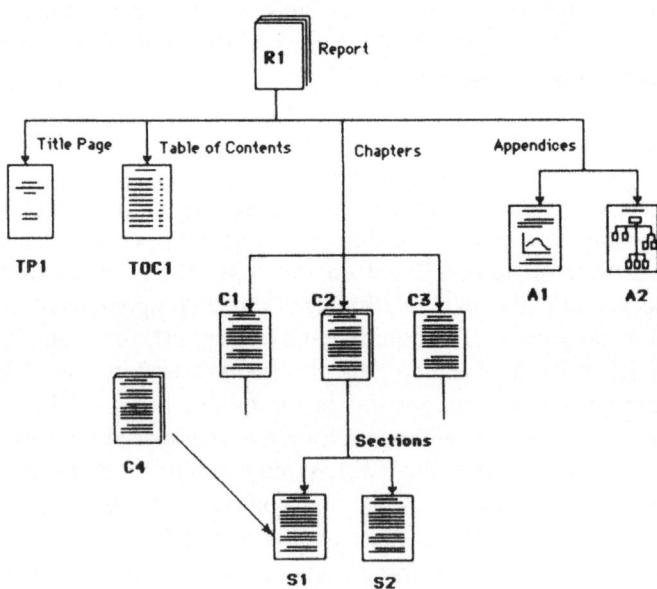

FIGURE 6. Example of component structure.

ifies the content of an object, a new version of that object is created. This implies that all objects with content are conceptual objects.

Attributes are different from content in that their values can change without causing a new version to be created. Attributes simply make statements about things. They do not affect the identity of an object. Of course, it is always possible to create explicitly a new version of a conceptual object at any time so that one can force an attribute change to be reflected in a new version. It is also possible to designate an attribute as one for which changes will force the system to create a new version of the object. We will call this later kind of attribute a *version sensitive attribute*.

3.5. References

A *reference* is the way in which one names or refers to other objects in the OMS database. When a component of an object is set to a value by one of the *set-component* operations for that object type, a reference to that value is constructed and held by the object. A reference is just another object type. As such, it has associated operations. The most fundamental operation that must be supported for all objects of type reference is the *Dereference* operation. *Dereference* returns the object or set of objects to which the reference refers. We will call the objects that are returned the *referent*. Users may create new subtypes of the type *Reference* as long as they provide the code for the *Create-Reference* and the *Dereference* operations.

There are two types of references, static references and dynamic references. A static reference refers to a particular object or set of objects for all time. Dereferencing a static reference always yields the object from which the reference was constructed. The referent of a dynamic reference can vary with time. Dereferencing it today can yield a different result from what was returned yesterday. An example of this type of reference is the *latest-version-reference*. This type of reference refers to the latest version of some conceptual object. The referent will obviously change as new versions are created.

There are many other interesting examples of dynamic references. Many of them resemble database queries. Consider the *longest-version-reference* subtype that refers to the version of a conceptual report that has the greatest number of pages. The program that implements the Dereference operation for this reference subtype looks through all versions of the conceptual object counting pages and returns the appropriate version.

As another example, consider a file folder that contains documentation on accounts that have an outstanding balance. The file folder object would contain a dynamic reference to the set of all dunning letters sent to delinquent customers regarding accounts that have not as yet been paid. If objects in the class *Dunning-letters* have an attribute named *account* that has as a value the offending account record, then the program to dereference a reference of type *Current-dunning-letters* would be as follows:

> **Define** current-dunning-letters$dereference (r: reference)
> **Select** [Dunning-letters,
>   lambda (dl)
>   not [dl.account.paid]]

The *Select* operation takes two arguments, a set and a predicate. The result is all those members of the set that satisfy the predicate. This code assumes the existence of a class Dunning-letters that contains all dunning letters in the database. It also assumes that each account record has an attribute named *paid* that has a value of either *true* or *false*. Notice that as accounts become paid, the corresponding dunning letters no longer are referents and, therefore, automatically "leave" the folder.

All references have a special property called its *definition*. This property is a part of the internal representation for the reference type. The value of this property is the object that is evaluated by the *dereference* operation. A value for this property is normally supplied by one of the arguments to the *Create-reference* operation. For example, the value of this property for a *latest-version-reference* might be a static reference to a version set object. The dereference code, then, finds that version set and returns the latest version.

3.6. Version Propagation

Creating new versions of low-level components of an object can create new versions of each of the higher-level containing objects. This can best be seen by an example. Suppose that report R contains chapters C_1 and C_2, chapter C_1 contains paragraphs P_1 and P_2, and chapter C_2 contains paragraphs P_3 and P_4. If a new version of paragraph P_2 is created, its containing chapter C_1 is also changed since it now contains a different paragraph. The effect, then, is to create a new version of C_1. Of course by a similar argument, this change could also be propagated to R, the containing report, by generating a new version of it.

It is not always desirable to propagate all changes at a lower level to all the higher containing levels. Conceptually, propagation of versions always makes sense. What is lost if we do not always force propagation? In the example above, we could simply collect all changes to paragraphs in their respective version sets and never propagate any of these new versions to the containing chapters and reports. If we follow this course, information is lost. The intermediate states of the chapters and the report are not recorded and could only be reconstructed if we stored some additional information (e.g., as an attribute) such as a timestamp that indicates when the version was written.

A simple example using the report structure outlined above will illustrate this. If P_1 and P_3 change as a result of one transaction, and paragraphs P_2 and P_4 change as a result of another transaction, each of the version sets for the four paragraphs will contain a new version. If these transactions do not cause their changes to be propagated, the information about the intermediate state created by the first transaction is lost. The paragraphs that it created can be seen, but the time sequencing of how these changes were made with respect to the second transaction has not been recorded. Propagation of the changes would solve this problem. The changes caused by the first transaction could cause a new version of the chapter and, likewise, the report to be created. This new report version is a snapshot of the intermediate state.

The decision of what changes should be propagated lies with the database schema designer. The definition of the containing object specifies which components should have their changes propagated. Within the object schema for the *Chapters* class, one might find the following component definition:

Paragraphs (version-sensitive): **Set of** Paragraphs

This would cause any change to any paragraph in the Paragraphs component of a chapter to cause a new version of that chapter to be created by propagating the changed paragraphs to the chapter level.

A more complicated case arises when a version sensitive component of an object contains a particular component that is represented by a branching version set (i.e., an object that has alternatives). In this case, the system must be careful about how it propagates new versions. A new version along one alternative path must be propagated to the correct alternative at the parent level.

Assume that an object of type S contains a version sensitive component of type T. If an alternative exists for a T-object the creation of the alternative would have propagated a similar fork to the S-object level. If a new version of one of the T-object alternatives is created, to which S-object alternative is this change propagated? The system must ensure that it is connected to the S-object alternative that has the same transaction as a value for *Derived-by* as the previous version on its alternative path.

If report R_1 contains two chapters C_1 and C_2 and two alternatives to

C_1 are created, C_{1x} and C_{1y}, then two alternatives to R_1, R_{1x} and R_{1y}, will be created by version propagation. Further changes to R_{1x} will propagate to the path that contains R_{1x} since Derived-by (C_{1x}) = Derived-by (R_{1x}).

This is complicated still further if C_2 also generates alternatives. Call these alternatives C_{2p} and C_{2q}. Since both R_{1x} and R_{1y} contain C_2 as a version sensitive component, the change is propagated to both alternative reports. This creates two alternatives to R_{1x} (i.e., R_{1xp} and R_{1xq}) and two alternatives to R_{1y} (i.e., R_{1yp} and R_{1yq}).

3.7. Inheritance

The type hierarchy found in most object-oriented programming languages allows for the *inheritance* of operations by types from their supertypes. Any operation that is defined for a type T is also defined for all subtypes of T. In data models like the SDM (and the OMS), attributes of a class are inherited from its supertypes. Inheriting an attribute amounts to inheriting the operations to get and set its value.

In the OMS, it is also possible to inherit attributes via the content hierarchy. An object can inherit attributes from its containing objects, although it should be understood that this type of inheritance must be controlled very carefully. It only happens in cases that have been clearly specified by the schema designer.

A couple of examples will illustrate the usefulness of this idea. Suppose that lines are objects and that lines are defined to have an attribute called *point-size* that indicates the number of points of height that the line is to occupy. Lines contain characters. If one were to ask for the point size of a character, it would be convenient to inherit that attribute from the line that contains them, assuming that a point size attribute were not defined explicitly for individual characters.

This situation is illustrated in Figure 7. In this case, the schema for the type *Line* would contain the following definitions:

> **Define Type** Line
> **Content**
>   Chars: **Set of** Characters
> **Attributes**
>   Point-size: Integer
>   **Extent via** Chars **is through** Char

Line offers its *Point-size* attribute to all components of the line that are connected "via" the *Chars* component. This offer is terminated at the point at which the component path encounters an object of type *Char*.

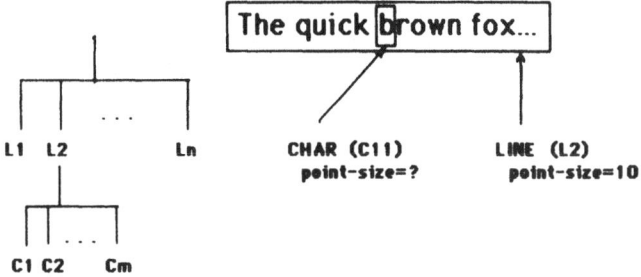

FIGURE 7. Example of component inheritance.

As another example, suppose that objects in the class *Reports* have an attribute called *author*. Since reports contain chapters, and chapters contain paragraphs, we would like to be able to say that the author of a paragraph is actually just the author of the report. Even if an *author* attribute was not defined for paragraphs, a paragraph in the context of a containing report should inherit this attribute from that report.

This type of inheritance is specified in the object schema by an *extent expression* attached to the attribute. An extent expression indicates that the given attribute is to be made available to some contained objects. It indicates how far this inheritance is to extend by indicating a class name that will terminate the inheritance path. If an object that is a member of the given class is encountered in moving down from the offering object, then the extent ends there. That is, objects below that object in the containment hierarchy cannot inherit the attribute.

The following schema fragment illustrates how the author attribute for the Reports example given above might look:

> **Define Type** Reports
> **Content**
> Chapters: **Set of** Chapters
> . . .
> **Attributes**
> . . .
> Author: Person
> **Extent via** Chapters **is through** Paragraphs

This would allow any object between a report and a paragraph in the content hierarchy to inherit the attribute (and its value) from the report. Of course, if one of the types for the intervening objects defines its own *author* attribute, that attribute definition will prevail. The phrase *via Chapters* indicates that this inheritance is only available to components that are ultimately connected to the report by means of the Chapters component (i.e., the one component that is specified in the Content section above).

4. Example

The following is an example of a fragment from an OMS schema. It contains two object schemas that define the types *MailMessage* and *Paragraph* and the corresponding classes *MailMessages* and *Paragraphs:*

Define Type MailMessage
Associated Class: MailMessages
Description: This class contains all sendable objects.

Content
 Paragraphs: **Set of** Paragraphs

Attributes
 Subject: Strings
  extent via Paragraphs **is through** Paragraphs
 Sender: Users
 Receivers: **Set of** Users
  extent via Paragraphs **is through** Paragraphs
 Cc: **Set of** Users
 Number-of-words: Integers
  derived-by: Sum [Image [self.paragraphs, lambda (p) p.number-of-words]]

Define Type Paragraph
Associated Class: Paragraphs
Description: This class contains all paragraphs.

Content
 Words: **Set of** Strings

Attributes
 Number-of-words: Integer
 Point-size: Integer
  extent via Words **is through** Strings

Given these definitions it would be possible to create instances of the type *MailMessage.* Each of these instances would be collected in the class named *MailMessages.* It would, then, be possible to retrieve messages from the OMS database based on some of the attributes defined in these schemas. For example, one could ask for all mail messages that have more than 500 words, that were sent to me by Smith, and that were also copied to Jones. This retrieval would conceptually require iterating through the members of the class *MailMessages.* Of course, if additional access structures were defined (e.g., indices on these attributes), the system could process this request more directly.

Notice that in order to process the above query, the system would have to make use of the *number-of-words* attribute on mail messages. The value of this attribute is derived by an expression that directly follows the attribute definition in the schema. This code is evaluated at retrieval time to determine the number of words in a given message. (N.b., The *Sum* function takes a set of integers as an argument and returns the sum of the members as a result. The *Image* function takes a set *A* of *n* elements and a function *f* as arguments and returns a set *B* of *n* elements such that the *i*th member of *B* is the function *f* applied to the *i*th member of *A* for all *i* between 1 and *n.*)

It is not possible to set the values of *Paragraphs* of a *MailMessage* or *Words* of a *Paragraph.* Since they are defined as the content of the respective objects, changing them would require the creation of a new version of the *MailMessage* or the *Paragraph* objects.

The attribute *Point-size* of a paragraph will be inherited by each of the words that make up the content of the paragraph because of the extent specification that accompanies the definition of that attribute in the paragraph object schema. The attribute *Number-of-words* is not inherited by the individual words because that would not make sense.

Similarly, the attributes *Subject* and *Receivers* are inherited by the paragraphs that are contained in the MailMessage. Even though the paragraph inherits these attributes, they are not passed on to the lower-level components (i.e., the words).

Notice that the attributes *Senders* and *Receivers* have *Users* as a value class. *Users* is a special class that contains an entry for every person that uses the office system. This class is used by the system to perform some of its bookkeeping functions (e.g., security checking).

If user *U* wanted to have the receivers of a message *M* be his bosses, he might use the following code:

```
r: = Bosses-ref$create (U)
Set-attribute-value (M, Receivers, r)
```

Bosses-ref is a reference type with a given user as the value for its definition attribute. *Bosses-ref$create* names the create operation that is associated with the type *Bosses-ref.* All references have an attribute called *Definition* that stores the information required to define it (see Section 3.5). Notice that the create operation for the type *Bosses-ref* takes a definition as an argument (in this case, the object *U*). The dereference operation for the *Bosses-ref* reference type is as follows:

> Define Bosses-ref$dereference (r:reference)
> Get-attribute-value(
> Get-attribute-value(r, Definition), Works-for)

A *Bosses-ref* reference type will always evaluate to the current set of bosses of the user who has been bound to the *Definition* attribute of the reference. If that user's position in the company changes by acquiring or losing bosses, the value of this reference will change automatically. This is because the reference is defined in terms of the *Works-for* attribute of a user object. Sending the message *M* to the Receivers will have different effects at different times.

A derived attribute and a dynamic reference are similar in some respects in that they both compute a value at the time the reference is evaluated. However, they are different in that a derived attribute requires the specification of the expression at schema definition time, whereas a dynamic reference allows the selection of the expression at run time.

5. Summary

The above discussion has given an overview of a few of the more fundamental aspects of our object management system, ENCORE. We feel that this type of system can be used effectively as a tool in supporting integrated office applications. A simple prototype of ENCORE has been implemented, and a more complete and usable version is currently being constructed.

We also feel that ENCORE addresses many of the differences between office applications and data processing applications that were cited in Section 2. More specifically, it addresses the following:

1. Integration of Many Types. The OMS is designed to accommodate objects of arbitrary type. As long as a type manager (i.e., the set of operations and attribute definitions) for objects of type *T* has been provided, the OMS can handle instances of *T* in the same way that it would handle any other object.

2. Highly Structured Objects. The content portion of an object is reflective of this structure. Our data model allows one to build a single object out of many others. One can then treat this as a single object or access its pieces.

3. Change. The version set mechanism is a technique for handling change. Since it is a primitive construct, the system can assist in the management of these versions. The propagation of versions is an example of this.

4. Flexibility. The basic notions of type hierarchies that can handle instances of multiple types and the notion of inheritance through this hierarchy as well as through the content hierarchy helps to make the system more flexible and consistent.

5. Incremental Development. It is possible to add new types and class definitions to the OMS without causing any disruptions to what has already been created. Furthermore, using the version mechanism to manage modifications to the type defining objects supports the inevitable evolution of programs.

Lessons that have been learned over the last 15 years from database technology can be effectively applied to the new field of Office Information Systems. The technology of an OMS adapts some of these database ideas to better fit the office environment.

Acknowledgments

The author wishes to thank Michael Hammer, Irene Grief, and Dan Carnese for their helpful suggestions during the early stages of this work. The author would also like to acknowledge the contributions of Tom Atwood and Gordon Landis.

This research was supported in part by the Office of Naval Research and the Defense Advanced Research Projects Agency under contract N00014-83-K-0146 and ARPA order no. 4786.

References

1. H. BILLER and E. J. NEUHOLD, Semantics of databases: The semantics of data models, *Inf. Syst.* **3**, 11–30 (1978).
2. C. L. CESAR *et al.*, An integrated office system, Technical Report RC 9659, IBM Thomas J. Watson Research Center, October, 1982.
3. P. P. S. CHEN, The entity-relationship model: Towards a unified view of data, *ACM TODS* **1**, 1 (March 1976).
4. E. F. CODD, A relational model for large shared data banks. *Commun. ACM* **13**(6), 377–387 (June 1970).
5. E. F. CODD, Extending the database relational model to capture more meaning, *ACM Trans. Database Syst.* **4**(4) 397–434 (December 1979).
6. G. CURRY, L. BAER, D. LIPKIE and B. LEE, Traits: An approach to multiple-inheritance subclassing, Proceedings of the Conference on Office Information Systems, ACM/SIGOA, University of Pennsylvania, Philadelphia, Pennsylvania, June, 1982.

7. C. A. ELLIS and M. BERNAL, OfficeTalk-D: An experimental office information system. Proceedings of the Conference on Office Information Systems, ACM/SIGOA, University of Pennsylvania, Philadelphia, Pennsylvania, June, 1982.

8. S. GIBBS, Office information models and the representation of office objects, Proceedings of the Conference on Office Information Systems, ACM/SIGOA, University of Pennsylvania, Philadelphia, June, 1982.

9. A. GOLDBERG, Introducing the Smalltalk-80 system. *Byte* August, 14–26 (1981).

10. A. GOLDBERG and D. ROBSON, *Smalltalk-80: The Language and its Implementation,* Addison-Wesley, Reading, Massachusetts, 1983.

11. M. HAMMER and M. ZISMAN, Design and implementation of office information systems. Proceedings of the NYU Symposium on Automated Office Systems, New York University Graduate School of Business Administration, May, 1979, pp. 13–24.

12. M. HAMMER and D. MacLEOD, Database description with SDM: A semantic database model, *ACM Trans. Database Syst.* September, 351–386 (1981).

13. W. HORAK and G. KRONERT, An object-oriented document architecture model for processing and interchange of documents, Proceedings of the Conference on Office Information Systems, ACM/SIGOA, University of Toronto, Toronto, Canada, June, 1984.

14. R. H. KATZ and T. J. LEHMAN, Storage structures for versions and alternatives, Computer Sciences Technical Report No. 479, Computer Sciences Department, University of Wisconsin-Madison, July, 1982.

15. R. LORIE and W. PLOUFFE, Complex objects and their use in design transactions, Technical Report RJ 3706, IBM Research Laboratory, San Jose, California, December, 1982.

16. J. MYLOPOULOS, P. A. BERNSTEIN, and H. K. T. WONG, A language facility for designing database-intensive applications, *ACM Trans. Database Syst.* 5(2) 185–207 (1980).

17. D. REINER, M. BRODIE, G. BROWN, M. CHILENSKAS, M. FRIEDELL, D. KRAMLICH, J. LEHMAN, and A. ROSENTHAL, A database design and evaluation workbench: Preliminary report, Procedings of the International Conference on Systems Development and Requirements Specification, Gothenburg, Sweden, August, 1984.

18. D. W. SHIPMAN The functional data model and the data language DAPLEX, *ACM TODS* 6(1), 140–173 (1981).

19. J. M. SMITH and D. C. P. SMITH, Database abstractions: Aggregation, *CACM* **20,** 6 (1977).

20. D. C. TSICHRITZIS and F. H. LOCHOVSKY, *Data Models,* Prentice-Hall, Englewood Cliffs, New Jersey 1982.

21. S. B. ZDONIK, Object management system concepts: Supporting integrated office workstation applications, Ph.D. dissertation, Massachusetts Institute of Technology, May, 1983.

22. S. B. ZDONIK, Object management system concepts, Proceedings of the Conference on Office Information Systems, ACM/SIGOA, University of Toronto, Toronto, Canada, June, 1984.

23. M. M. ZLOOF, Office-by-Example: A business language that unifies data and word processing and electrtonic mail, *IBM Syst. J.* **21,**(3) 272–304 (1982).

11

TM—AN OBJECT-ORIENTED LANGUAGE FOR CAD AND REQUIRED DATABASE CAPABILITIES

J. Miguel Gerzso and
Alejandro P. Buchmann

1. Introduction

In recent years there has been growing interest in handling structured objects as a means for establishing more natural interfaces to sophisticated application systems, such as integrated CAD systems and the databases underlying expert systems, i.e, knowledge bases. At the same time language designers have recognized the advantage of programming with objects which allow the encapsulation of procedure and data definitions. This property leads to modular code which is easier to maintain. As a growing number of object-oriented languages appear and mature[16,21,22,23,18,9] it becomes more interesting to explore their potential as vehicles in expert systems and application systems that have to deal primarily with structured objects.

The main reason for the interest in object representation in CAD has been the desire of the designer to handle structured objects that are meaningful to him and not amorphous collections of records which have to be retrieved individually. This requirement has led to extensions to relational systems,[6,31,24] as well as new implementations of network-type DBMSs,[19] all of them still in an experimental stage. In all these cases, however, there is a lack of uniformity in the handling of the objects in the database and their use in application programs which are typically written in a language which is not designed for object handling. Using an object-oriented lan-

J. MIGUEL GERZSO and ALEJANDRO P. BUCHMANN • IIMAS, National University of Mexico, Mexico D.F. 01000, Mexico

guage both as interface to a DBMS (which has the capability of handling objects) and as the programming language of the application programs will enhance the ability to manipulate them in a consistent and easy way. A framework for classifying objects has been presented in Ref. 4. There the notion of molecular objects is introduced. A molecular object is an aggregate of records that can be of the same or of different record types. Molecular objects can be classified into disjoint/nondisjoint and recursive/nonrecursive objects. Disjoint objects do not share records with other objects, while nondisjoint ones do. Recursive objects can be described recursively using the same record type(s), while nonrecursive objects cannot.

The main advantages, besides ease of use, of object-oriented languages are the facilities for class and type definitions. This has as its immediate effect that an object which belongs to a class will inherit automatically all the properties of the class. The facility of type definition is definitely useful when handling data of heterogeneous structure, such as numeric and alphanumeric data, unstructured text, drawings, raw images, etc. Existing DBMSs support only a very limited set of types. Finally, the extensibility of some object-oriented languages with their capability for encapsulation can be used effectively for integrity constraint handling in a database environment.

This paper introduces the object-oriented language TM,[11] whose genealogy is shown in Figure 1, and presents the necessary extensions to make TM a suitable programming environment that includes full database capabilities required in a CAD environment. Section 2 of the paper justifies the design of a new language, such as TM, whose main characteristicsare presented in Section 3. Section 4 analyzes what extensions are required to make TM into a programming environment that includes a full-fledged DBMS. Section 5 discusses the issue of molecular objects in a CAD environment and how these can be incorporated in TM, while Section 6 addresses the issue of constraint handling. Finally, the current status and future work are discussed.

2. Justification

Designing a new language these days almost requires an apology or at least a good justification. We feel there are enough reasons to warrant the design and implementation of TM. Original motivation and interest in designing TM came as a response to studying SMALLTALK in 1978.[18] But its design was also very much influenced by other ideas.[1,8,15,16,17,26,32,33] However, the present version of TM was a consequence of studying SMALLTALK-80[12] and was implemented as a response to the difficulties in acquiring its

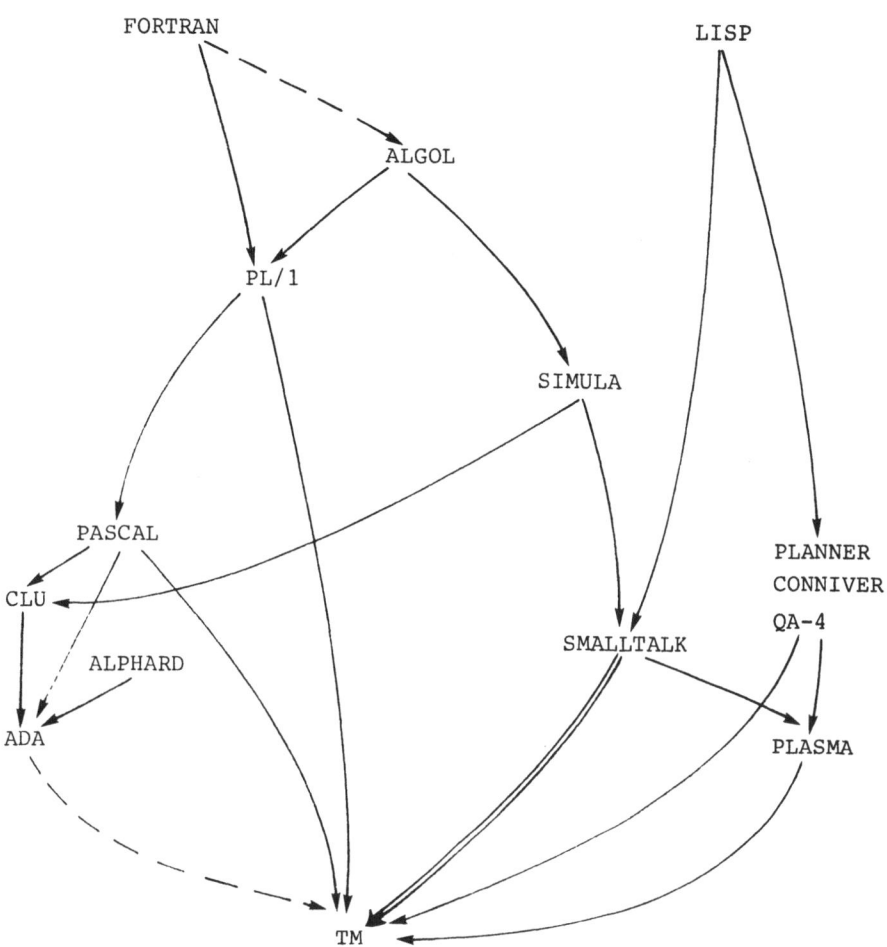

FROM:

PLASMA	OBJECT BASED, MESSAGE PASSING, SYNTAX
PL/1	CONDITIONS, SYNTAX
LISP	EXTENSIBILITY
SMALLTALK	OBJECT BASED, MESSAGE PASSING, SYNTAX
QA-4	LITERALS
PLANNER	FAILURES
PASCAL	TYPING, SYNTAX
ADA	PUBLIC/PRIVATE VIEWING.

FIGURE 1. The genealogy of TM.

source code. The idea was to design a language that was appropriate for computer-aided architectural design and would run on standard computers instead of the special purpose hardware built at Xerox PARC. It also had to be economic, like C and PASCAL vs. PL/I and perhaps ADA, but at the same time it should be extensible, like LISP. TM attempts to clear up some of SMALLTALK's ambiguities, such as the distinction between passive and active objects, and its syntax provides a clear visualization of who is the receiver and who is the sender of a message. Finally, we should point out that some of the newer object-oriented languages, ADA for example, do not appear in TM's genealogy because design and implementation of TM occurred in parallel with those languages.

However, they had some influence during refinement of the language. We are sure that solutions to CAD problems obtained using TM will be applicable to other object-oriented languages.

3. Basic Concepts of TM

3.1. Objects

Objects are the basic entities that can be defined and manipulated in TM, that is, TM is a collection of objects. However, not all the objects have the same characteristics, but they can be classified into two groups: active and passive objects.

3.1.1. Passive Objects. Passive objects are objects that cannot be executed, that is, they can do nothing by themselves but can be accessed, processed, or otherwise acted upon. Most other languages have built-in passive objects, such as numbers, character strings, sets, etc. In TM any given passive object is said to be a member or instance of a class. When an instance is created we say that the object is instantiated. For example, we declare variable c as a member of the class of coordinate triplets "coords" in a procedure by writing

dcl c:coords. temporary

In this case c will have all the characteristics of a coordinate triplet. "Temporary means it is allocated dynamically in a procedure when it is invoked.

Whenever a class is defined one can specify also other classes or metaclasses the objects belong to. This has the effect that "lower ranking" objects can inherit the properties of another class. Property inheritance is further discussed in a later paragraph. Passive objects can only be thought of as data objects.

3.1.2. Active Objects or Administrators. Active objects are those objects that can operate on other objects. In a computational system we are not only interested in classifying objects based on their characteristics, but also in terms of what operations are performed on them. Therefore, associated with each class of passive objects we have codes that describe the properties of the class and its implementation, as well as the operations that can be performed,[12,13] much like abstract data types. Thus another metaphor for class-code is the notion of an administrator who is responsible for processing the instances of the class. The administrator is also considered an object, although an active one.

The administrator, itself an object, is not a member of the class it administers, but is associated with the class. Whenever a request for alteration or instantiation or any other kind of information about the object is needed, the request is directed at the administrator, which handles it transparently to the user.

Once we have defined the notion of adminstrator we can define a class consisting of a collection of passive objects that have the same characteristics and the same administrator to act upon them.

Since administrators are themselves objects they can also be grouped by common attributes and can be acted upon by another, higher-ranking administrator. A collection of active objects and their associated administrator is a metaclass in TM. The actions of the administrator of a metaclass differ from the actions of the administrator of a class, since the elements of a metaclass are themselves active objects which can be modified, created dynamically, linked together, or executed. To illustrate the concept of administrators let us borrow an example from LISP. The function EVAL is a code that executes a code of functions written by a user in LISP. However, EVAL itself can be written in LISP.

3.2. Effects of Having Administrators

The idea of having instances of objects administered by class code has some practical consequences. Some are directly related to the software development process, its maintenance and reliability. Others, such as the possibility of adding responses to an administrator, will be the basis for constraint checking. We will discuss the former here, while discussion of the latter is delayed until Section 7.

In designing and developing software in non-object-oriented languages it is common to define program modules around processing functions. In these functions there may be a code that deals with processing one or more objects. If the system expands and more objects are defined, then all of the functions that affect these objects have to be updated. As the system grows, more arbitrary interconnections are established until the moment when modifying one part would bring the whole system to its knees. Given that the maintenance cost is one of the major costs during the useful life of software, it becomes important to localize any change. Object-oriented languages achieve this through partitioning, modularity, public vs. private definitions, and object communication with the outside.

3.2.1. Partitioning. Partitioning means that the implementation of all functions needed to operate on a particular class or metaclass are implemented within that particular administrator, thus localizing any changes that might be necessary during system development. This is particularly useful in very large applications, such as sophisticated CAD systems which integrate a variety of complex applications and evolve over a long period of time. Since only the administrator is allowed to alter an object within a class, all possible causes of error are concentrated in one place.

3.2.2. Extensibility of TM. TM is considered as a collection of objects, each belonging to a class or metaclass. It is anticipated that the majority of classes and metaclasses will be defined by the users of TM (application programmers, database administrators, etc.) and not by TM's implementors. The way this is accomplished is by means of the extensibility of the system. TM provides a basic collection of classes, such as numbers, strings, lists, etc. and their administrators, as well as metaclasses, such as the virtual machines (as in abstract machines) of numbers, strings, etc. Users will implement other classes or metaclasses by defining a new class or metaclass after the system is loaded by coding its administrator. Then, after the administrator has been accepted by the system, the system is considered extended in that the language includes a new class. From the point of view of the user, there is no difference between sending a message to an instance of a built-in or a user-defined class. There is no need to recompile the TM system in order to include a new administrator. The notion of extensibility was primarily influenced by LISP.

3.3. Representation of Objects

3.3.1. Literals. Any instance of an object can be written as a literal. While the internal representation of instances of an object is hidden from the user, every instance of a class also has an external representation. The general format of the external representation of the object is

$$[.\langle \text{ name of class or metaclass } \rangle \langle \text{constituent parts } \rangle]$$

The square brackets mean that it is a literal. The name of the class or metaclass of the instance appears first, and then the elements that make up the instance. The entire expression enclosed in the square brackets is considered the instance. Simple examples of literals are

```
[.coords 10 1112 4]
[.list[.coords 10 1112 4] [ .coords 23 44 56]]
[.stream | number. : = 'S15'| flowrate. : = 2000| temperature. : = 68]
[.array | rows. : = 24 | columns. : = 45 | 1 1: = 0 | 1 2: = 3]
[.pc 〈 code of an adminstrator 〉]
```

The first two literals have constituent elements which are accessed by position. The third uses names to access its fields (separated by a vertical bar), and the fourth uses indexes, where the third and fourth subexpressions are the initialization. The last is the code of an administrator that has been compiled into pseudocode.

Literals of very common objects (numbers, character, and TM source code) have an abbreviated form to facilitate writing code and making them more readable. Specifically,

```
[.fix 5] as 5
[.float 5 7] as 5.7
[.string John Smith] as 'John Smith'
[.tm source 〈 the code 〉] as {〈 the code 〉}
```

3.3.2. Names of Objects: Variables. In TM a variable is an identifier which can have as a value an instance of a class or of a metaclass. We can distinguish variables because they always have the dot to the right of the identifier:

〈 identifier of object 〉.

A variable is made known at the time it is declared. At this time, the variable is typed in the sense that the values it can take on must be members of a particular class or metaclass. However, the user does have the option to leave the variable untyped by means of the "anything" clause. Consider the following examples:

```
declare c:coords. p:pc. temporary
c. : = [coords 10 10 34];
p. : = [ .pc 〈the code of administrator coords 〉];
```

3.3.3. Names of Objects: Administrators. There is also a need for names of classes and administrator which results from the necessity to send a message to the administrator itself, instead of to an instance of the class. For example, in the event that an instance must be created, a message to the effect "create an instance" must be sent to the administrator. It is impossible to send a message to an instance that does not exist.

The syntax of an administrator name is very similar to that of a variable name, the only difference being that the dot appears to the left of the name, instead of right:

$$.\langle \text{ administrator name } \rangle$$

Examples are

  .fix the administrator of fixed numbers
  .if an administrator that acts as the TM if statement
  .pc administrator that executes code of other administrators
  .tm the administrator of the entire system

3.4. Internal Organization of Objects

So far, we have discussed objects in TM by viewing them from the outside, regardless of their internal organization. Hiding the internal organization allows greater flexibility in the implementation of both passive and active objects.

The simplest object is one that has only one field, such as a number. An object that has more than one field can be accessed via different methods: position, name, and index.

Since the implementation of the instances of a class is hidden from the user, the so-called "instances" may not really exist as passive data objects. They may be implemented algorithmically (also known as virtual data) and the administrator computes the value whenever needed.

3.5. Interaction Between Objects in TM: Message Passing

Without some mechanism of interaction between all objects in the system, there would be no way of processing objects or performing computations. Thus, an object has to send another object a message requesting a particular action. The action can cause the receiving object to return a message with or without a result of some computation, or to perform some side effect. The general format of a message is

$$\langle \text{ receptor object } \rangle \Leftarrow \langle \text{ message pattern } \rangle;$$

The receptor object can be any object known to the system. The representation of the object can either be a literal, a variable, a variable in the field of an object, or an object which is the result of a message expression.

In contrast to SMALLTALK-80 and PLASMA, syntax of the message always requires the use of the arrow ⇐. It is a simple syntax that clarifies the precedence in a message expression. The arrow in the expression above points to the left, which means that it is being "sent out" to another administrator or another invocation of the same administrator. If the arrow points to the right, it means that a message is "coming in" to the present invocation of the administor. The message pattern is made up of a selector followed by keywords:

⟨ selector ⟩ ⟨ parameters or keywords or both ⟩

The selector is the name of the operation that the administrator is requested to carry out and is followed by a pattern of variables, literals, and keywords. A keyword is like a selector but always appears after the selector qualifying the operation being requested. The pattern of parameters serves as the mechanism of invocation of a procedure (response) in an administrator. This mechanism differs from the usual invocation mechanism in most languages where invocation is realized by naming the desired procedure. A simple message expression

pipeno. ⇐ get pnumber. ⇐ concat diameter. ⇐ concat material.;

In this example, a variable pipeno is sent a message to get a pipelinenumber and this number is concatenated to the diameter and the material to form a pipeline descriptor. This expression has the keyword "pnumber."

Parameters are evaluated before they are sent to the receiver object. In the case of a message expression, the expression is evaluated and the resulting object is sent to the receiver object. In the case of a variable, a value is sent. In the case of a literal or a keyword, the literal or keyword itself is sent.

Messages are sent to an administrator not only to create new instances, but also to administrators who function as control-classes, such as "if" and "do." Let us take the example of the "if" administrator, which is the TM version of the "if" statement in other languages. Here, the selector is either "true" or "false," and depending upon which it is, the control is passed to code following the keywords "then" and "else." The "else" part is optional.

. if ⟨ boolean expr ⟩ { ⟨ TM statements ⟩}
    else { ⟨ TM statements ⟩ };

3.6. Responses and Response Adding

In the previous paragraphs we have discussed the mechanism of message passing. Let us look at the internal organization of an administrator to see how an administrator responds to it. The general form of an administrator is

```
{ admi_mod
⟨ administrator and class name ⟩
⟨ field, public and class declarations, if any ⟩

to_itself
⟨ response procedure 1 ⟩
⟨ response procedure 2 ⟩
⟨ response procedure n ⟩
end_it

to_instance
⟨ response procedure 1 ⟩
⟨ response procedure 2 ⟩
⟨ response procedure n ⟩
end_inst

end_mod }
```

This is an oversimplification but sufficient for the purposes of this discussion, the point being that with an "admi_mod" (or administrator module) an administrator and a class are defined for the system. A simplified example of an admi_mod is the extract of a heat exchanger definition given below:

```
{ admi_mod
    equipment class
    instance
        equipment_number:string.
        equipment_descriptor:string.
        location:coords. fields
to_itself
    ⟨ one or more responses ⟩
end_it
```

```
to_instance
  ⟨ one or more responses ⟩
end_inst
end_mod }
{ admi_mod
    heat_exchanger class
    .equipment super_class
    instance
      heat_duty:float.
      number_of_shells:fix.
      type:string. fields
to_itself
  ⟨ one or more responses ⟩
end_it
to_instance
  calc_LMTD temp_shell_in:float. temp_shell_out:float.
    temp_tube_in:float. temp_tube_out:float.⟹
←calc_LMTD temp_shell_in. temp_shell_out.
    temp_tube_in. temp_tube_out;
  end⟹
  set_type type:string.⟹
←instance type.:=type.;
  end⟹
  ⟨ more responses ⟩
end_inst
end_mod }
```

In the "admi_mod" all the public (global), field and class variables are declared. All of these variables are accessible to all of the responses in the administrator. The public variables must have been declared previously at the TM system level. In the equipment and "heat_exchanger" examples we have shown how administrators are defined and how a class can be declared as member of a superclass. We also illustrated how a response procedure can be invoked in the calculation of the logarithmic mean temperature difference in a heat exchanger and how fields are assigned values.

In real systems the code for an administrator can be very large. The problem then becomes: how can the code be broken down into manageable pieces in order to facilitate code development and maintenance? The answer is by means of an "add_resp" (add response module) which has the effect of adding one or more responses to an administrator defined in an "admi_mod" module. All of the gobal, field, and class variables defined in the "admi_mod" are available to the code in the "add_resp" module.

3.7. Failure Procedure

TM includes, as part of a response, the necessary failure procedures. The idea of locating failure code near the response is, first, for clarity of the code and second, to avoid that the calling procedure has to check the failure code and take the remedial action as it is done in most other languages. The failure procedure is similar to condition code in PL/I. If the result of any response contains a failure, TM will take over automatically and transfer control to the failure code. The programmer does not have to code the transfer. In other languages the programmer not only has to code the control transfer, but after taking remedial action has to return and retry the code which originated the failure. The TM failure procedure attempts to repair the damage and then allows execution to resume either at the next statement after the one that originated the failure or to retry execution of the same statement. This feature is attractive when doing database accesses in which one does not want to abort a transaction because of certain failures and it is often difficult to resume execution, since the user has no insight over the internal processing of the database.

3.8. Property Inheritance

Property inheritance has two characteristics: function inheritance and attribute inheritance. Assume a class is defined and its administrator is coded and debugged. If a new type of object is required that has most of the functions of the already defined class, we can define a subclass that has the functionality of the existing class *except* those features we want to redefine in the new administrator. The effect of this relation is that the new object is said to be a member of the subclass of the class to which the old object belongs. In addition, in the new administrator, the programmer declares that the present administrator is responsible for a class which has as its superclass the old class.

The same idea can now be extended beyond mere programming advantages. If we define an object which describes, for example, a pump with attributes such as impeller-diameter and capacity, and another object describing a heat exchanger with heat-transfer-area and number of tubes, we still have to define for both common attributes, such as equipment number, service description, and location, which are common to all pieces of equipment. By defining pumps and heat exchangers as subclasses of equipment, we can define the equipment attributes once and inherit them to the subclasses when needed or we can deal with the superclass without bothering about unncessary detail. This is the notion of generalization.[29]

Property inheritance also affects the way the system processes messages. If a message is sent to an object, the system determines first if the administrator has the necessary response code. If it does not, then the system attempts to determine which administrator has the response code. Thus, if a response is written once and never replaced by a subadministrator, it is the only response needed and maintained.

4. Extending TM for Database Management

For years DBMSs supported more complex structures than the host languages in which the applications were written. Recently developed object-oriented languages have inverted this situation and now commercially available DBMSs are unable to support objects or abstract data types handled by new languages. At the same time it has become clear that most object-oriented languages and the languages developed for artificial intelligence seldom provide extensive data-handling capabilities. However, this is a necessary condition for object-oriented languages to become useful as a CAD environment. Therefore, it is of practical interest to expand an object-oriented language into a programming environment that provides full database capabilities and allows homogeneous handling of data at the database and the application levels.

Since the requirements posed by CAD applications on a DBMS are different from those required by administrative applications we will first analyze the shortcomings of existing DBMSs and will then discuss the necessary additions to TM.

- Off-the-shelf systems cannot handle adequately complex engineering objects which may consist of hundreds or even thousands of records of different types.
- Existing concurrency control mechanisms are tuned for short transactions (1 or 2 sec) while engineering applications require long interactive transactions often spanning hours or even days between consistent states of the database.
- Commercial systems do not provide versioning and update propagation support.
- Most existing DBMSs break down when required to handle in an integrated form the typical mix of unstructured text, numeric data, and graphic data of various kinds which is common to engineering applications.
- Business-oriented DBMSs provide only minimal capabilities for integrity checking.

TM was developed as a language for CAD applications. The main motivation for expanding TM into a full DBMS lies in the features of TM that

that will allow solving some of the major problems stated above: the handling of molecular objects, handling of integrity and consistency constraints, and integrated handling of multiformatted data. As a subproduct of being able do define molecular objects we expect to handle versioning and update propagation in a more elegant manner than is currently possible with existing DBMSs.

Disk access mechanisms will be provided as a first approach by a B+ tree indexing mechanism and will be used both for primary keys and for secondary indexes. Records receive a unique identifier that is used for all the necessary relationships among records through the use of a binary relation of unique internal sequence numbers, also to be managed by the same B+ tree mechanism. This mechanism allows easy linking of database records into molecular objects, even after loading the database. Future expansions will be concerned also with the handling of different graphics capabilities and spatial indexes.

TM also needs schema definition capabilities, which can be provided in an integrated fashion, since TM is particularly well suited for the definition of classes and their administrators. Defining a schema through a Data Definition Language (DDL) is nothing more than defining the logical structure of a database by defining record and item types and, depending on the underlying model, relationships among records. Each relationship between records can also be modeled as an object. Each class has an administrator and the metaclass of all the administrators forms the schema definition and manipulation capabilities of the DBMS. The responses needed for object processing within an application program not using the database features of TM can be separated from the responses required for database operations. However, the class will have all its characteristicsdefined uniformly and the administrator will be able to use all the responses if required.

Commercial DBMSs have to provide special Data Manipulation Languages (DML) whose statements are interspersed in the application program written in the host language. Special DML processors are required that substitute the DML statements for calls to subroutines of functions. This will not be necessary in TM since TM is extensible and the necessary DML commands are just added to the language. The TM compiler can treat DML commands as any other commands. The main advantage is the homogeneous treatment of molecular objects (to be explained in the next section) at the database and the application level, since both levels rely on TM, thereby avoiding unnecessary mappings in which part of the semantics is lost or distorted.

Extended TM is viewed by us as a DBMS for CAD and, therefore, the concurrency control mechanisms will satisfy only CAD requirements. This means that no generalized concurrency control mechanism will be provided and that a check-out mechanism based on the library paradigm will be enforced.[20,15]

Recovery mechanisms are necessary for any DBMS, regardless of the realm of application. However, the recovery mechanism in a CAD-DBMS should double in the update propagation mechanism and workarea reconciliation strategies. This is still an open research topic and no decision has been made as to the mechanics of the recovery system.

5. Abstract Data Types and Molecular Objects in TM

Recent research has shown the need for use of molecular objects in CAD.[4,14,19,24,31] A molecular object is defined as an object that can be viewed at different levels of abstraction. While it may appear to one application as a single tuple of a relation it consists really of several tuples from possibly heterogeneous relations. For example, a pipeline is a molecular object consisting of pieces of pipe, valves, flanges, etc. Each of the components is represented by at least one record or tuple, but the pipeline as a whole is also an object of interest that can either be addressed at a higher level of abstraction or may in turn be a component of another object, such as a plant. The notion of abstract data types, consisting of a data structure and operators defined on it, has been applied by Stonebraker et al. to arrays and other structures embedded in a single tuple of a relation.[31,25] It is, however, necessary to extend the notion beyond that to include molecular objects as defined above. These "molecular ADTs" consist of a multirecord data structure and the necessary operators defined on it. TM's encapsulation facilities are ideally suited for the definition of molecular ADTs.

Restricted notions of molecular objects have been implemented both in relational systems and CODASYL network type systems.[24,19] Neither model offers the necessary DBMS support. In the relational model molecular objects have been formed through foreign keys. An extension to System-R accomplishes it through use of system-defined foreign keys, but unfortunately restricts itself to handling only disjoint molecular objects.

This is not sufficient for practical purposes as shown by Batory et al.[4] In a CODASYL network environment[7] one can use CODASYL sets for the definition of molecular objects. The problem here is the lack of DBMS support for the operations on molecular objects. The semantics are buried in the application programs and their specification is left to the user.

In TM each record type can be viewed as the definition of an object. Each object definition consisting of atomic fields can be viewed as the intention of a relation. The instances of the object are equivalent to tuples. It has been shown in Section 3 how fields can be updated in an object. An object can be defined as consisting of several objects, thereby defining multirecord structures. While the conventional relational operators can be defined for single-record objects, additional operators can be defined for the multirecord structures. We will call these multirecord structures and the operators defined on them "molecular objects." Examples of the operations which are defined on molecular objects are extraction, deletion, 3-D transformations for some graphical molecular objects, etc. Current research is underway to determine the set of necessary operators. In this paper we will only exemplify the kind of operations under discussion using a small subset of molecular operators.

Deletion as used in the context of molecular objects is different from the notion of a single tuple. In the case of disjoint molecular objects, i.e. objects that do not share tuples with other objects, deletion is simply the elimination of the instances of the component objects. This is not possible in the case of nondisjoint molecular objects, in which deletion is equivalent to unlinking a component object from the higher level object and actual deletion only occurs whenever an object is not a component of any other object and has no further use as a simple object in itself.

Extraction is the action of creating a copy of the instance of a molecular object together with the context necessary to operate on it. The context may consist of other objects or different indexing required for an application.

The discussion of transformation operators forces us to analyze the kind of data handled in a complex CAD environment. In the graphics domain a CAD database has to deal with two-dimensional schematics, isometrics, orthographic projections, three-dimensional wire-drawings, three-dimensional solid drawings, etc. Typically some drawings are generated and kept as vector graphics and others as pixels. Most DBMSs cannot interface adequately with graphics packages since basic incompatibilitiesexist between current DMLs and graphic command languages. These incompatibilities worsen as the internal representations change from vector graphics to pixels or quad-trees. TM offers two features which alleviate the problem. First, through the encapsulation mechanism it is possible to present an object to the user as a regular TM object, regardless of the underlying implementation. This means that an object can have internally any data structure and can be implemented either in TM or in any other language or assembler. Second, operator overloading allows us to define graphic operations independently of its internal representation. The user can then apply the operator regardless of whether the data are stored as vectors or pixels.

Molecular objects are particularly useful when dealing with graphic data. Typical examples of molecular objects involving graphics are engineering schematics, such as process flowsheets (PFS) and piping and instrumentation diagrams (P and ID). A complex P and ID can consist of several hundred objects if one counts major equipment, instruments, valves, pipelines, etc.

Common engineering practice calls for version control at the document level, and this in turn requires knowledge of where an object is used. A molecular object of type drawing will consist of all the objects that appear on that drawing, and whenever a change is made to one component of the drawing, all the other drawings that use that particular component can be easily flagged. This notion can be extended to any other molecular object and its components and will thus be the basis of a version control mechanism. We have implemented such a mechanism in a conventional DBMS using sets and relying on the Data Dictionary for it. Extended TM would provide the necessary framework for an integral implementation of this feature.

Molecular objects are also essential for the concurrency control mechanism.

This mechanism relies on the extraction operator. A CAD-DB is actually a confederation of databases consisting of a project-wide database with approved data and workarea databases in which the designer does his preliminary designs. After approval by the supervisor and/or project manager the data from the work-area DB are integrated into the project-wide approved database where they are accessible to other designers working on the project. Modifications are not allowed directly in the approved database. Any modification there is subject to a revision process which is documented in the same database. The reason for this is twofold: CAD transactions are long and can take hours and even days to take the database from one consistent state to another and the changes have to be approved by the affected users. It is unreasonable to treat these data with a conventional concurrency control mechanism which would lock out other users for long periods. The solution consists in making extracts of the project-wide database, an operation that may require the access and copy of hundreds or even thousands of tuples or records to set up the correct context for a transaction. Doing this either manually or mechanically without the definition of high-level objects is a tedious and time-consuming process.

Extended TM will use the molecular objects defined in it for extraction of workarea databases and will suport this mechanism for concurrency control. Extracted objects are flagged in the process-wide database, but they are still accessible. Users checking out flagged objects are recorded for notification of changes. This process is reminiscent of intention locks where a transaction records its intention to carry out an update but does not acquire an exclusive lock until it is ready to execute the update. The impact analysis that has to be performed whenever data are checked back will be discussed in the next section since it involves the handling of constraints.

7. Handling Constraints in Extended TM

The difference between a sophisticated DBMS and an expert system is becoming smaller. DBMSs are incorporating more "intelligence" through definition and enforcement of constraints and expert systems are handling increasingly higher volumes of data. Let us, therefore, identify the types of integrity constraints a CAD-DBMS has to handle. The most important are type checking, range checking, value matching, property inheritance, and consistency constraint handling.

Type checking is performed by most languages and DBMSs. In many cases there exist discrepancies among the types supported by the DBMS and those supported by the host-languages with the consequent, often awkward, mappings. A language like TM both simplifies and complicates the issue. Type discrepancies between the application language and the DML and DDL are eliminated. On the other hand, the facility for defining ADTs, and particularly molecular objects, requires an extension of the notion of type checking to multirecord structures. Fortunately, TM provides through the class definition for extensive type checking, not only on each individual field in an object but also for component objects.

Range checking consists in checking that a database value falls within preestablished ranges—for example, no negative absolute temperatures, mole fractions between 0 and 1, etc. Currently available commercial systems offer little help in range checking and several patches can be used: through hardcoded database procedures (if the system has that option), or storing range values in the Data Dictionary and checking against them. Hardcoded databases procedures are always invoked and cannot be easily changed without recompiling the schema. Using metadata stored in the Data Dictionary is only a feasible option if the DBMS offers triggers for database procedures and suffers some shortcomings similar to the first solution. An extension of TM can use responses that are defined for a class or specific field (equivalent to hard-coded database procedures) using variable parameters. This is important when dealing with a variety of units and conversion factors. In addition, while hardcoded procedures are always invoked, responses can be selectively loaded whenever the design process reaches a stage in which checking is required, and not for every update operation. This is important when the percentage of errors is small or a large portion of the data is input from simulation programs or similar software. Checking all updates or insertions would degrade the system.

Value matching is similar to range checking, but here the values to be input to the database have to assume one value out of a set of possible values. Built-in objects, such as lists and sets, are essential for solving value matching easily. No additional problem is therefore expected when implementing extended TM.

Property inheritance pretends to ensure that all objects in a subclass inherit the properties of another class eliminating inconsistencies that arise whenever attributes are repeated. Most DBMSs do not provide this capability. TM offers attribute inheritance as an integral part of its class definition, and since no mapping is required between DBMS and application language the problem is virtually solved.

Consistency constraint handling depends most on the particular application. One example is logical sequence checking, which refers to the proper position of engineering elements in a molecular object. Examples are the correct sequence of block and check valves in a pipeline, or the correct connection of instruments in an instrument loop. Other examples are the compliance with current codes, such as TEMA for heat exchangers or ASME codes for pressure vessels. These constraints correspond to engineering knowledge, and existing CAD systems embed them exclusively in application programs. However, when changes are made to the database, these may affect other data and their users. Previously we concentrated on the extraction problem for concurrency control. When a workcopy is checked back, it is necessary to trace the impact any change has on other portions of the database. Current engineering practice requires that this is done through inspection of design documents by experienced designers. Constraints defined on the database can be of great help to the engineer. But it is unrealistic to assume that a complete set of constraints can be defined before start-up of a CAD system. Therefore, it is important that a CAD system can "learn" through the addition of new constraints. In extended TM, constraints can be implemented as responses of an administrator and new responses can be added to an administrator as they are needed. As a CAD system is used, criteria are added, deleted, or modified, thus allowing also for changes in engineering codes or design practice.

8. Current Status and Future Work

At present two implementations of TM have been completed.[3,10] The first one was implemented in PASCAL on a VAX-780. The main contribution of this work was an implementation design which specified the internal data structures of the system, the characteristics of the compiler, etc. This design was used as a basis for the second implementation, also in PASCAL but under TENEX on a Foonly-F2. During these implementations the main concern was to develop first a thoroughly thought-out design and not to worry about performance at this stage. Accordingly, since transportability was another issue of concern, the TM compiler produces a pseudocode that is processed by a so-called virtual machine. Future implementations may eventually produce machine code. Finally, since the user has to be aware only of the valid message patterns, the implementor has entire freedom to implement an administrator in any language or to incorporate off-the-shelf software.

Currently, the basic language definition and its compiler are stable enough so that we can think now about the proposed extensions. While none of the database extensions has been implemented in TM, we have experimented extensively with existing DBMSs in real-life CAD systems for process plant design and architectural CAD systems. From these implementations and the pitfalls suffered applying conventional DBMSs to CAD systems we have extrapolated the requirements of an extended TM. Future research will concentrate on run-time support required for molecular objects, expansion of TM for handling heterogeneous data, and implementation of constraint handling capabilities.

9. Conclusions

An object-oriented language developed for CAD applications was presented. The language offers various characteristics that make it well suited for CAD applications. The most important are object-oriented approach, encapsulation, extensibility, attribute inheritance, public and private definitions, communication among objects through message passing, and response adding.

The previous characteristics make TM a candidate for extension into a programming environment that includes full database capabilities, which are needed for sophisticated CAD systems. The extension is natural and smooth since the language supports all the necessary constructs. While a DBMS that grows out of a programming language offers naturally all the constructs present in the language and the schema definition is essentially the same as the object definition for the applications, it still is necessary to design the schema carefully, and for that purpose we need a data-model which is not provided by extended programming languages. It is this point where purely language-based approaches would flounder. On the other hand, DBMS capabilities have to be extended to match the capabilities of object-oriented languages. Pairing the developments in language design with the advances made in database design and implementation within the context of a demanding real-life application, such as CAD, appears to be a promising direction for future research.

Acknowledgments

The most important modifications of the several design versions of TM were a consequence of the comments of the graduate students at IIMAS: Salvador Barra, Fernando Fraustro, Victor Valadez, and to some extent, Alfonso San Miguel. Comments from Renato Barrera, Jeannie Becerra, and Hector Calvario of IIMAS are also appreciated. Finally, thanks to Fred Moses of Computervision for his interest in TM and his penetrating observations, and to Don Batory and Hank Korth of the University of Texas, Austin, for many interesting discussions of the database issues and helpful comments.

References

1. J. ALLEN, *Anatomy of LISP*, McGraw-Hill, New York, 1978
2. M. M. ASTRAHAN *et al.* System R: Relational approach to database management, *ACM Trans. Database Syst.* **1**(2), June (1976), pp. 97–137.
3. S. BARRA, Diseño detallado de la implementación de TM, M. S. Thesis, IIMAS, in preparation.
4. D. S. BATORY and A. P. BUCHMANN, Molecular objects, abstract data types, and data models: A framework, 10th International Conference on Very Large Data Bases, Singapore, August, 1984, pp. 172–184.
5. A. P. BUCHMANN, Current trends in CAD databases, *CAD*, May (1984), **16**(3), pp. 123–126.
6. E. F. CODD, Extending the database relational model to capture more meaning, *ACM Trans. Database Syst.* **4**(4), December (1979), pp. 379–434.
7. CODASYL Data Base Task Group, Report, April, 1971.
8. M. L. DERTOUZOS, DELPHI: A time-shared system for students of MIT subject 6.031, structure and interpretation of languages (LISP), MIT, Cambridge, Massachusetts, 1974.
9. Department of Defense: The programming language ADA reference manual, *Lecture Notes in Computer Science*, Springer Verlag, New York, 1981.
10. F. FRAUSTRO, Implementación de TM en la Máquina Foonly-F2, M.S. Thesis, IIMAS, in preparation.
11. J. M. GERZSO, Report on the language TM: Its design and definition, Instituto de Investigaciones en Matemáticas Aplicadas y en Sistemas, Universidad Nacional Autonoma de Mexico, Mexico (report forthcoming).
12. A. GOLDBERG *et al.*, The SMALLTALK-80 system, *Byte Mag.* **6**(8) August (1981), pp. 14–26.
13. A. GOLDBERG and D. ROBSON, SMALLTALK-80, the Language and its Implementation, Addison-Wesley, Reading, Massachusetts, 1983.

14. R. HASKIN and R. LORIE, On extending the functions of a relational database system, Proc. 1982 ACM SIGMOD Intl. Conf. on Management of Data, Gainesville, Fla., pp. 207–212.
15. C. HEWITT, The description and theoretic analysis (using schemas) of PLANNER: A language for proving theorems and manipulating models in a robot, Ph.D. thesis, MIT, Cambridge, Massachusetts, 1971.
16. C. HEWITT and B. SMITH, A PLASMA Primer, AI Lab Working Paper 92, MIT, Cambridge, Massachusetts, 1975.
17. J. HUGHS, PL/I Structured Programming, Wiley, New York, 1979.
18. D. INGALLS, The SMALLTALK-76 programming system, 5th Annual Symposium on Principles of Programming Languages, Tucson, Arizona, January, 1978, ACM-SIGACT-SIG-PLAN Report.
19. R. JOHNSON, J. E. SCHWEITZER, and E. R. WARKENSTINE, A DBMS for handling structured engineering entities, Proc. 1983 ACM Engineering Applications, San Jose, California, May, 1983, pp. 3–12.
20. R. H. KATZ, DAVID: Design aids for VLSI using integrated databases, IEEE TC Database Eng. Bull. 5(2), June (1982), pp. 29–32.
21. B. H. LISKOV, An introduction to CLU, Computation Structures Group Memo 136, Laboratory for Computer Science, MIT, Cambridge Massachusetts, February, 1976.
22. B. H. LISKOV et al., Abstraction mechanisms in CLU, Computation Structure Group Memo 144-1, Laboratory for Computer Science, MIT, Cambridge, Massachusetts, January, 1977.
23. B. H. LISKOV et al., CLU reference manual, Lecture Notes in Computer Science, Springer Verlag, New York, 1981.
24. R. LORIE and W. PLOUFFE, Complex objects and their use in design transactions, Proc. 1983 ADM Engineering Applications, San Jose, California, May, 1983, pp. 115–121.
25. J. ONG, D. FOGG, and M. STONEBRAKER, Implementation of data abstraction in the relational database system INGRES, ACM SIGMOD Rec. 14(1), March (1984), pp. 1–14.
26. J. F. RULIFSON, J. A. DERKSEN, and R. J. WALDINGER, QA-4: A procedural calculus for intuitive reasoning, AI Center Technical Report, 73 Stanford Research Institute, Menlo Park, California, November, 1972.
27. J. W. SCHMIDT, Some high level language constructs for data of type relation, ACM Trans. Database Syst. 2(3), September (1977), pp. 247–261.
28. M. SHAW, ALPHARD: Form and Content, Springer Verlag, New York, 1981.
29. J. M. SMITH and D. C. P. SMITH, Database abstraction: Aggregation and generalization, ACM Trans. Database Syst. 2(2), June (1977), pp. 105–133.
30. M. STONEBRAKER, E. WONG, and P. KREPS, The design and implementation of INGRES, ACM Trans. Database Syst. 1(3), September (1976), pp. 189–222.
31. M. STONEBRAKER, B. RUBENSTEIN and A. GUTTMAN, Application of abstract data types and abstract indices to CAD databases, Proc. 1983 ACM Engineering Design Applications, San Jose, California, May, 1983, pp. 107–113.
32. N. WIRTH and K. JENSEN, PASCAL user manual and report, Lecture Notes in Computer Science, Springer Verlag, New York, 1974.

12

A CAD/CAM DATABASE MANAGEMENT SYSTEM AND ITS QUERY LANGUAGE

Y. C. LEE† AND K. S. FU

1. Introduction

In general, functions performed during a design/manufacturing cycle include sales and marketing, product design and analysis, production planning and control, manufacturing, quality control, as well as shipping and inventory control.[1] These functions apparently have to share a large amount of data that are collected from many different sources. As a consequence, a centralized database is needed in order to prevent data from being repeatedly or inconsistently stored. However, a database composed of nothing but data files with different formats can never be adequate in supporting all these design/manufacturing functions owing to the expensive efforts required by data preparation and the programming of the interfaces between file formats. Therefore, the prevailing database management techniques, which have been successfully used in business, are possibly of essential importance for such a database to be accurate, useful, and convenient.

A database management system (DBMS) can be considered as a software/hardware system through which only a conceptual database, i.e., abstract representation of the physical store, is presented to the user.[2] Two significant characteristics of DBMS are *data independence* and *data consistency or integrity*. The abstract (conceptual) database can be represented

†Y. C. Lee is now with the Department of Electrical Engineering and Computer Science, University of Michigan, Ann Arbor, Michigan 48109-1109.

Y. C. LEE and K. S. FU • School of Electrical Engineering, Purdue University, West Lafayette, Indiana 47907

by one of the three most influential data models including relational, network, and hierarchical. Both the network and hierarchical models have been widely implemented as commercial packages, partly because of their relatively cheaper software cost and higher efficiency. Nevertheless, the relational model based on the set-theoretic concept has had its simplicity and high descriptive power commonly approved.[3] In the proposed approach, the relational model is chosen primarily based on the belief that the productivity of the application programs needed in a design and manufacturing environment can be enhanced, and that data models related to different design/manufacturing functions can be flexibly integrated together.

The first step in the design of a relational model is to conceptually identify entity sets and the relationships among them. Also needed are the properties (attributes) associated with entities and relationships. This is actually a semantics capturing task assumed by the responsibility of the so-called database designer. However, it is not trivial at all to map the result of the conceptualization into a data scheme, considering the variety of the requirements in each environment. In fact, it has been emphasized by many researchers[3-7] that an ill-defined data scheme, which fails to incorporate some of the important semantic information about the real world, would result in the poor performance of a database system. As a consequence, these researchers have proposed their own models or schemes. It is not the purpose of this paper to comment on these various approaches. Among them, however, the generic scheme proposed by Smith and Smith[7] is preferred because of the close match between the generic scheme and the BNF grammar,[8] which is considered as a conceptualization tool in the proposed database design method. It appears that the dynamic nature resulting from the recursiveness and nondeterminism in grammar rules can best be characterized by the data abstractions based on aggregation and generalization.

The idea of employing BNF grammar as a conceptualization tool stems from the experience accrued during an effort to store the constructive solid geometry (CSG) trees in a relational database.[9] The CSG scheme, which is one of the most commonly used solid modeling representations,[10] can be well specified by a set of BNF grammar rules. In converting this grammar into a generic scheme, it is realized that most of the important steps can be formulated as a systematic procedure. Accordingly, the proposed database design approach, called a syntactic method, is as follows:

1. As a formulation scheme, the grammar rules which concisely depict the fundamental structure embedded in the underlying data will be considered as a conceptualization tool.

2. Following a systematic conversion procedure, the grammar rules are then transformed into a more rigorous generic scheme.

To support the resulting generic scheme, that is, to map the generic scheme into relational model, the language SEQUEL[11-13] has been modified, at the syntax level, and been implemented to define, control, and manipulate the flat relations which represent the highly structural generic scheme.

In Section 2, the syntactic method is described along with the basic notations used for BNF grammar and generic scheme. Following in Section 3 are generic schemes resulting from the syntactic method. The scheme for CSG representations is an example where the grammar already exists. The other two generic schemes, one for machinability data and the other one for production information, are two exmaples where grammar can be first created to conceptualize the world model. Together, these two sections demonstrate an effort to achieve a uniform relational treatment of design/manufacturing data items. In Section 4, example queries are shown in order to discuss the key issues related to the augmented query language. The syntax rules of major concern are listed in the Appendix. Section 5 describes the interface between the CSG-based geometric database and one of its applications—a ray casting solid modeler. Using queries, the geometric data for a mechanical part is retrieved from the database and organized in the required format for the solid modeler. Section 6 summarizes the main themes of this paper.

2. Conversion from BNF Grammar into Generic Scheme

In this section, the definition of BNF grammar is first reviewed, followed by the additional descriptions which are needed for grammar rules to become a conceptualization tool. The grammar depicting the CSG scheme will be used as an example. Then the reason why the generic scheme is considered as an adequate model for data that are structured by grammar rules is discussed. Lastly, the procedure for converting grammar rules into a generic scheme is introduced. A procedure slightly resembling this one has been introduced to construct relations from a BNF grammar.[14] In the proposed approach, however, the underlying generic relational model requires quite different and much more complicated treatment.

2.1. Backus–Naur Form Grammar

A BNF grammar[8] can be defined as a four-tuple $G = (V_N, V_T, P, S)$, where

V_N: set of *nonterminals;*

V_T: set of *terminals;*

$S \in V_N$: *start symbol,*

and P is a set of production rules of the form

$$X_0 ::= \alpha 1 \mid \alpha 2 \mid \ldots \mid \alpha i \mid \ldots \mid \alpha m$$

where α_i is the *i*th *alternate* for some nonterminal X_0 and is represented as

$$X_{i1}X_{i2} \cdots X_{ij} \cdots X_{im_i}, \quad \text{where } X_{ij} \in (V_T \cup V_N)$$

The BNF grammar is the same as the context-free grammar defined in Ref. 15. Being commonly used in specifying syntax for programming languages, it is of course very much involved in designing language syntax when descriptive power, precedence relation, parsing efficiency, and ease of use are concerned. However, this should not concern the database designer when the grammar rules are used only for the purpose of conceptualizing the world model.

For example, the CSG scheme can be represented as a BNF grammar $G_{CSG} = (V_N, V_T, P, S)$, where $V_N = \{\langle \text{mech\_part} \rangle, \langle \text{object} \rangle, \langle \text{primitive} \rangle, \langle \text{set\_operator} \rangle\}$, $V_T = \{\text{movement, cube, cylinder, cone, sphere, torus, union, intersect, difference}\}$, $S = \langle \text{mech\_part} \rangle$, and P:

```
⟨mech part⟩ :: = ⟨object⟩
   ⟨object⟩ :: = ⟨primitive⟩ |
                 ⟨object⟩ ⟨set_operator⟩ ⟨object⟩ |
                 ⟨object⟩ movement
⟨primitive⟩ :: = cube | cylinder | cone | sphere | torus
⟨set_operator⟩ :: = union | intersect | difference
```

In addition to the foregoing definition, some a priori knowledge is assumed that allows the database designer to partition terminals into two categories:

- *Atomic terminals*, which correspond to atomic (indecomposable) values, and
- *Entity terminals*, which correspond to entity sets composed of entities that are distinguishable by associated attribute values.

Also, the meta symbol "$::=$" will be replaced by "$\xrightarrow{x}$" where x is a character string indicating the specific information being considered. This notation is needed because the same nonterminals (usually the starting symbol) may also be used for some other information. For instance, the rule $\langle$mech_part$\rangle \xrightarrow{\text{geometric}} \langle$object$\rangle$ is associated with the geometric information while $\langle$mech_part$\rangle \xrightarrow{\text{production}} \langle$part$\rangle$ may deal with the production information. This facilitates flexible merging of two or more separately defined individual data schemes.

Note that terminals are those symbols which never appear on the left-hand side of any rule. In the above G_{CSG}, the motion argument can be either a set of six translation/rotation arguments or a set of 16 entries of its associated homogeneous transformation matrix. The cube, cylinder, cone, sphere, and torus all require different sets of size parameters. Yet the union, intersect, and difference only carry different procedural information that requires no parameters at all. Therefore, it is clear that all the terminals, except the union, intersect, and difference, can be classified as *entity terminals*.

It can be easily observed that different constructive trees will be derived, based on the CSG grammar, to represent different mechanical parts. Theoretically, the dynamic nature is attributed to the recursiveness and nondeterminism embedded in the grammar rules.[15] Notice that in the rules associated with the left-hand side (LHS) symbol $\langle$object$\rangle$ and $\langle$primitive$\rangle$, there are different alternatives that imply different tree structures such as binary or unary operations. Also, in some of these alternatives, the LHS symbol is repeated. Both the recursiveness and nondeterministism require a data scheme that allows a uniform treatment of these symbols at a higher level while providing their individual details at a lower level.

2.2. Generic Scheme

The idea of aggregation and generalization[7] seems to be a good approach to characterizing the dynamic nature of a grammar. Aggregation is an abstraction in which a relationship is regarded as a higher-level object, while generalization refers to an abstraction in which a set of objects is generalized as a generic object. The generic schemes based on such data abstractions are capable of capturing the structural information and of being represented as relations. Conceptually, both grammar rules and generic scheme are based on the decomposition of information. The grammar, with its more natural and easier formulation, can be used as a tool to relieve a database designer of more tedious database design considerations. On the other hand, the generic scheme serves as an intermediate model between the grammar and the relational model. Therefore, it is of significant importance to devise a systematic procedure capable of identifying the phenomena of aggregation and generalization from a set of grammar rules. As the third alternative shown in Figure 1, the proposed database design methodology is divided into two phases:

1. BNF grammar rules are used by the designer to conceptualize the world model;
2. Grammar rules are then converted into a generic scheme according to the following systematic procedure.

Before we introduce the conversion procedure, the formal definition of a generic structure is given below[7]:

Var R: generic

$$S_{k1} = (R_{11}, \ldots, R_{1p_1});$$
$$\cdots$$
$$S_{km} = (R_{m1}, \ldots, R_{mp_m})$$

of
aggregate[keylist]

$$S_1:\{key\}R_1;$$
$$\cdots$$
$$S_n:\{key\}R_n$$

end

where

1. Each S_{ki}, $(i = 1,m)$, is a selector name, and is the same as some S_j $(j = 1,n)$;
2. R_{ij}, $(i = 1,m, j = 1,p_i)$, is a generic identifier whose key domain is the same as that of R. The range $(R_{i1}, \ldots, R_{ip_i})$ forms the image domain for S_{ki};
3. Keylist is a sequence of S_i's $(i = 1,n)$;
4. S_i, $(i = 1,n)$, is an attribute name. If S_i is the same as S_{kj}, then the type "$\{key\}R_i$" is the range $(R_{j1}, \ldots, R_{jp_j})$;
5. R_i $(i = 1,n)$ is either a generic identifier or a type identifier (in which case "key" must not appear).

As indicated in Figure 2, the generic structure R simultaneously specifies two abstractions. It specifies R as an aggregation of a relationship between objects R_1 through R_n, and as a generalization of a class containing objects R_{11} through R_{mp_m}. The most important feature of a generic scheme is that all the generalization descendant objects must have the same key domains as their ascendant object. This requirement allows individual objects to be uniformly referred to regardless of the generic class in which they appear.

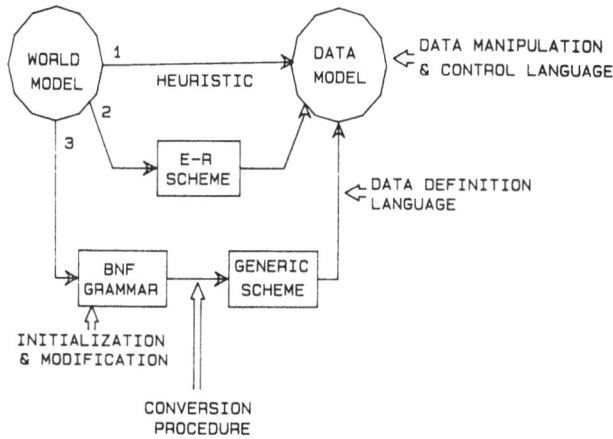

FIGURE 1. Alternatives for database design.

2.3. Conversion from Grammar to Generic Scheme

To be emphasized is that a complete database design process can hardly be automated, especially the highly subjective naming process.[6] The procedure described below should not be understood as an automatic procedure that can be programmed. Instead, the procedure is a set of straightforward steps which can be followed by the database designer to map the grammar rules into a generic scheme.

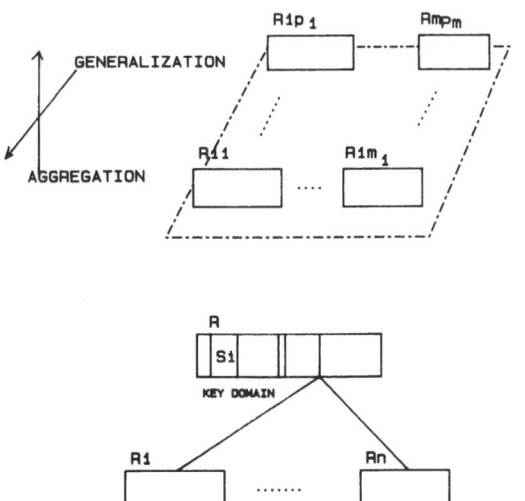

FIGURE 2. A graphical notation for the generic structure.

Based on the a priori knowledge about the partition of terminals, non-terminals and alternates are classified according to the following definitions so as to facilitate identifying the phenomena of generalization and aggregation.

- *Nonentity Nonterminal.* An alternate composed of atomic terminals and/or nonentity nonterminals is defined as a *nonentity alternate*. A *nonentity nonterminal* is a nonterminal associated with only *nonentity alternates*. According to the number and composition of alternates, nonentity nonterminals can be further classified into four different types as shown in Table 1, where a simple example is shown.
- *Entity Nonterminal.* This is a nonterminal of which every alternate includes at least one entity terminal or entity nonterminal. Clearly, this type of alternate is defined as *entity alternate*. Entity nonterminals can also be partitioned as shown in Table 1.

Obviously, the above two nonterminal types do not cover all kinds of nonterminals that are associated with different possible combinations of alternates. However, an assumption to be made is the *homogeneity among alternates*, that is, all the alternates for a certain nonterminal must be either

TABLE 1
Classification of Nonterminals[a]

Nonterminal	Example	No. of alternates	Composition of alternate
Nonentity nonterminal			**Nonentity alternate**
Single-valued	E	$=1$	Single atomic terminal
Multivalued	C	>1	Single atomic terminal
Single-aggregate	D	$=1$	Multiple atomic terminals and/or nonentity nonterminals
Multiaggregate	F	>1	Multiple atomic terminals and/or nonentity nonterminals
Entity nonterminal			**Entity alternate**
Individual entity nonterminal	G	$=1$	Including at least one entity or entity terminal
Generic entity nonterminal	S,A,B	>1	Including at least one entity nonterminal or entity terminal

[a]The grammar used as an example in this table is $G = \{V_N, V_T, P, S\}$, where $V_N = \{S, A, B, C, D, E, F, G\}$, $V_T = \{a, b, c, d, e, f, g, h\}$, $S = S$, and P:
$S::= A \mid B, A::= a \mid b \mid c, B::= BCB \mid BDE \mid AaF \mid BG \mid a, C::= e \mid f \mid g \mid h, D::= efg, E::= e, F::= fgh \mid gfE \mid Cgf, G::= d$, assuming that a, b, c, and d are entity terminals; e, f, g, and h are atomic terminals.

entity alternates or nonentity alternates. This assumption is reasonable in practice since a nonterminal represents either an entity or an attribute value while different alternates simply indicate various appearances of the nonterminal. Furthermore, we do not expect any occurrence of multiaggregate nonterminal. The reason is that, while a generic entity may have various representations, a value of an attribute should be specified in a unique format.

As indicated by the close match between the grammar rules and data abstractions, to convert the CSG grammar into a generic scheme is conceptually easy. But it is specificationwise rather tedious since the generic scheme calls for rigorous definitions such as the requirement of unified key domains at different levels of generalization. The conversion procedure described below includes three phases for entity terminals, individual entity nonterminals, and generic entity nonterminals, respectively.

Phase 1. For each entity terminal (ET), X, define the generic structure as

Var X:

> generic
> of
> aggregate [$X\#$]
> > $X\#$:identification number;
> > .... (other attributes). ...
> end

Phase 2. For each individual entity nonterminal (IEN), X_0, with its single alternate designated as $X_1 X_2 \cdots X_i \cdots X_n$, define the generic identifier following the following procedures.

1. Define the fundamental part of the identifier,
   Var X_0:

   > generic
   > of
   > aggregate [$X_0\#$]
   > > $X_0\#$:identification number;
   > > .... (other attributes). ...
   > end

2. For each X_i, ($i = 1, n$), which is an entity terminal (ET) or entity nonterminal (EN), add an attribute $X_i\#$ to the definition of X_0.
3. For each X_i which is an atomic terminal (AT) or a single-valued

nonentity nonterminal (SVNN), no explicit action needs to be taken.

4. For each X_i which is a multivalued nonentity nonterminal (MVNN), add to the definition of X_0 an attribute X_i with as its domain the set composed of all the alternates.

5. For each X_i which is a single-aggregate nonentity nonterminal (SANN), add an attribute $X_i\#$ to the definition of X_0. In addition, define an identifier for X_i as,
   Var X_i:

```
        generic
        of
        aggregate [X_i#]
                X_i#:identification number;
                .... (other attributes). ...
        end
```

and for each y_j, $(j = 1,r)$, which appears in the alternate of X_i and is either a multivalued nonentity nonterminal or a single-aggregate one, follow step 4 if y_j is a MVNN, or otherwise, follow step 5. If $r = 0$, ignore the whole step 5, and treat X_i as in step 3.

Phase 3. For each generic entity nonterminal (GEN), X_0, with its multiple alternates designated as α_1, α_2, ... , α_n, define the following identifiers:

1. the first one, related to the GEN X_0, is defined as
   Var X_0:

```
        generic
                CT = category of alternates;
        of
        aggregate [X_0#]
                X_0#:identification number;
                .... (other attributes). ...
                CT:category of alternates;
        end
```

2. for each alternate α_i, if α_i is simply an entity nonterminal or an entity terminal, X,
   then

    i. insert the name X into the set CT, that is, include X as a subcategory of X_0.

    ii. modify the definition of the identifier X by substituting the key attribute of X_0 for that of X. This step should be done recursively if X is also a GEN.

else

    i. insert into CT an appropriate name for α_i, say X.

    ii. define the fundamental part of the identifier X as Var X:

> generic
> of
> aggregate $[X_0\#]$
> $X_0\#$:identification number;
> end

then follow the steps 2, 3, 4, and 5 in phase 2.

The result of the above procedure is a generic scheme only with the fundamental definitions which can be extracted from the grammar. The following tasks must be performed in order to complete the description of the generic scheme.

1. Add individual names to those aggregate relations and modify others, if necessary. Again, this is inevitable because natural naming is rather subjective.
2. Identify and add additional attributes when needed, especially those that carry less structural information.
3. Check functional dependency and other design issues. Brief discussion on functional dependency with respect to generalization and aggregation has been mentioned along with the generic scheme in the original paper.[7]

3. The CSG-Based Geometric Database and Two Examples of Manufacturing Databases

For the fast-growing CAD/CAM technologies, geometric solid modeling plays one of the most important roles.[10,16-17] Solid modeling, which focuses on complete representations of solids,[10] is undoubtedly a key to allow an integrated CAD/CAM database management system to facilitate the calculation of well-defined geometric properties of any stored object. Although none of the existing solid representation schemes are suitable for all applications, we have chosen the constructive solid geometry (CSG) as the basis for the geometric database model, primarily because of its conciseness, unambiguity, simple validity checking, ease of creation, and high descriptive power.[18-20]

Based on the BNF grammar in Section 2, the generic scheme for CSG representations can be designed following the conversion procedure described in the previous section. Owing to the space limitation, only a graphical representation of this generic scheme is shown in Figure 3. Basically, aggregation occurs up the page while generalization occurs, orthogonally, out of the page. More detailed notation can be found in the paper of generic scheme.[7]

In this scheme, two additional concerns have been included. First, the relation *primitive* is added with an attribute MV# based on the belief that except the starting primitive, other primitives are often created with an initial movement with respect to the reference point of the starting primitive. Second, since the database is not used for a single object, the key attribute of the relation *object* is replaced by a pair of key attributes, MP# and O#, and so are those of its generalization children relations such as *primitive*, etc.

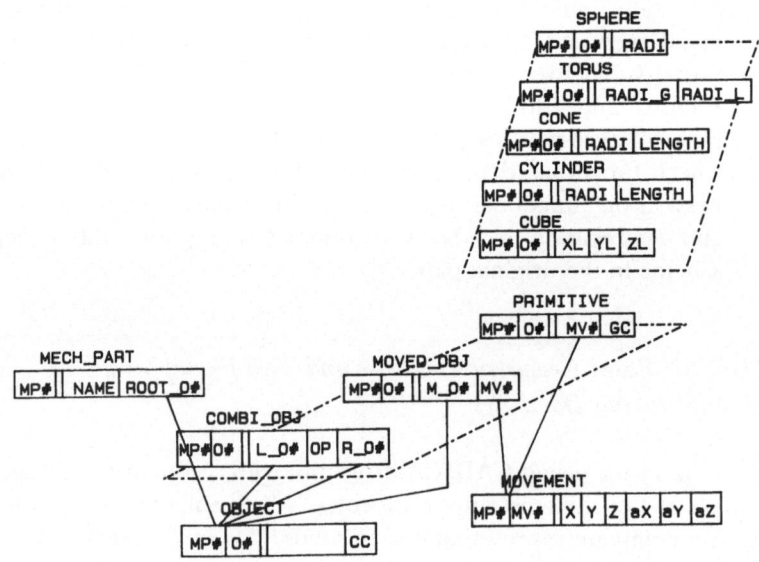

FIGURE 3. Generic scheme for CSG representations.

Compared with the generic scheme in Fig. 3, the following set of relations, which can be designed in an ad hoc fashion, is simpler:

> OBJECT (ID,L-Child-Object, R-Child-Object,
> Operation, Origin, Angles-Wrt-Axis)
> CUBE (ID, Length, Width, Height)
> CYLINDER (ID, Height, Radius)
> CONE (ID, Height, Radius)
> SPHERE (ID, Radius)
> TORUS (ID, R_global, R_local)

However, this scheme has at least two disadvantages: First, NULL value has to be defined for some attributes. For example, the associated value of the attribute R-Child-Object is NULL for primitive or moved objects. NULL value complicates the implementation of relational operators. Second, this scheme lacks the relationships that are required to facilitate data retrieval. For instance, how would a user know a L-Child-Object of a combined object is a cube and be able to retrieve the Length of the cube in the relation CUBE?

After dealing with the geometric information of mechanical parts, it is important to examine the adequateness of the syntactic method by applying it to other kinds of manufacturing data. Following in this section are two examples of manufacturing database: one for production data and the other one for machinability data. As these two examples are intended to demonstrate how the generic structure embedded in each example can be characterized, they are not complete as far as the practical detailedness is concerned.

Process operations include those manufacturing activities that transform a work part from one state of completion into a more advanced state of completion.[1] Among them, basic processes like casting give the work material its initial form. Machining processes, such as turning, drilling, and milling, follow the basic processes and give the work part its desired final geometry. Finally, finishing processes can improve the appearance or provide a protective coating. As for assembly, the distinguishing feature is that components are joined together. The assembly operations may be mechanical fastening operations or joining processes like weldings.

For simplicity, a detailed description of each production operation is not specified in the following grammar rules. However, it is obvious that these operations may require different sets of parameters and thus cannot be stored together in a single relation. This is a problem that can best be solved by the generic model, which provides the flexibility to allow structural dissimilarities. The production grammar can be defined as $G_{production}$

$= (V_N, V_T, S, P)$, where $V_N = \{\langle\text{product}\rangle, \langle\text{assembly}\rangle, \langle\text{subassembly}\rangle,$ $\langle\text{assembly\_primitive}\rangle, \langle\text{mech\_part}\rangle, \langle\text{part}\rangle, \langle\text{initial\_part}\rangle, \langle\text{basic\_opera-}$ $\text{tion}\rangle, \langle\text{machined\_part}\rangle, \langle\text{finishing\_part}\rangle\}$, $V_T = \{\text{assembly\_operation},$ vendor_part, machining_operation, raw_material, finishing_operation, casting, molding$\}$, $S = \langle\text{product}\rangle$, and P:

$$\langle\text{product}\rangle \xrightarrow{\text{production}} \langle\text{assembly}\rangle$$

$$\langle\text{assembly}\rangle \xrightarrow{\text{production}} \langle\text{subassembly}\rangle \mid \langle\text{assembly\_primitive}\rangle$$

$$\langle\text{subassembly}\rangle \xrightarrow{\text{production}} \langle\text{assembly}\rangle \text{ assembly\_operation } \langle\text{assembly}\rangle$$

$$\langle\text{assembly\_primitive}\rangle \xrightarrow{\text{production}} \langle\text{mechanical\_part}\rangle$$

$$\langle\text{mechanical\_part}\rangle \xrightarrow{\text{production}} \langle\text{part}\rangle$$

$$\langle\text{part}\rangle \xrightarrow{\text{production}} \langle\text{initial\_part}\rangle \mid \langle\text{machined\_part}\rangle \mid$$
$$\langle\text{finishing\_part}\rangle \mid \text{vendor\_part}$$

$$\langle\text{initial\_part}\rangle \xrightarrow{\text{production}} \langle\text{basic\_operation}\rangle \text{ raw\_material}$$

$$\langle\text{machined\_part}\rangle \xrightarrow{\text{production}} \text{machining\_operation } \langle\text{part}\rangle$$

$$\langle\text{finishing\_part}\rangle \xrightarrow{\text{production}} \text{finishing\_operation } \langle\text{part}\rangle$$

$$\langle\text{basic\_operation}\rangle \xrightarrow{\text{production}} \text{casting} \mid \text{molding}$$

Assume that *vendor_part* is an entity nonterminal, while *assembly_operation, raw_material, machining_operation, finishing_operation, casting,* and *molding* are atomic terminals. These rules can be transformed into the generic scheme shown graphically in Figure 4, in which key attribute sets PROD#, (PROD#, AS#), MP#, and (MP#,PT#) are associated with the relations PRODUCT, ASSEMBLY, MECH_PART, and PART, respectively. Other attributes are named as the abbreviations of their corresponding terminals or nonterminals in the above grammar rules. Note that the two databases of the production information and the part geometry can be combined by defining the relation mechanical_part as mechanical_part(MP#, name, PT#, root_O#). The attribute root_O# is related to the root of the geometric constructive tree while PT# leads to the production list.

The second example is concerned with machinability data. Machinability is defined as the relative ease with which a metal can be machined by an appropriate cutting tool. Usually, after the machine tool, cutting tool, and depth of cut have been selected, success in the operation depends on the proper choice of cutting conditions—cut speed and feed rate. Machinability data systems are basically intended to support this decision problem.

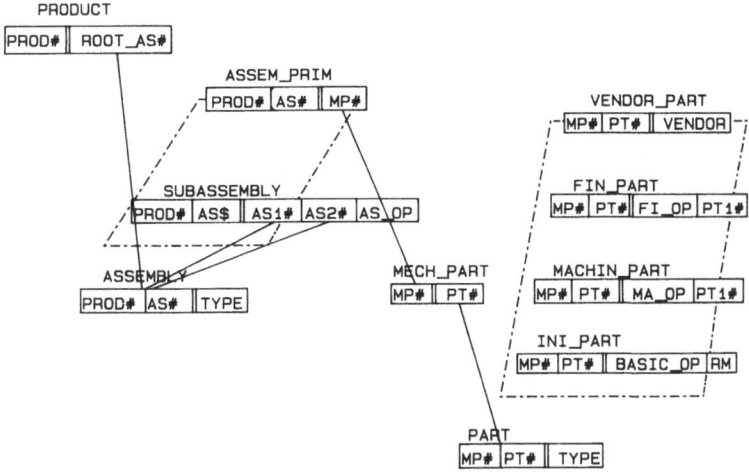

FIGURE 4. Generic scheme for production information.

Tabular formats are commonly used for machinability data concerning drilling, face-milling, turning, etc.[21] However, these tables are not purely two-dimensional tables. Reorganization of them are needed before being put into relational tables. The grammar intended to characterize the structure about each operation can be defined as $G_{machinability} = (V_N, V_T, P, S)$, $V_N = \{\langle\text{machinability\_data}\rangle, \langle\text{part\_material}\rangle, \langle\text{operation}\rangle, \langle\text{drill}\rangle, \langle\text{face\_milling}\rangle, \langle\text{turning}\rangle, \langle\text{tool\_type}\rangle\}$, $V_T = \{\text{material\_type, hardnessH, hardnessL, condition, hole\_diameter, speed, feed, tool\_material, depth\_of\_cut, high\_speed\_steel, cast\_alloy, carbide\_brazed, carbide\_throwaway}\}$, $S = \langle\text{machinability\_data}\rangle$, and P:

$\langle\text{machinability\_data}\rangle \xrightarrow{\text{machinability}} \langle\text{part\_material}\rangle \langle\text{operation}\rangle$

$\langle\text{part\_material}\rangle \xrightarrow{\text{machinability}}$ material_type hardnessH hardnessL condition

$\langle\text{operation}\rangle \xrightarrow{\text{machinability}} \langle\text{drill}\rangle \mid \langle\text{face\_milling}\rangle \mid \langle\text{turning}\rangle$

$\langle\text{drill}\rangle \xrightarrow{\text{machinability}}$ hole_diameter speed feed tool_material

$\langle\text{face\_milling}\rangle \xrightarrow{\text{machinability}}$ depth_of_cut speed feed $\langle\text{tool\_type}\rangle$ tool_material

$\langle\text{turning}\rangle \xrightarrow{\text{machinability}}$ depth_of_cut speed feed $\langle\text{tool\_type}\rangle$ tool_material

$\langle\text{tool\_type}\rangle \xrightarrow{\text{machinability}}$ high_speed_steel $\mid$ cast alloy $\mid$ carbide_brazed $\mid$ carbide_throwaway

Assume that all terminals are atomic terminals; the associated generic scheme is shown in Figure 5. The above rules are not complete by any means. However, it suggests that in order to accommodate the multitude of different parameters that characterize various operations, a structural organization of the machinability data is needed and can possibly be done. Furthermore, it allows more flexibility to combine with other databases. In the next section, example queries posed against these databases will be shown.

4. The SEQUEL-Augmented Query Language

A generic relational model is a relational data model that represents the highly structural generic scheme in terms of flat tables. The significance of the generic relational model is a uniform relational treatment for all kinds of identifiers—individual, aggregate, and generic. To achieve this, consistent relationships in both aggregation and generalization domains must be maintained during the use of the database. In this section, key issues about the modification of the query language SEQUEL are briefly discussed. Also emphasized is the additional facility for defining procedural relations, which are aimed at reducing design or programming efforts. Syntax rules needed for these modifications are listed in the Appendix.

For data definition, the syntax is modified according to the definition of generic identifier. It is used to set up data structures in both data directory and data storage. Associated with each relation, it is important to collect all four kinds of relations that characterize the generic structure. As shown in Figure 6, they are aggregation parent (PaR), aggregation children (CaR), generalization parent (PgR), and generalization children

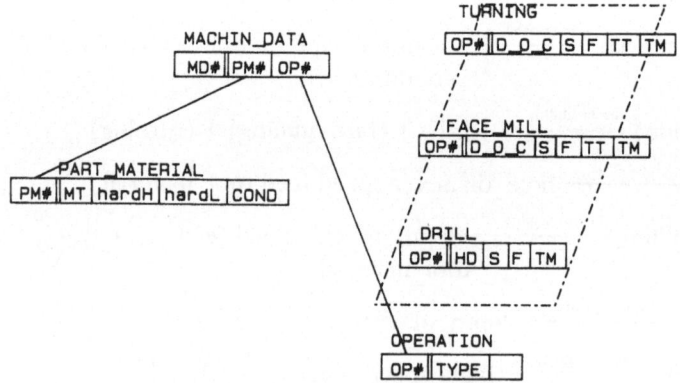

FIGURE 5. Generic scheme for machinability data.

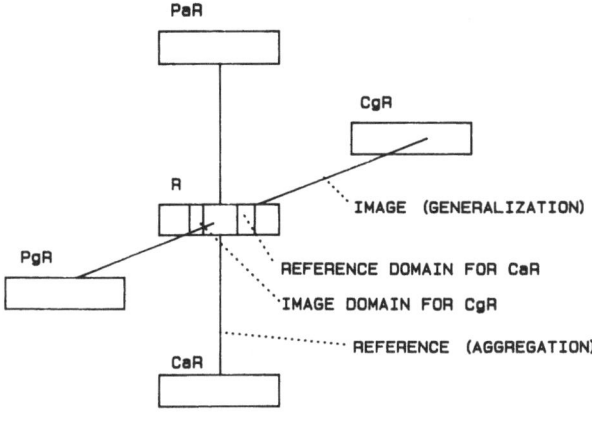

PaR — AGGREGATION PARENT
CaR — AGGREGATION CHILD
PgR — GENERALIZATION PARENT
CgR — GENERALIZATION CHILD

FIGURE 6. A generic relation and its parents and children along generalization and aggregation.

(CgR) relations. These relations present the data semantics by which a user knows how to write queries, and also from which the relational invariant constraints can be enforced.

To create new relations, the definition of a generic scheme is preceded by the command CREATE. For example:

Q1. Create the relation 'moved_object':

> CREATE
> VAR moved_obj:
>   GENERIC
>   OF
>   AGGREGATE[MP#,O#]
>   MP#:INT;
>   O#:INT;
>   M_O#:KEY object;
>   MV#:KEY movement
>   END

Recall that the newly created relation, moved_obj, is associated with a pair of key attributes, namely, MP# and O#. They are used for inter-

and intramechanical-part distinction purposes, respectively. Both attributes are of type INT (integer). The moved-obj relation has no children relation along generalization but two along aggregation, namely, relation object and movement.

By entering the complete generic scheme defined in Section 2 through the use of CREATE, all the relations needed for the CSG scheme are registered in the database directory. Associated with them are the parent–children relationships along both aggregation and generalization planes.

Relational invariants that must be maintained between these relations address nontrivial data control issues. We have reduced the size of considerations in Ref. 7 so as to explicitly specify these invariant constraints in terms of data control statements. Assume that the attributes of each identifier include four parts: key attributes, aggregate attributes, generic attributes, and the rest, and that no overlaps between different kinds of attributes are allowed except between key and aggregate attributes as well as between key and generic attributes. Also assume that the update only modifies nonkey attributes while a modification of key values is considered as a deletion followed by an insertion. A simplified table of possible violations against the relational invariants can be found in Ref. 9. Following the syntax rules for data control, these violations can be stated as integrity assertions, and thus, the relational invariants can be enforced. For instance:

Q2. Relational invariants assertion on relation 'moved_object':

```
ASSERT ON INSERT OF moved_obj:
(((⟨MP#, O#⟩ IN SET(object.MP#, object.O#))
 AND
 (⟨MP#,M_O#⟩ IN SET(object.MP#, object.O#))
 AND
 (⟨MP#, MV#⟩ IN SET (movement.MP#, movement.MV#)))
ASSERT ON DELETE OF moved_obj:
 (⟨MP#, O#⟩ NOT IN SET (object.MP#, object.O#))
ASSERT ON UPDATE OF moved_obj:
(((⟨MP#, NEW M_O#⟩ IN SET (object.MP#, object.O#))
 AND
 (⟨MP#, NEW MV#⟩ IN SET (movement.MP#, movement.MV#)))
```

The foregoing integrity assertions clearly state what relational invariants might be violated and should be prevented. For example, the first statement asserts that the two key values, ⟨MP#, O#⟩, of the inserted moved_object should be in the key set of the object relation, which is the PgR of the moved_obj relation. The next two assertions are related, respectively, to the two CaRs, namely, object and movement.

After relations are defined and relational constraints asserted, relational tuples can be inserted, deleted, and updated. To create an object in the database, a sequence of insertions is performed in the same way as a CSG tree is established. The following query is a simple example of insertion:

Q3. Combine primitive#101 and primitive#102 to form the unioned object#103 for the mech_part#101:

> INSERT INTO object:
> ⟨101, 103, "combi_obj"⟩
> INSERT INTO combi_obj:
> ⟨101, 103, 101, "union", 102⟩

Essentially, a basic purpose of database management system is to support the retrieval of database content. SEQUEL is a keyword-oriented query language and provides convenient ways to specify predicates. A typical example is given as follows:

Q4. List all the *movements* incurred in composing the object named "shaft":

```
SELECT M.MV#, M.X, M.Y, M.Z, M.aX, M.aY, M.aZ
FROM   mech_part, moved_obj MO, movement M
WHERE  mech_part.name = "shaft"
AND    mech_part.MP# = MO.MP#
AND    M.MP# = MO.MP#
AND    M.MV# = MO.MV#
```

As shown above, the SELECT clause lists the attributes to be returned. The WHERE clause may contain any collection of predicates defined on the attributes of the relations that are listed in the FROM clause in terms of relation names or variables.

A second retrieval example is a query posed against the generic scheme for machinability data, defined earlier in Section 3. A table-lookup into a machinability handbook can now be replaced by a query of the following form:

Q5. Find out the cut speed and feed rate for a turning operation using a high_speed_steel tool of *M*2 material. The depth of cut is 0.15 in. and the part material is of Cold_Drawn B111 type with hardness between 120 and 130.

```
SELECT Y.S, Y.F
FROM    machinability_data X, turning Y
WHERE Y.TT = "high_speed_steel"
AND     Y.TM = "M2"
AND     Y.D_O_C = 0.15
AND     Y.OP# = X.OP#
AND     X.PM# =
        SELECT PM#
        FROM   part_material
        WHERE MT = "B111"
        AND   hardH ) = 130
        AND   hardL ( = 120
        AND   COND = "Cold_Drawn"
```

In addition to the above facilities, two important features of SEQUEL, view and trigger, are used to support other useful definitions. *View* is defined as a relation derived from one or more other relations while being used in the same way as a base relation. As virtually nonexistent relations, views can be used to filter out unrelated and/or unauthorized data, and to modify data representations for a specific application. As the generic structure sometimes appears to be rather complicated for certain specific application, a view definition can always reorganize the generic structure into a different abstraction hierarchy.

In the CSG model, scale information is stored with individual primitives such as cones and cylinders. Suppose that a user or an application programmer prefers a view in which the generic *primitive* is of the lowest generalization level; then he can do the following:

Q6. Define a view "prim" in which all the geometric information about the primitives for the mechanical part "link" is included:

```
DEFINE VIEW prim (O#,type,XL,YL,ZL,X,Y,Z,aX,aY,aZ)
AS
((    SELECT PR.O#, PR.GC, CO.length, CO.radi, CO.radi,
             M.X, M.Y, M.Z, M.aX, M.aY, M.aZ
      FROM   mech_part, primitive PR, movement M, cone CO
      WHERE PR.GC = "cone"
      AND    mech_part.name = "link"
      AND    mech_part.MP# = PR.MP#
```

```
        AND     M.MP#  =  PR.MP#
        AND     M.MV#  =  PR.MV#
        AND     CO.MP#  =  PR.MP#
        AND     CO.O#  =  PR.O#  )
UNION
(       SELECT PR.O#, PR.GC, CU.XL, CU.YL, CU.ZL,
                M.X, M.Y, M.Z, M.aX, M.aY, M.aZ
        FROM    mech_part, primitive PR, movement M, cube CU
        WHERE PR.GC  =  "cube"
        AND     mech_part.name  =  "link"
        AND     mech_part.MP#  =  PR.MP#
        AND     M.MP#  =  PR.MP#
        AND     M.MV#  =  PR.MV#
        AND     CU.MP#  =  PR.MP#
        AND     CU.O#  =  PR.O#  )
UNION
(       SELECT PR.O#, PR.GC, CY.length, CY.radi, CY.radi,
                M.X, M.Y, M.Z, M.aX, M.aY, M.aZ
        FROM    mech_part, primitive PR, movement M, cylinder CY
        WHERE PR.GC  =  "cylinder"
        AND     mech_part.name  =  "link"
        AND     mech_part.MP#  =  PR.MP#
        AND     M.MP#  =  PR.MP#
        AND     M.MV#  =  PR.MV#
        AND     CY.MP#  =  PR.MP#
        AND     CY.O#  =  PR.O#  )
UNION
(       SELECT PR.O#, PR.GC, SP.radi, SP.radi, SP.radi,
                M.X, M.Y, M.Z, M.aX, M.aY, M.aZ
        FROM    mech_part, primitive PR, movement M, sphere SP
        WHERE PR.GC  =  "sphere"
        AND     mech_part.name  =  "link"
        AND     mech_part.MP#  =  PR.MP#
        AND     M.MP#  =  PR.MP#
        AND     M.MV#  =  PR.MV#
        AND     SP.MP#  =  PR.MP#
        AND     SP.O#  =  PR.O#  )
UNION
(       SELECT PR.O#, PR.GC, TOR.radi_G, TOR.radi_L, TOR.radi_L,
                M.X, M.Y, M.Z, M.aX, M.aY, M.aZ
        FROM    mech_part, primitive PR, movement M, torus TOR
        WHERE PR.GC  =  "torus"
```

AND	mech_part.name = "link"
AND	mech_part.MP# = PR.MP#
AND	M.MP# = PR.MP#
AND	M.MV# = PR.MV#
AND	TOR.MP# = PR.MP#
AND	TOR.O# = PR.O#))

Basically, the user intends to collect all location, orientation, and scale information within a unified relation *primitive*. The tradeoff is that he has to allow multiple interpretations of certain attributes for different kinds of primitives.

The following view is another example of view definition showing how the content of the aforementioned production database can be conveniently retrieved in order to provide the information required by a Bill-of-Material (BOM)[1,22]:

Q7. Define a view that lists the quantity and vendor of each vendor-part needed for each product:

DEFINE VIEW vending (PROD#, PT#, vendor, quantity)
AS(
   SELECT A.PROD#, Z.PT#, Z.vendor, COUNT(C.AS#)
   FROM   product A, assem_prim C, vendor_part Z
   WHEREA.PROD# = C.PROD#
   AND   C.MP# = Z.MP#)

where COUNT is a built-in function. Similarly, other views can be posed to find out how much raw material is needed for each product, or what kinds of machining/finishing operations are needed for a part.

In contrast with the passive role played by *view, procedure* is included as an active data definition facility. The need for a procedure definition arises from the frequent use of procedural design. The idea behind it is to use a single update to start a set of prespecified relation updates. A procedure relation is therefore defined as a relation whose name is the same as that of the procedure and whose attributes are labeled by their associated parameter names. The sequence of prespecified relation updates is defined in terms of the *trigger* statement, which is used in SEQUEL as a data control facility to automatically update some data when other related data are updated. By calling a procedure relation with a tuple specifying its associated parameters, a sequence of queries is executed and new tuples are inserted.

In the following example, a procedural object is defined. Note that

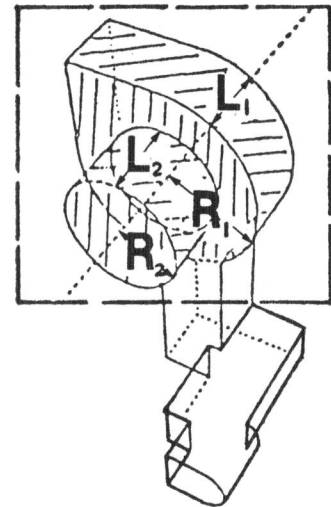

FIGURE 7. A procedural object "head."

procedural objects are defined and stored simply for the ease of design while the complete geometric information is stored in base relations.

Q8. Define a procedural object, as shown in Figure 7, of which the four parameters are R1, R2, L1, and L2:

```
DEFINE PROCEDURE head:[MP#,head#]
    MP#:INT;
    head#:INT;
    R1,R2,L1,L2:REAL;
    root.O#:KEY object;
    MV#:KEY movement
DEFINE TRIGGER add_head
    ON INSERT OF head:
    ( INSERT INTO object:
        ((MP#, O#,   "combi_obj"),
         (MP#, O# + 1, "combi_obj"),
         (MP#, O# + 2, "primitive"),
         (MP#, O# + 3, "primitive"),
         (MP#, O# + 4, "primitive"));
      INSERT INTO primitive:
        ((MP#, O# + 2, MV#,      "cube"),
         (MP#, O# + 3, 100,   "cylinder"),
         (MP#, O# + 4, MV# + 1, "cylinder"));
      INSERT INTO combi_obj:
        ((MP#, O#,   O# + 1, "union", O# + 2),
         (MP#, O# + 1, O# + 3, "union", O# + 4));
```

INSERT INTO movement:
  ($\langle$MP#, MV#, -R1,0.,0.,0.,0.,0.$\rangle$,
  $\langle$MP#, MV# + 1, 0.,0.,L1,0.,0.,0.$\rangle$));
INSERT INTO cylinder:
  ($\langle$MP#, O# + 3, R1, L1$\rangle$,
  $\langle$MP#, O# + 4, R2, L2$\rangle$));
INSERT INTO cube:
  $\langle$MP#, O# + 2, R1, R1, L1$\rangle$))

It is obvious that the *trigger* add-head contains nothing but a sequence of creation steps with all identification numbers being replaced by parameters. The deletion part is omitted for simplicity. Although the definition of procedure seems to be very tedious, it needs to be defined only once.

5. Interfacing a Ray Casting Solid Modeler via Queries

An experimental prototype of the augmented-SEQUEL language has been implemented in "C" language on the UNIX† operating system at the Engineering Computer Network of Purdue University. The main purpose of this implementation is to verify the modified syntax and to provide a language for actual interfaces with applications. Except at the syntax level, no attempt has been made to follow the implementation of the language SEQUEL or its supporting DBMS, the system *R*.

Shown in Figures 8 and 9 are two mechanical parts drawn by a solid modeler based on the algorithm of ray casting.[23] The solid modeler, which is also locally implemented, can handle all the primitives mentioned in Section 2. It requires three tabular input files:

1. *Object File.* This includes four fields, namely, the object_number (O#), operation (OP), left_child_object (L_O#), and right_child_object (R_O#). The R_O# field has multiple meanings: it indicates the right_child_object for a combined object; stores the movement_number for a moved object; and is redundant for primitives.
2. *Primitive File.* It contains the type of the primitive as well as the parameters characterizing the size and transformation. Similarly,

---

†UNIX is a Trademark of AT&T Bell Laboratories.

FIGURE 8. A mechanical part "link."

the three size fields have to be multiply defined due to the variety of size parameters.

3. *Movement File.* It corresponds to those movement arguments needed to move objects.

It is obvious that for some applications, the application program must allow fields in its input format to be multiply defined so as to reduce the number of input files. In such cases, the correctness of the input data is sometimes very doubtful. With a well-defined database and a convenient query language, however, data preparation can be easily and accurately accomplished. For the above solid modeler, as an example, the *movement file* can be retrieved using the query Q4 in Section 4 while Q7 is responsible for the *primitive file*. The *object file* can also be prepared using a query similar to Q7. The two mechanical parts, link and clevis, are stored with many others in the geometric database organized according to the CSG scheme shown in Fig. 3.

FIGURE 9. A mechanical part "clevis."

6. Conclusions

Aiming at a uniform relational treatment of data items that are related to various aspects in design and manufacturing, this research emphasizes convenient facilities for defining relational data schemes, manipulating relational data, and incorporating with application uses.

Based on BNF grammar and generic scheme, we have proposed a syntactic method to facilitate scheme definition. The significance of this approach is magnified by the augmented-SEQUEL query language which supports the relational database built from a generic scheme. Application interface via queries is shown to be very useful with the implementation of a prototype of this language. Future researches include a more thorough examination of practical data and the enhancement of the query language both in its syntax and in its implementation.

Acknowledgment

This work was supported by CIDMAC, a research unit of Purdue University, sponsored by Purdue, Cincinnati Milicron Corporation, Control Data Corporation, Cummins Engine Company, Ransburg Corporation and TRW.

Appendix: Modified Syntax of The Augmented SEQUEL Language

In the following syntax, square brackets "[" and "]" are used to indicate optional parameters. The true square brackets are denoted by "[[" and "]]." All the grammar symbols that are not defined here are comparable to those in Refs. 11 and 12.

Creation of Relation

⟨create-relation⟩::=	CREATE ⟨generic-identifier⟩
⟨generic-identifier⟩::=	VAR ⟨relation-name⟩:⟨generic-part⟩ OF ⟨aggregate-part⟩ END
⟨generic-part⟩::=	GENERIC ⟨cluster-list⟩ | GENERIC
⟨cluster-list⟩::=	⟨cluster⟩ | ⟨cluster-list⟩ ; ⟨cluster⟩
⟨cluster⟩::=	⟨cluster-name⟩ = (⟨object-list⟩)
⟨object-list⟩::=	⟨relation-name⟩ | ⟨object-list⟩ , ⟨relation-name⟩
⟨relation-name⟩::=	identifier
⟨cluster-name⟩::=	identifier
⟨aggregate-part⟩::=	AGGREGATE[[⟨keylist⟩]]⟨attri-defn-list⟩
⟨keylist⟩::=	⟨attri-name⟩ | ⟨keylist⟩ , ⟨attri-name⟩
⟨attri-name⟩::=	identifier
⟨attri-defn-list⟩::=	⟨attri-name⟩:⟨attri-domain⟩ | ⟨attri-defn-list⟩ ; ⟨attri-name⟩:⟨attri-domain⟩
⟨attri-domain⟩::=	KEY ⟨relation-name⟩ | ⟨type⟩
⟨type⟩::=	CHAR(⟨integer⟩) | CHAR(*) | INTEGER | SHORTINT |

	REAL(⟨integer⟩ , ⟨integer⟩) |
	FLOAT |
	identifier
⟨integer⟩::=	number

View, Trigger, and Procedure

⟨define-view⟩::=	DEFINE VIEW ⟨relation-name⟩ [(⟨attri-name-list⟩)] AS ⟨query⟩
⟨define-trigger⟩::=	DEFINE TRIGGER ⟨trig-name⟩ ON ⟨trig-condition⟩:(⟨statement-list⟩)
⟨trig-name⟩::=	⟨name⟩
⟨trig-condition⟩::=	⟨action⟩
⟨statement-list⟩::=	⟨statement⟩ |
	⟨statement-list⟩ ; ⟨statement⟩
⟨action⟩::=	INSERTION OF ⟨relation-name⟩ [⟨var-name⟩]
	DELETION OF ⟨relation-name⟩ [⟨var-name⟩]
	UPDATE OF ⟨relation-name⟩ [⟨var-name⟩][⟨attri-name-list⟩]
⟨define-procedure⟩ ::=	DEFINE PROCEDURE ⟨relation-name⟩ :[[⟨keylist⟩]]⟨attribute-defn-list⟩ ⟨define-trigger⟩

Integrity Assertion

⟨asrt-statement⟩::=	ASSERT [⟨asrt-name⟩][IMMEDIATE][ON ⟨asrt-condition⟩]:⟨boolean⟩
⟨asrt-condition⟩::=	⟨action-list⟩ |
	⟨relation-name⟩[⟨var-name⟩]
⟨action-list⟩::=	⟨action⟩ |
	⟨action-list⟩, ⟨action⟩

References

1. M. P. GROOVER, *Automation, Production Systems, and Computer-Aided Manufacturing,* Prentice-Hall, Englewood Cliffs, New Jersey, 1980.
2. J. D. ULLMAN, *Principles of Database Systems,* Computer Science Press, Rockville, Maryland, 1980.
3. E. F. CODD, Relational database: A practical foundation for productivity, *Commun. ACM* **25,**(2), 109–117 (1982).
4. S. A. BORKIN, *Data Models: A Semantic Approach for Database Systems,* MIT Press, Cambridge, Massachusetts, 1982.

5. P. P. Chen, The entity-relationship model—Toward a unified view of data, *ACM Trans. Database Syst.* **1**(1), 9–36 (1976).

6. N. Roussopoulos and R. T. Yeh, An adaptable methodology for database design, *Computer* **17**(5), 64–80 (1984).

7. J. M. Smith and D. C. P. Smith, Database abstractions: Aggregation and generalization, *ACM Trans. Database Syst.* **2**(2) 105–133 (1977).

8. P. Naur (Ed.), Revised report on the algorithmic language ALGOL 60, *Commun. ACM* **6**(1), 1–17 (1963).

9. Y. C. Lee and K. S. Fu, A CSG based DBMS for CAD/CAM and its supporting query language, Proc. ACM-IEEE SIGMOD 83 (Engineering Databases), San Jose, California, May 23–26, 1983, pp. 123–130.

10. A. A. G. Requicha, Representations for rigid solids: Theory, methods, and systems, *Comput. Surv.* **12**(4), 437–464 (1980).

11. M. M. Astrahan *et al.*, System R: Relational approach to database management, *ACM Trans. Database Syst.* **1**(6), 93–133 (1976).

12. D. D. Chamberlin *et al.*, SEQUEL 2: A unified approach to data definition, manipulation, and control, *IBM J. Res. Dev.*, **20**(6), 560–575 (1976).

13. D. D. Chamberlin *et al.*, A history and evaluation of system R, *Commun. ACM* **24**(10), 632–646 (1981).

14. S. S. Yau and P. C. Grabow, A model for representing programs using hierarchical graphs, *IEEE Trans. Software Eng.*, **SE-7**(6), 556–574 (1981).

15. A. V. Aho and J. D. Ullman, *The Theory of Parsing, Translation, and Compiling*, Prentice-Hall, Englewood Cliffs, New Jersey, 1972.

16. W. Beeby, The future of integrated CAD/CAM systems: The Boeing perspective, *Comput. Graph. Appl.* **2**(1), 51–56 (1982).

17. M. A. Wesley, Construction and use of geometric models, in *Computer Aided Design, Modeling, Systems Engineering, CAD-Systems*, J. Encarnacao (Ed.), Springer Verlag, New York, 1980, pp. 79–136.

18. J. W. Boyse and J. E. Gilchrist, GMSolid: Interactive modeling for design and analysis of solids, *Comput. Graph. Appl.* **2**(2), 27–40 (1982).

19. C. M. Brown, PADL-2: A technical summary, *Comput. Graph. Appl.* **2**(2), 69–84 (1982).

20. A. A. G. Requicha and H. B. Voelcker, Solid modeling: A historical summary and contemporary assessment, *Comput. Graph. Appl.* **2**(2), 9–24 (1982).

21. Metcut Research Associates Inc., *Machining Data Handbook*, Cincinnati, Ohio, 1966.

22. R. J. Tersine, *Principles of Inventory and Materials Management*, Second Edition, North-Holland, New York, 1982.

23. S. D. Roth, Ray casting for modeling solids, *Comput. Graph. Image Process.* **18**, 109–144 (1982).

13

SELF-DESCRIBING AND SELF-DOCUMENTING DATABASE SYSTEMS

Nick Roussopoulos and Leo Mark

1. Introduction

The ANSI/SPARC DBSSG[1] took a major step forward for the database community when it identified the need for a conceptual schema in the context of a three-schema framework for database systems. The three-schema framework allows a clear separation of the conceptual schema from the external schema and the internal physical schema, resulting in databases which are flexible and adaptable to changes. A new step forward needs to be taken if the database system framework is to allow databases to be flexible and adaptable to changes of the conceptual schema.

To accommodate this flexibility two new notions, that of a self-describing database and that of a self-documenting database, are introduced. A self-describing database contains an explicitly stored, self-describing metaschema which models and controls the conceptual schema of the database. A self-documenting database contains a "documentation" of the evolution of the conceptual schema of the database.

This chapter also presents a simple database system framework supporting self-describing and self-documenting (SD/D) databases. It includes a single data language DL which facilitates access, change, and documentation of the conceptual schema evolution through the metaschema. The data language also facilitates access and the manipulation of all levels including the metaschema, the data dictionary, the conceptual schema, and the database extension. This uniformity allows the production of plug compatible data management tools like DDL processors, DML processors, data editors, report generators, etc. for a variety of data models.

NICK ROUSSOPOULOS and LEO MARK • Department of Computer Science, University of Maryland, College Park, Maryland 20742

An SD/D Database System supports two *orthogonal* dimensions of data description:

- the point-of-view dimension; and
- the intension–extension dimension.

The *point-of-view dimension* is the three levels of data description proposed by ANSI/SPARC,[1] which consist of several *external schemata*, a *conceptual schema*, and an *internal schema*. The *intension–extension dimension* has four levels of data description: the first contains information about the data model in use and is modeled in the *metaschema*. The second contains information about the management and usage of data and is modeled in the *data dictionary schema*. The third contains information about the database applications and is modeled in the *application schema* (or schemata). The fourth level is the *applications data* stored in the database. Each of these levels in this intension–extension dimension is the intension of the next level and the extension of the previous one. The only exception to this is that the metaschema contains its own intension that is, it is *self-describing*.[14]

The idea of a metaschema, or a schema for schemata, has been mentioned in some previous efforts. But, to our knowledge, it has never been fully developed. Most recently, the ISO Working Group on Concepts and Terminology for the Conceptual Schema and Information Base[4] pointed out the need for representing and changing both schema and data in a database and briefly listed some capabilities of a schema facility supporting this. In Nijssen,[16] an architecture of a system which does some management of both schema and data is outlined. In a report prepared by the Computer Corporation of America for the National Bureau of Standards,[15] the concept of a schema for schemata is briefly discussed, and it is mentioned that this schema should be accessible through the same interface as user data, and that it therefore would be convenient to represent the metaschema using the same data model as for the schema. In Kent,[8] some of the peculiarities of a separation of schema and data are highlighted, calling for a unified way of representing and manipulating both. In System R[2] and in QBE[25] a relational catalog represents base relations and integrates views and indexes built upon these.

Some of our previous work is related to the present one. In Ref. 9 we called for an explicit metaschema to model and control the changes of a database schema. In Ref. 19 the notions of self-description and self-documentation were defined and the latter was thoroughly analyzed. In Refs. 10 and 11, the concept of self-description was used for a binary relational model and a data language supporting schema changes was defined for the data model. In Ref. 20, a very detailed and precise self-describing metaschema for an extended relational data model was presented. In Refs. 12 and 14, we concentrate on the architecture of an SD/D Database System.

Section 2 of this paper introduces the concept of self-description and self-documentation. Section 3 defines the new framework for self-describing/self-documenting database systems. Section 4 analyzes the elementary and compound functions required by this new database framework. The conclusions are in Section 5.

2. On Self-Description and Self-Documentation

"I WISH MY WISH WOULD NOT BE GRANTED."
Achille's typeless wish to the Genie in Ref. 6.

"SELF-DESCRIPTION" IS A BUZZ-WORD.
"BUZZ-WORD" IS A BUZZ-WORD.
""BUZZ-WORLD" IS A BUZZ-WORD" IS SELF-DESCRIBING.
The Authors

2.1. Self-Description

Our position is that the metaschema should not only be made *explicit,* but also be made *self-describing.* That is, the metaschema describing the class of acceptable schemata for the used data model should itself be represented in terms of the same data model. In other words, if the relational model is the data model in use, the class of all acceptable relational schemata must be described in terms of relational primitives, i.e., domains, relations, attributes, etc. Self-description is necessary to make explicit the model's semantics and the way the schemata are allowed to evolve.

All existing database systems actually do have a metaschema, but it is implicitly represented in the definition and implementation of the DDL and DML. The ISO Working Group on the Conceptual Schema and Information Base[4] stresses explicitness in its 100% principle: "All relevant general static and dynamic aspects, i.e., all rules, laws, etc., of the universe of discourse should be described in the conceptual schemas." The class of acceptable schemata is a very relevant aspect because changing the schema of a database reflects changes in the general classifications, rules, and laws of the universe of discourse and is therefore clearly subject to the 100% principle. Following this principle, we create an explicit model for the metaschema, which models the class of schemata allowed to be defined in the data model in use. The explicit model of the metaschema gives us control in managing the schema changes, which is our main objective.

Having an explicit metaschema does not imply self-description. In fact, we could have an explicit metaschema which is not self-describing. There is, however, a justified and appealing urge to use the same data model at all layers of the system, and the same data language, for manipulating them.

One advantage of a database with a self-describing metaschema is that it allows users with some basic knowledge about the model to browse and find out all the information needed to use the database. The basic required knowledge is minimal, dependent on the data model itself, but independent of the database content. This will enhance information exchange. With the recent expansion in network technology, databases are often being used by new groups of users all over the world—users for which the database was not initially designed. The need for SD/D databases in this context is illustrated by ongoing research on Standard Format Data Units for exchanging scientific data among the various space agencies.[24]

Another advantage of having a self-describing metaschema is that knowledge of only one set of primitives of one data model and one DL is needed for any user to be able to work with the database system. This uniformity enhances communication among users working with the system in different roles, application programmers, end-users, database designers, database administrators, etc. From an implementation point of view, this is also advantageous because only one data language will have to be implemented.

A third advantage of a self-describing metaschema is that the database designers and administrators will be able to ask questions, not only about what the current application schemata are, but also about what an acceptable application schema looks like, and what the rules for and consequences of changing an application schema are. The latter kind of questions cannot be answered using the same language if the metaschema is not self-describing.

It is natural to ask questions like: How do we design a self-describing metaschema? And, just how self-describing can a self-describing metaschema be? These questions are not easy to answer. Designing a schema for a database involves a detailed analysis of the universe of discourse. The same should be done with the metaschema where the universe of discourse is the very data model itself—or the class of schemata which can be defined in it. We can therefore only provide subjective arguments on the "goodness" of a self-describing metaschema based on our experiences.[10,11,13,19,20] "Good" self-describing metaschemata only come from data models that are well defined, that support aggregation and generalization,[23] that use entity-based rather than name-based representation,[3-5] and that have primitives for explicitly representing a rich variety of constraints. When we design a metaschema, we must strive to model basic facts, be precise, and be explicit; we must forget about physical representation and efficiency, and about how adequate the metaschema is from the users' point of view, and only concentrate on modeling the universe of discourse at hand.

The question of how self-describing a self-describing metaschema can be cannot be answered yet. Current data models still lack generally accepted primitives for expressing constraints. We are not aware of other theoretical work on self-description or self-reference which can be applied to our problem. None of the results on recursive functions or logic seem to be directly applicable.

2.2. Self-Documentation

The main purpose of self-documentation is to keep a record of what the database has become. It is based on "what it was" and "what has happened to it." Since all objects (schema and data objects) are derived via some data language whose semantics are controlled by the SD/D model, the meaning of any derived object in the database is documented by the meaning of the operand objects and the semantics of the operators used in the derivation, without a need for additional knowledge external to the database. Self-documentation is essential in a database system to allow a smooth and natural evolution.

Conventional database systems have mostly dealt with the management of the database extension. They simply retrieve, insert, modify, delete, or restructure (derive views of) data instances. This includes maintenance of the views during updates of the base files, view indexing techniques for storage savings and performance improvement, etc. On the contrary, the schema, the *database intension,* is considered static or "relatively stable over time," and no operators analogous to the extensional ones have been defined which modify the intensional part of a database.

However, when we define a derived object in a database, the intension of this object comprises the derived objects name, the names of its attributes, the inherited properties, the inherited constraints, the inherited security and authorization properties, etc. The relational model is the only one that allows the definition of derived data objects, but unfortunately, it has no rules for generating the intension of the derived objects. It only provides rules for generating the extension of the derived objects. In other words, it has *extensional* but no *intensional semantics.* The first attempt to define the intensional semantics of the relational model was done in Ref. 19. In the above paper a discipline was imposed on the relational algebra in order to guarantee the meaningfulness of the results. Operators that dynamically generate new relation types, such as joins, Cartesian products, and projections, were disciplined to create only temporary displays which are not self-documenting. For the remaining operators, which were defining existing types of relations, the names of the attributes and all properties of the results are inferred from the operands.

Self-documentation is done by keeping a record of all derivations of data objects. By examining the derivations, a user can fully understand a derived object's meaning provided that he understands the basic data objects it was derived from. The names used for the derivations, and the deriving operators, are stored in the extension of the data dictionary schema.

3. The SD/D Database Systems

This section defines a new framework for database systems. This framework supports multiple schema dimensions and evolution of the schemata.

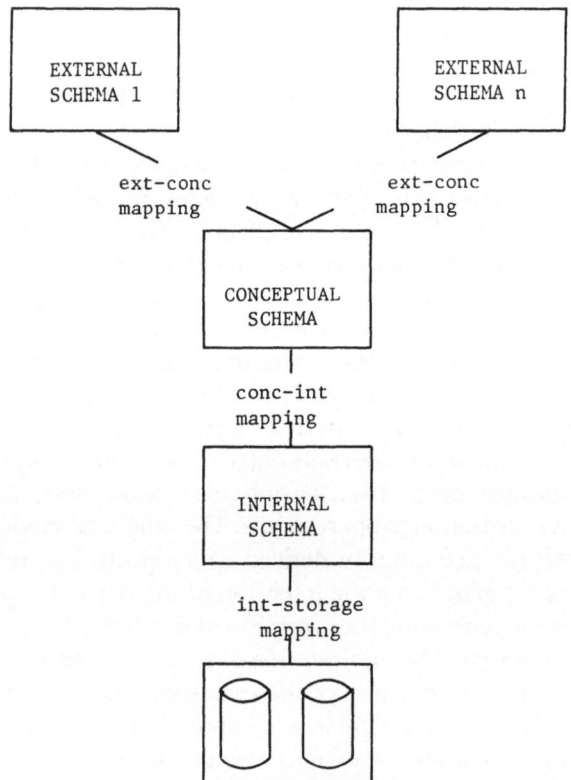

FIGURE 1. Point-of-view dimension.

3.1. Dimensions of Data Description

Data can be viewed from two orthogonal dimensions of description:

- the point-of-view dimension; and
- the intension–extension dimension.

The *point-of-view dimension* has three levels of data description resulting in databases with a logical architecture illustrated in Figure 1.

The three-schema logical database architecture[1] allows a clear separation of the information meaning, described in the *conceptual* schema, from the *external* data representation and from the *internal* physical data structure layout. This results in databases which are flexible and adaptable to changes in the way users view the data and in the way data are stored. This flexibility and adaptability is usually called *data independence*. The three-schema logical architecture is based on the assumption that the meaning of data, that is, the conceptual schema, is relatively stable over time compared to the external and internal schemata.

The *intension–extension dimension* has four levels of data description resulting in databases with a logical architecture illustrated in Figure 2.[14]

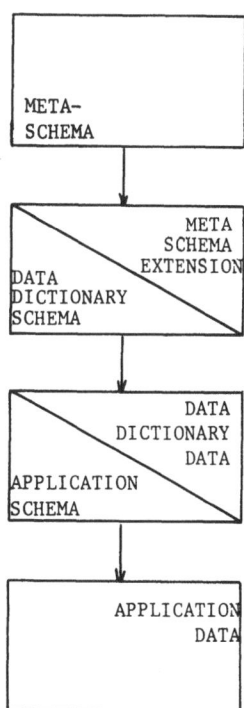

FIGURE 2. Intension–extension dimension.

Each level of data description is the *extension* (the data) of the level above, and, at the same time it is the intension (the schema) for the level below. The four-level logical data description allows a clear separation of information about the *data model,* described in the *metaschema;* information about the *management and use* of databases, described in the *data dictionary schema;* information about specific *applications,* described in *application schemata;* and application data stored in the database.[7,14]

The four-level intension–extension dimension is based on the assumption that the data model in use does not change, i.e., the *metaschema* intension is static. It does, however, support evolution and change of the meaning of application databases and of their management and use; that is, data dictionary schema and application schemata can change. It also facilitates explicit *meta-data-management,*[13] which is essential to developers of plug-compatible Data Management Tools.

3.2. The Point-of-View Dimension

A *conceptual schema* describes all *relevant general static* and *dynamic aspects,* i.e., all rules, laws, etc., of the universe of discourse. It only describes *conceptually relevant aspects,* excluding all aspects of data representation, physical data organization, and access.[4] *All static* and *dynamic aspects* of general nature must be described in the conceptual schema. If any of them were to be described in application programs, a very strict programming discipline would have to be enforced to control, verify, and maintain the multiple copies of the same rules. On the other hand, if all the rules are described in the conceptual schema it is much easier to extend or modify the rules, and it is much easier to verify one set of rules which completely controls all operations on the data. *Only relevant general aspects* should be described; that is, classes, types, and variables rather than individual instances, and rules and constraints having wide influence on the behavior of the universe of discourse rather than narrow influence. *Only conceptually relevant* aspects should be in the conceptual schema, excluding any aspect which refers to representation of data in the user's views or aspects related to machine efficiency and physical data organization. Focusing on conceptaully relevant aspects not only simplifies the conceptual schema design process, it also makes the conceptual schema insensitive to changes in users' views on data and to changes in the way data are physically stored.

An *external schema* describes parts of the information in the conceptual schema in forms convenient in a particular user-group's view including the manipulation aspects of these forms. The information described in an external schema can only be a subset of the information described in the conceptual schema. This means that no new information whatsoever can be produced by any mapping from conceptual schema to external schema.

The *internal schema* describe how all the information described in the conceptual schema is physically represented within the computer. It concentrates on which forms give the most efficient access with respect to storage media, control of concurrent use, recovery, etc. The internal schema must be designed to provide the overall best physical representation of all information in the conceptual schema. The internal schema design cannot be done without information about the external schemas and the way they are accessed: weight of importance, access frequencies, etc. An *access path schema* is necessary for global optimization, indexing, and reorganization of the physical structures.[17,18]

3.3. The Intension–Extension Dimension

The universe of discourse of the *metaschema* is a *data model*. The metaschema *describes* and *controls all operations* on the class of schemata which can be defined by means of the data model. This means that the metaschema is itself a member of the class of schemata it describes—it is *self-describing*. Self-description allows a copy of the metaschema to be explicitly stored as part of its own extension.

For example, for the relational model, the self-describing metaschema is based on the concepts of relation, domain, attribute, and the relational algebra, and it models these concepts, that is, how they can be put together to form a schema, and which operations are allowed to the schema. A part of such a metaschema and the initial extension explicitly describing it are shown in Figure 3. The "headers" of the tables constitute the metaschema and the tables hold the initial extension of the metaschema. The metaschema in Figure 3 defines four relations, namely, RELATIONS, DOMAINS, REL_ATT_DOM, and CONSTRAINTS. The initial extension, boxes labeled A, describes these four relations, their name and type, their attributes and domains, and the type of their domains. The relation CONSTRAINTS stores constraints defined for each relation. The tuple (RELATIONS, r1) in CONSTRAINTS indicates that there is a set r1 of intra- and interrelational constraints involving RELATIONS. r1 comprises all constraints needed on RELATIONS to model an acceptable class of relational schemata and to control the operations on them. An explicit and detailed representation of the constraints can be found in Ref. 13.

The full metaschema extension describes the *data dictionary schema*. The universe of discourse of the data dictionary schema is *the management and use* of the database system.

RELATIONS

RELNAME:NAME	RELTYPE:RTYPE
RELATIONS	MS
DOMAINS	MS
REL_ATT_DOM	MS
CONSTRAINTS	MS

A

DOMAINS

DOM_NAME:NAME	DOMTYPE:DTYPE
NAME	ALPHA
RTYPE	{MS, DDS, AS}
DTYPE	{ALPHA, INTEG, ...}
RULE	DML_STMT

A

REL_ATT_DOM

REL_NAME:NAME	ATT_NAME:NAME	DOM_NAME:NAME
RELATIONS	RELNAME	NAME
RELATIONS	RELTYPE	RTYPE
DOMAINS	DOMNAME	NAME
DOMAINS	DOMTYPE	DTYPE
REL_ATT_DOM	RELNAME	NAME
REL_ATT_DOM	ATTNAME	NAME
REL_ATT_DOM	DOM_NAME	NAME
CONSTRAINTS	RELNAME	NAME
CONSTRAINTS	RULE	RULE

A

CONSTRAINTS

REL_NAME:NAME	RULE:RULE
RELATIONS	r1
REL_ATT_DOM	r2
DOMAINS	r3
CONSTRAINTS	r4

A

FIGURE 3. Data model schema and its initial extension.

In addition to the metaschema copy, the data dictionary schema defines concepts like authorization, derivation, program, schema, etc., and contains the static and dynamic rules which determine who may use the database and how.

In Figure 4, boxes A and B, we illustrate how the definition of the data dictionary schema is stored in the extension of the metaschema.

The example data dictionary schema of Figure 4 illustrates two of the important relations one must include in a data dictionary. Relation DE-RIVATIONS defines the record of relation derivations, thus supporting parts of the system's capability to document its own evolution. This evolution, which is achieved via self-documentations as described in Section 2.2, is most important when it comes to schema objects which control other schema or data objects. For this reason the meaning of the operations on schema objects is critical. Relation REL_AUTH_ZATION defines access rights to relations. It should be pointed out that these two relations provide the minimal requirements for accounting about security and evolution. In more sophisticated systems each of the above relations is replaced by a set of relations modeling more complex security and authorization mechanisms and evolution management.[21]

In general, the data dictionary schema models information on the management and usage of the complete database. It contains the relations

RELATIONS

RELNAME:NAME	RELTYPE:RTYPE
RELATIONS	MS
DOMAINS	MS
REL_ATT_DOM	MS
CONSTRAINTS	MS
REL_AUTH_ZATION	DDS
DERIVATIONS	DDS

(A = rows 1–4, B = rows 5–6)

DOMAINS

DOM_NAME:NAME	DOMTYPE:DTYPE
NAME	ALPHA
RTYPE	{MS, DDS, AS}
DTYPE	{ALPHA, INTEG, ...}
RULE	DML_STMT
AUTH_NO	INTEGER
OP	RELATIONAL_OP
EXPRESSION	RELATIONAL_EX

(A = rows 1–4, B = rows 5–7)

REL_ATT_DOM

REL_NAME:NAME	ATT_NAME:NAME	DOM_NAME:NAME
RELATIONS	RELNAME	NAME
RELATIONS	RELTYPE	RTYPE
DOMAINS	DOMNAME	NAME
DOMAINS	DOMTYPE	DTYPE
REL_ATT_DOM	RELNAME	NAME
REL_ATT_DOM	ATTNAME	NAME
REL_ATT_DOM	DOM_NAME	NAME
CONSTRAINTS	RELNAME	NAME
CONSTRAINTS	RULE	RULE
REL_AUTH_ZATION	RELNAME	NAME
REL_AUTH_ZATION	AUTH_CODE	AUTH_NO
DERIVATIONS	RELNAME	NAME
DERIVATIONS	OPERAND1	NAME
DERIVATIONS	OPERAND2	NAME
DERIVATIONS	OPERATOR	OP
DERIVATIONS	EXPRESSION	EXPRESSION

(A = rows 1–9, B = rows 10–16)

CONSTRAINTS

REL_NAME:NAME	RULE:RULE
RELATIONS	r1
REL_ATT_DOM	r2
DOMAINS	r3
CONSTRAINTS	r4
REL_AUTH.ZATION	R5
DERIVATIONS	R6

(A = rows 1–4, B = rows 5–6)

FIGURE 4. Data dictionary schema.

needed to document schema evolution. It is important to create these relations as early as possible because we may want to document the evolution of the rest of the data dictionary schema as well. By far the most needed relations for documenting schema evolution are provided by the meta-schema part of the data dictionary schema, simply because the meta-schema—except for the relation DERIVATIONS—describes the whole class of schemata which can be defined in the data model.

The data dictionary schema also models application schemata and information about application programs, such as their names, the input–output data structures of programs, how programs are related to each other, and how programs use database schemata. Thus, the data dictionary schema should include relations for representing information about users, their access rights, whom they may grant access to, information about the physical representation of the database, usage statistics, etc. It is not, however, the purpose of this paper and the examples shown here to provide a complete data dictionary schema.

The extension of the data dictionary schema naturally stores the application schema. That is, the application schema is part of the data dictionary extension. The universe of discourse of an *application schema* is a *"real world" application*. Also, as part of the data dictionary extension, we have information about how specific programs use application schemata, how specific users are authorized to access data through specific application schemata, etc. All these data dictionary data should, of course, be properly described by data structures in the data dictionary schema as discussed above.

The data dictionary extension contains all the specific data used to manage application schemata and their use, and there are no real reasons why we should not store, as part of the data dictionary extension, specific data on how we manage and use the data dictionary schema itself. We must, for example, store the information about who is allowed to access the database through the data dictionary schema. The data dictionary schema, is however not part of the data dictionary extension, and we, therefore, cannot record information about its specific use, unless, of course, we use the same principle once again, and store a description of the data dictionary schema as part of the data dictionary data. And this is precisely what we do.

Figure 5, boxes A, B, and C, illustrates part of the extension of the data dictionary schema, including the definition of two relations, SUPPLIERS and PARTS, in the application schema.

It is desirable to keep the SD/D implementation independent of individual management strategies of organizations using a database system. In our example database we could insert extra constraints in the relation CONSTRAINTS enforcing the rule that any time a user accesses a relation, he can only do so if his authorization code matches the AUTH_CODE in REL_AUTH_ZATION for the particular relation he is trying to access. This way, the SD/D system only needs to know that it must always enforce those constraints found in relation CONSTRAINTS, which is a metaschema concept and therefore known by the SD/D system.

Finally, the *application data* is the extension of that part of the data dictionary extension which constitutes the application schema, that is, relations SUPPLIERS and PARTS.

We can summarize our description of the intension–extension dimension in Figure 6. The arrows on the left-hand side of Figure 6 lead from intension to extension. The arrows on the right-hand side show how intensions are explicitly stored as part of their extension. The two black boxes

RELATIONS

RELNAME:NAME	RELTYPE:RTYPE
RELATIONS	MS
DOMAINS	MS
REL_ATT_DOM	MS
CONSTRAINTS	MS
REL_AUTH_ZATION	DDS
DERIVATIONS	DDS
SUPPLIERS	AS
PARTS	AS

(row groups labeled A, B, C)

DOMAINS

DOM_NAME:NAME	DOMTYPE:DTYPE
NAME	ALPHA
RTYPE	{MS, DDS, AS}
DTYPE	{ALPHA, INTEG, ...}
RULE	DML_STMT
AUTH_NO	INTEGER
OP	RELATIONAL-OP
EXPRESSION	RELATIONAL-EX
S#	INTEGER
S_NAME	ALPHA
CITY	ALPHA
P#	INTEGER
P_NAME	ALPHA

(row groups labeled A, B, C)

REL_ATT_DOM

REL_NAME:NAME	ATT_NAME:NAME	DOM_NAME:NAME
RELATIONS	RELNAME	NAME
RELATIONS	RELTYPE	RTYPE
DOMAINS	DOMNAME	NAME
DOMAINS	DOMTYPE	DTYPE
REL_ATT_DOM	RELNAME	NAME
REL-ATT_DOM	ATTNAME	NAME
REL_ATT_DOM	DOM_NAME	NAME
CONSTRAINTS	RELNAME	NAME
CONSTRAINTS	RULE	RULE
REL_AUTH_ZATION	REL	NAME
REL_AUTH_ZATION	AUTH_CODE	AUTH_NO
DERIVATIONS	RELNAME	NAME
DERIVATIONS	OPERAND1	NAME
DERIVATIONS	OPERAND2	NAME
DERIVATIONS	OPERATOR	OP
DERIVATIONS	EXPRESSIONS	EXPRESSIONS
SUPPLIERS	S_NO	S#
SUPPLIERS	S_NAME	S_NAME
SUPPLIERS	S_LOC	CITY
PARTS	P_NO	P#
PARTS	P_NAME	P_NAME

(row groups labeled A, B, C)

CONSTRAINTS

REL_NAME:NAME	RULE:RULE
RELATIONS	r1
REL_ATT_DOM	r2
DOMAINS	r3
CONSTRAINTS	r4
REL_AUTH.ZATION	R5
DERIVATIONS	R6
SUPPLIERS	R7
PARTS	R8
RELATIONS	R9
DOMAINS	R10
REL_ATT_DOM	R11
CONSTRAINTS	R12
DERIVATIONS	R13
REL_AUTH_ZATION	R14
SUPPLIERS	R15

(row groups labeled A, B, C)

REL_AUTH_ZATION

REL:NAME	AUTH_CODE:AUTH_NO
RELATIONS	a1
DOMAINS	a2
REL_ATT_DOM	a3
CONSTRAINTS	a4
DERIVATIONS	a5
REL_AUTH_ZATION	a6
SUPPLIERS	a7

(row groups labeled A, B, C)

DERIVATIONS

RELNAME	OPERAND 1	OPERAND 2	OPERATOR	EXPRESSION

FIGURE 5. Data dictionary extension.

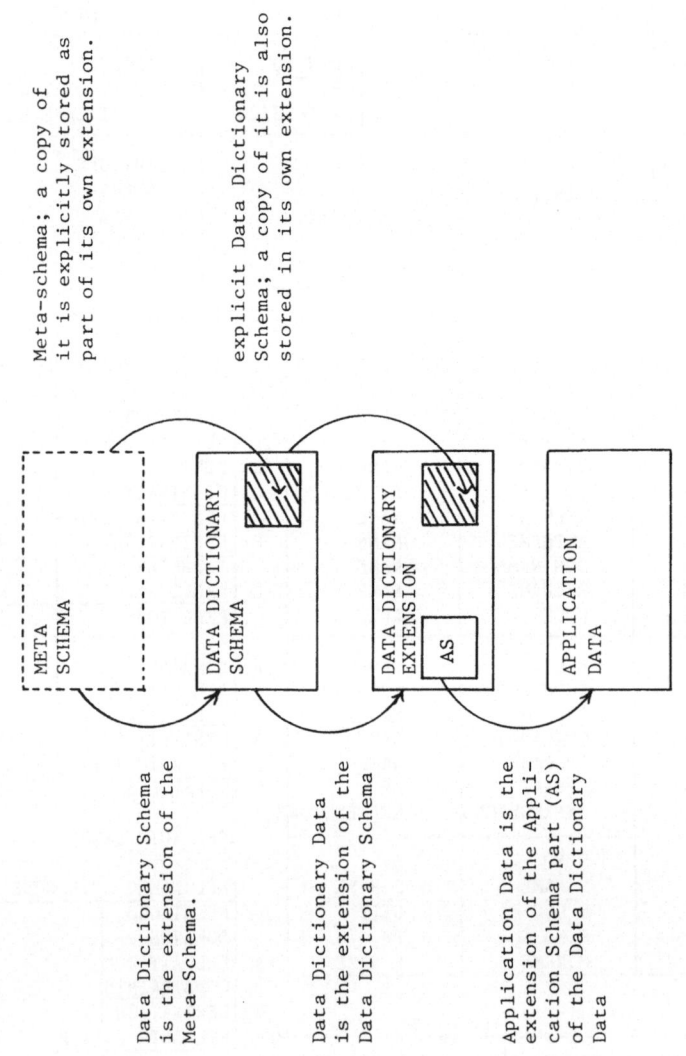

FIGURE 6. Logical self-describing DB.

indicate data that cannot be changed directly by any user or data management tool acting on his behalf. The black box in DATA DICTIONARY EXTENSION is an explicitly stored copy of the DATA DICTIONARY SCHEMA. If the data dictionary schema is to be changed, this cannot be done by changing the copy kept as part of the data dictionary extension. Instead it must be done through the metaschema. These changes, will, of course, be propagated down to the copy kept in the black box in the data dictionary extension.

The black box in the data dictionary schema contains the description of the metaschema which cannot be changed. Any system in this new architecture is born with relations to hold the metaschema extension; that is, the data dictionary schema, and the black box of the DATA DICTIONARY SCHEMA is *filled in*. It is also born with *part* of the relations to hold the DATA DICTIONARY EXTENSION which will have the part that corresponds to the black box *partly filled-in*. The filled-in part corresponds to the metaschema extension.

It is worth noting that only the data model is fixed in the system (the metaschema cannot change) whereas the DATA DICTIONARY SCHEMA can be designed to support the database management strategies of any particular organization.

4. Function Analysis

In this section we discuss two kinds of SD/D system functions:

- elementary functions, and
- compound functions built by elementary/compound functions.

We shall base our discussion on the data dimensions of the previous section.

4.1. Elementary SD/D System Functions

We shall limit our discussion of *elementary functions* to the *conceptual level* of the *point-of-view dimension,* and we shall *classify* the elementary functions with respect to the *intension–extension dimension.*

The classification is illustrated in Figure 7. In clarifying these functions we use a simple model of the SD/D system supporting the intension–extension dimension. This is illustrated in Figure 8.

Reference to any data object in Fig. 8 must, strictly speaking, be done through its intension. Reference to an application data object must be done through the application schema. Reference to a data dictionary

function \ data	META SCHEMA	DATA DICTIONARY SCHEMA	DATA DICTIONARY EXTENSION	APPLICATION DATA
deletion	///////////			
creation	///////////			
reference				

FIGURE 7. Elementary functions. Intension–extension dimension.

extension object must be done through the data dictionary schema. Reference to a data dictionary schema object must be done through the metaschema. The intension, which a reference to some data must go through, is itself a data object which is referenced. In effect, each reference goes through the metaschema, the data dictionary schema, etc. until it reaches the required data object. This strict method for referencing does not, of course, apply after compilation.

The DML processor has the metaschema built in, i.e., it can access the data dictionary schema stored in the metaschema extension.

Deletion and *creation* both require a *reference*. Aside from this the two functions are described as follows.

Deletion of any data object must be done through its intension, which describes all rules and laws, static as well as dynamic, that must be obeyed at the data level. As part of these rules and laws there may be *update dependencies* describing how the deletion of one data object propagates to other data objects at the same level. This kind of propagation is termed *intralevel propagation*.[22] Deletion of a data object which is part of a set of data objects interpreted as describing the intension of another data level may propagate to this other data level. This kind of propagation is termed *interlevel propagation*.[22]

For example, if we delete a relation from an application schema we must delete all its tuples from application data. If we delete a key from an application schema, no interlevel propagation occurs. If we delete the superdomain of an "is-a" constraint from the application schema, the data dictionary schema would probably enforce the deletion of the subdomain and the "is-a" constraint too. This is intralevel propagation caused by an update dependency in the data dictionary schema. The deletion of the tuple (Smith, B) from a relation USER_AUTHORIZATION, defined in the data dictionary schema, causes no interlevel propagation, because the tuple is not interpreted as part of the intension for another data level.

Creation of any data object must be done through its intension, which describes all rules and laws, static as well as dynamic, that must be obeyed at the data level. As part of these rules and laws there may be update dependencies causing intralevel propagation. For example, if we insert in the application data the tuple (Smith, $37,000) we may have to insert the tuple (Smith, project-alpha) too, because of the application schema update dependency PERSON_PAID [NAME] ⊆ PERSON_WORKING [NAME]. This example illustrates intralevel propagation.

Interlevel propagation of a creation of a data object occurs when the data object is interpreted as intension for another level. For example, if we insert a key in the data dictionary schema we may have to delete some tuples in the data dictionary extension: USER_AUTHORIZATION [USER_NAME, *AUTH_ZATION_CODE*]. If we create a new relation on some level, we have to set up a relational structure to hold its extension at the next level.

As part of deleting or creating data objects which are used as part of the intension for another level, one may experience the problem that the

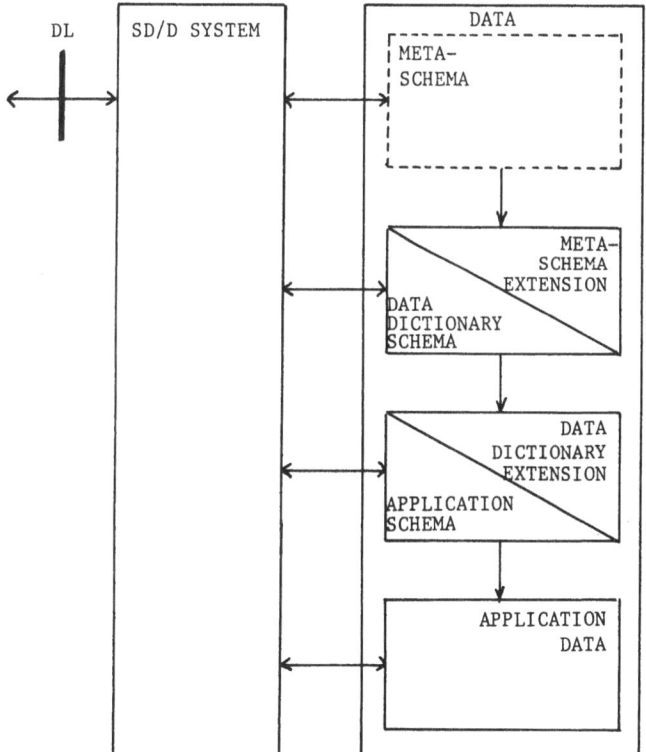

FIGURE 8. Logical self-describing DB architecture.

system knowledge is not sufficient to allow all propagations to be carried out *automatically.*[11] In this situation some of the extensions must be marked as not fully specified and the user is responsible for providing additional information. For example, the creation of a relation derived from other relations by means of the data language (e.g., relational algebra) can be handled automatically by the system. The creation of a new relation with empty extension can be handled automatically by the system. The creation of a new constraint PERSON_PAID [NAME] ⊆ PERSON_WORKING [NAME] cannot be handled automatically by the system in the case where there are violations of this constraint already.

We finally note that because each level of description in the intension–extension dimension is the extension of another level and because the metaschema is self-describing there is no need to distinguish between data definition and data manipulation languages (DDL and DML). We can simply discard the DDL. There is just one syntax for the DML and a mechanism for enforcing the update propagation.

4.2. Compound SD/D System Functions

The compound SD/D functions must be built upon elementary ones. They must support and enforce the point-of-view dimension and the intension–extension dimension of data. The compound SD/D system functions must be a supplement to and should not overlap or replace functions which, from a database point-of-view, should be provided by an operating system. We have identified the following essential compound functions to be supported by an SD/D system:

- data language processing;
- integrity control;
- authorization;
- supporting the point-of-view dimension of data;
- supporting the intension–extension dimension of data;
- concurrency;
- performance control;
- physical access.

Data-Language Processing. The SD/D system must be able to accept syntactically correct DL statements issued from a user or a Data Management Tool. And, it must return retrieved data through the DL interface.

Integrity Control. The SD/D system must know that a set of data objects in the database must be interpreted as rules and laws for *what* we can store in the database and *how* it can be changed. The SD/D system must be capable of enforcing these rules and laws. This is closely related to enforcing the intension–extension dimension of data (below).

Authorization. The SD/D system must know that a set of data objects in the database must be interpreted as rules and laws for *who* can access *which* data, and *how*.

Support Point-of-View Dimension. The SD/D system must be able to transform data language statements at the external schema level into internal statements of the storage level.

Support Intension–Extension Dimension. The SD/D system must be able to enforce interlevel propagation of data object manipulation. And, it must support integrity control (above).

Concurrency. The SD/D system must be able to coordinate multiple user interaction "at the same time" through the DL interface. And, it must be able to coordinate multiple interaction "at the same time" to data objects at the storage level.

Performance Control. The SD/D system must collect statistics about the state and usage of the data object.

Physical Access. The SD/D system must be able to issue requests to the underlying operating system or file management system as a result of accepted data language statements which are properly checked, transformed, sequenced, etc. And, it must be able to accept data objects returned from the operating system.

5. Conclusions and Future Research

We have presented a framework for a new kind of database system, the SD/D database systems. The main purpose of the new framework is to allow database systems to support evolution of both data and schema. An SD/D database is different from a conventional database in that it has two dimensions of data description, the point-of-view and the intension–extension dimension, and it comprises a self-describing metaschema which models and controls the operation on the database schemata. An SD/D database system is different from a conventional database systems in that it only has one interface to its many users, a single data language.

An SD/D system can be thought of as a core DBMS based on one data model, but with the potential to support a multidata model DBMS.[14]

The SD/D system framework supports the familiar three-level data description dimension of Ref. 1 and the orthogonal intension–extension dimension. The latter one allows manipulation of data objects, sometimes as data objects, and sometimes as schema objects that have their own extension.

The SD/D framework approach to database management opens up a new area for research. One of the main topics for research is the schema change propagation rules. We have proposed two categories of rules for changing schemata: intralevel and interlevel. Both need to be explored with respect to their representation and their enforcing mechanisms. Ideally, these rules would be explicitly represented in the schemata, just as constraints on data are to be represented in the application's schemata. This is one more reason why research on data models must continue in order to incorporate the concept of constraints into the data models. When we have adequate means to represent the rules for schema change, we will be able to extend the propagation of schema changes to application programs using the modified schemata. These rules could be given in the data dictionary as constraints between schemata and the input–output part of the application programs using them.

References

1. D. TSICHRITZIS, and A. KLUG (Eds.), The ANSI/X3/SPARC DBMS framework, *Inf. Syst.* **3**(3), 173–191, (1978).
2. D. D. CHAMBERLIN *et al.*, SEQUEL 2: A unified approach to data definition, manipulation, and control, *IBM J. Res. Dev.* **20**(6), 560–575.
3. E. F. CODD, Extending the database relational model to capture more meaning, *ACM TODS* **4**(4), 397–434, (1979).
4. J. J. VAN GRIETHUYSEN (Ed.), ISO TC97/SC5/WG3: Concepts and terminology for the conceptual schema and information base, ANSI, 1982.
5. P. HALL, J. OWLETT, and S. TODD, Relations and entities, in G. M. Nijssen (Ed.) *Modeling in Data Base Management Systems,* North-Holland, Amsterdam, 1976.
6. D. HOFSTADTER, *Gödel, Escher & Bach,* Basic Books, New York, 1979.
7. D. JEFFERSON, Reference model priorities and definition, Note for ISO/TC97/SC5/WG5, No. 106, National Bureau of Standards, Washington, D.C., 1983.
8. W. KENT, *Data and Reality,* North-Holland, Amsterdam, 1978.
9. L. MARK and A. LEICK, Ideas for new conceptual schema languages, in H. Kanagassalo (Ed.), *First Scandinavian Research Seminar on Information Modeling and Data Base Management,* University of Tampere, Finland, 1982, Series B, Vol. 17.
10. L. MARK, What is the binary relationship approach?, in C. G. Davis (Ed.), *Entity-Relationship Approach to Software Engineering,* North-Holland, Amsterdam, 1983.
11. L. MARK and N. ROUSSOPOULOS, Integration of data, schema and meta-schema in the context of self-documenting data models, in C. G. Davis (Ed.), *Entity-Relationship Approach to Software Engineering,* North-Holland, Amsterdam, 1983.
12. L. MARK and N. ROUSSOPOULOS, Fall and rise of an ANSI/SPARC DBMS framework, Working Note for the ANSI/SPARC/X3 Database Architecture Framework Task Group, 1984.
13. L. MARK Self-describing database systems-formalization and realization, Ph.D. dissertation, Department of Computer Science, University of Maryland, College Park, Maryland, *April,* 1985.
14. L. MARK and N. ROUSSOPOULOS, The reference model, Working Note for the ANSI/SPARC/X3 Database Architecture Framework Task Group, 1984.

15. National Bureau of Standards, An architecture for database management standards, Prepared by Computer Corporation of America, Report No. NBS 500-86, 1982.
16. G. M. NIJSSEN, An architecture for knowledge base software, presented to the Australian Computer Society, Melbourne, July 30, 1981.
17. N. ROUSSOPOULOS, View indexing in relational database, *ACM Trans. Database Syst.* **7**(2), 258–290 (1982).
18. N. ROUSSOPOULOS, The logical access path schema of a database, *IEEE Trans. Software Eng.* **SE-8**(6), 563–573 (1982).
19. N. ROUSSOPOULOS, A self-documenting relational model, TR-1264, Computer Science Department, University of Maryland, College Park, April, 1983.
20. N. ROUSSOPOULOS and L. MARK, A self-describing meta-schema for the RM/T data model, In *IEEE Workshop on Languages for Automation*, IEEE Computer Society Press, New York, 1983.
21. N. ROUSSOPOULOS, C. BADER, and J. O'CONNER, ADMS—A self-describing and self-documenting relational database system, Department of Computer Science, University of Maryland, College Park, Maryland, August, 1984.
22. N. ROUSSOPOULOS and L. MARK, A framework for self-describing and self-documenting database systems, NBS Trends and Application Conference, 1984.
23. D. SMITH and J. SMITH, Database abstractions: Aggregation and generalization, *ACM TODS* **2**(2), 105–133, (1977).
24. W. TRUSZKOWSKI, Private communication, 1984.
25. M. M. ZLOOF, Query by Example: The innovation and definition of tables and forms, Proceedings on Very Large Database Conference, 1975.

COMMUNICATION
MANAGEMENT

MESSAGE MANAGEMENT IN OFFICE INFORMATION SYSTEMS

S. K. Chang, P. Chan, T. D. Donnadieu, L. Leung and F. Montenoise

1. Introduction

An office information system is a distributed system consisting of workstations and functionally specialized stations such as file servers and printer stations, interconnected by a communication network.[3,6] The workstations exchange messages. The contents of most messages can be interpreted as documents. A text document contains only character strings. A form document can be shown on an alphanumeric display device in a certain structured representation.[8] A graphic document consists of graphic objects (points, lines, etc.), which can be shown on a vector display device. An image document consists of an array of picture elements, which can be shown on a raster display device. A voice document contains digitized sound patterns, which can be reproduced by a voice synthesizer.[7]

Since there are many different types of documents, we can also distinguish many different types of messages. Messages containing graphic documents, image documents, and voice documents can be excessively lengthy. Messages containing text documents or form documents sometimes can also become lengthy. If such messages are sent without reasonable control, this could easily lead to congestion on the communication network, especially during peak office traffic hours.

On the other hand, not all messages are relevant to an office worker at a certain workstation. An office worker usually must dispose of a lot of "junk mail." Without reasonable control, the "electronic junk mail" in an

S. K. CHANG, P. CHAN, T. D. DONNADIEU, and F. MONTENOISE • Department of Electrical and Computer Engineering, Illinois Institute of Technology, Chicago, Illinois 60616   L. LEUNG • Gould Research Center, Rolling Meadows, Illinois 60008

office information system can become even more wasteful than the traditional "paper junk mail," because it is so easy to send or broadcast an electronic message.

We therefore propose to augment the message management system at each station of an office information system by a *message filter*. The message filter, based upon current network traffic conditions and knowledge about the sender, the receiver, and message contents, determines whether the full message is to be sent, or a short alert message is to be sent instead.

Another problem of the office information system is that the communication network may not be highly reliable in the sense that messages may get lost. To increase reliability, the message management system is further augmented by an *adaptive protocol handler*. The adaptive protocol handler can transmit multiple short messages several times, and the desirable number of multiple transmissions is determined adaptively. For long messages, the receiving station's protocol handler sends back multiple requests for message retransmission, and the number of such requests is determined dynamically.

Therefore, the message management system is *adaptive* in two senses: (a) the message filter can adapt to traffic conditions to send more long messages and less short messages, and vice versa; (b) the protocol handler can adapt to noisy channel conditions to send multiple messages and several retransmitted messages.

The message management system is illustrated in Figure 1. It consists of the message manager, the message filter, and the adaptive protocol han-

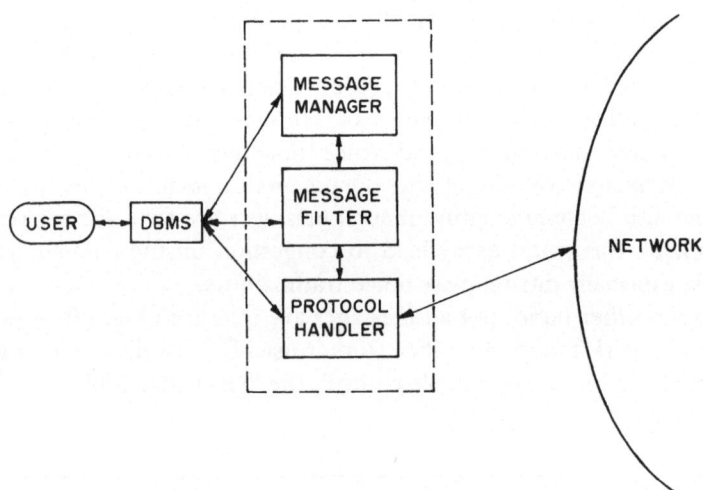

FIGURE 1. The message management system.

dler. A user can perform message management operations through a user interface to the message manager. Messages are stored in a relational database which is managed by a relational DBMS. All the control data are also stored as special relations in the relational database. Therefore, the DBMS performs data management functions for all the software modules.

From the viewpoint of implementation, the message manager serves also as the user interface. The message filter can be incorporated into the application layer of the network manager, and the protocol handler can be incorporated into the lower layers of the network manager.

In Section 2, we describe the message manager. The message filter design problem is stated in Section 3, where the message filter is designed based upon considerations of current network traffic condition, message type, and message length. Section 4 presents two filter design algorithms, and experimental results are described in Section 5. To design a more general *linguistic filter*, we should also consider message keywords, sender user profile, and receiver user profile. Design considerations of such linguistic filters are presented in Sections 6 and 7. In Section 8, the adaptive protocols are described. Applications and implementation status are discussed in Section 9. Extension to an intelligent filter is described in Section 10.

2. The Message Manager

The messages manager supports the following message management operations.

(a) Message Definition. Each message contains a *message-type* which can be defined by the user, a *message-identifier*, a set of *keywords* characterizing the contents of this message, and the *message-body*. The syntactic description of a message is given below:

⟨message⟩ :: = ⟨message-header⟩ ; ⟨message-body⟩
⟨message-header⟩ :: = (⟨m-type⟩ : ⟨m-value⟩) ;
       (⟨m-id⟩ : ⟨m-id-value⟩)
       (⟨keywords⟩ : ⟨m-value⟩)
⟨message-body⟩ :: = ⟨m-atom⟩ | ⟨m-atom⟩ ;
       ⟨message-body⟩
⟨m-atom⟩ :: = (⟨m-attribute-name⟩ : ⟨m-value⟩)

Each message consists of a message header and a list of self-descriptive m-atoms in the form of (m-attribute-name, m-value) pairs, where m-value may consist of a set of values. A *message type* is defined by the m-type and the list of m-attribute-names it consists of. Together with message definition, there is also an operation to remove an existing message definition.

(b) Message Sending. This is the operation to send message(s) from a sending station to the mailbox of a receiving station.

(c) Message Reception. This is the operation to receive message(s) from the mailbox into the message database (see below).

(d) Message Input. This is the operation to deposit a message in a *message file*, which is a relation in the relational database. For each message type, there is a separate message file. Therefore, at each station, there is a message database[9] consisting of message files (relations) and control relations (which are often invisible to the nonprivileged user). Each station also has a mailbox for temporary message storage during SEND and RECEIVE operations.

(e) Message Retrieval. For messages stored in the message database, the usual database retrieval operations can be performed.

(f) Message Deletion and Modification. This is to remove a message from a message file, or to modify an existing message already in the message file.

In our system, only the *system supervisor* can add new users or remove existing users. The current user names, passwords, and their privileges are stored in a control relation. A user with the privilege to create message types can define a new message file and, in so doing, become its *owner*.

Only the owner of a message file can send its contents from one station to another station. The SEND operation will remove the message(s) from a message file, and deposit them in the mailbox of the receiving station. Similarly, only the owner of a message file can receive message(s) from a mailbox into this message file. The RECEIVE operation will remove message(s) of the specified type from the mailbox, and deposit them in the message file.

A sender can leave a message to a receiver by using the INPUT command. The INPUT operation deposits a message in a message file owned by the receiver. The sender can also associate a message password with his message. This enables the same sender to delete and/or modify his message(s) later on, because only he knows his message password. Likewise, a sender can only retrieve messages sent by him, because he does not know other sender's message password. The owner of a message file, on the other hand, can retrieve all messages in this message file.

The owner of a message file can also define the access privileges of the users (senders) for this message file. For example, he may allow the sender to deposit at most three messages in this message file. He may allow the sender to retrieve and look at certain messages (with the correct message password, of course), but not delete or modify any message, etc. The DEFINE operation allows him to define sender's privileges, which are then stored in another control relation in the database.

The commands supported by the message manager are summarized below:

ADDUSER super-password name password
(system supervisor with "super-password" may add a user "name" and assign his "password")

RMUSER super-password name
(system supervisor with "super-password" may remove a user "name")

DEFINE message-type attributes owner-name password
(user whose name is "owner-name" and password is "password" may define a new "message-type")

REMOVE message-type owner-name password
(user whose name is "owner-name" and password is "password" can remove his own "message-type")

SEND message-type owner-name password
(send all messages in "message-type" to user whose name is "owner-name" and password is "password")

RECEIVE message-type name password
(user whose name is "owner-name" and password is "password" can receive all messages of type "message-type" from the mailbox)

INPUT message-type name message-password
(input a message having "message-type" from sender "name" and assign a unique "message-password" to this message)

DELETE message-type name message-password
(delete message from sender "name" having "message-type," and message must have password "message-password")

MODIFY message-type name message-password
(modify message from sender "name" having "message-type," and message must have password "message-password")

DISPLAY message-type name message-password
(retrieve all messages having "message-type" from sender "name," and messages must have message password "message-password")

DISPLAY message-type owner-name password
(retrieve all messages having "message-type" owned by "owner-name" whose password is "password")

LIST
(list all message types in message database)

HELP
(list all commands)

HELP command-name
(list command syntax for "command-name")

END
(exit from message manager)

In actual implementation, the message manager is implemented as a Shell process under Unix, so that it also serves as the user interface. The standard Unix mail functions have also been incorporated into the message manager. Since the message files as well as control tables are relations in the relational database, the message manager need only invoke the relational DBMS to perform most message management operations, thus simplifying its implementation.

3. Message Filter and Filter Design

As described in Section 2, a user can issue a SEND command to the message manager to transfer all messages in a message file from one station to the mailbox of another station. Upon reception of the SEND command, the message manager first invokes the message filter to decide whether indeed the SEND command should be obeyed literally.

The message filter may decide to send the full message, in which case nothing more need be said. It may decide to send a short *alert message*, which contains only its message type, message identifier, and keywords (if any). Conceptually, the alert message is similar to the messages created by database alerters.[1,2] The difference is that in the present case, the alert messages are determined by the message filters.

The receiver, upon reception of this alert message, can request the full message if he thinks it is relevant. Otherwise he can disregard the alert message. The message filter may also decide not to send any message at all, in which case the message still remains in the message database at the sending station, and the message filter will be invoked by the message manager to reevaluate how to handle this message at some later time.

The message filter is designed based upon considerations of current network traffic condition, message type, and message length. A more general *linguistic filter* need also consider message keywords, sender user-profile, and receiver user-profile. Such linguistic filters will be considered in Sections 6 and 7.

We will first present a mathematical formulation of the message filter design problem. An algorithm for message filter design will then be described.

The filter is specified by a *filter function f* and an *alert function g* as follows.

The filter function specifies for each message class k, at station i, with sender s, whether at station j, receiver r will receive the full message. In other words,

$$f_k(i,s,j,r) = \begin{cases} 1 & \text{if } r \text{ at station } j \text{ will receive} \\ & \text{full message from } s \text{ at station } i \\ \\ 0 & \text{otherwise} \end{cases}$$

Similarly, the alert function g specifies for each message class k, at statin i, with sender s, whether at station j, receiver r will receive an alert message. In other words,

$$g_k(i,s,j,r) = \begin{cases} 1 & \text{if } r \text{ at station } j \text{ will receive} \\ & \text{alert message from } s \text{ at station } i \\ \\ 0 & \text{otherwise} \end{cases}$$

Since we will not send both the full message and the alert message, $f_k(i,s,j,r)$ and $g_k(i,s,j,r)$ cannot both be 1.

As a simplification, we assume both filter function and alert function are invariant with respect to all s and r, and we can write $f_k(i,j)$ and $g_k(i,j)$, instead of $f_k(i,s,j,r)$ and $g_k(i,s,j,r)$.

Furthermore, we assume all messages in class k have identical length L_k, and all alert messages v have identical length L_o, and $L_k \geq L_o$. L_k can also be interpreted as the average length of class k messages, and L_o the average length of alert messages.

The *location function* $m(i,s)$ is 1 if user s is at station i, and 0 otherwise. In the restricted case, every user has a unique location. In general, a user may have multiple locations. This function can be expressed as a matrix $M = [m(i,s)]$, with row index i (stations) and column index s (users). If every user has a unique location, then there is exactly one 1 in each column of

M. If every station has some user, then there is at least one 1 in each row of *M*.

The *rate function* $q_k(s,r)$ is the rate of class *k* messages from user *s* to user *r*. This function can be expressed as a matrix $Q_k = [q_k(s,r)]$, with row index *s* (senders) and column index *r* (receivers).

Given *M* and Q_k, the rate of class *k* messages from station *i* to station *j*, $r_k(i,j)$, can be calculated as follows:

$$R_k = MQ_kM^t \qquad (1)$$

Let us define

$$c_k(i,j) = f_k(i,j)L_k + [1 - f_k(i,j)]g_k(i,j)L_0 \qquad (2)$$

where $c_k(i,j)$ is the effective length of class *k* message transmitted from station *i* to station *j*. Depending upon the values of $f_k(i,j)$ and $g_k(i,j)$, $c_k(i,j)$ can be L_k, L_0, or 0. We now define

$$t_k(i,j) = r_k(i,j)c_k(i,j) \qquad (3)$$

where $t_k(i,j)$ is the traffic of class *k* messages from station *i* to station *j*. The matrix $T_k = [t_k(i,j)]$, with row index *i* (sending stations) and column index *j* (receiving stations), expresses the traffic of class *k* messages in the network.

The *line traffic constraint function* $v(i,j)$ specifies the maximum allowable traffic from station *i* to station *j*. Therefore, we must have

$$\sum_k t_k(i,j) \le v(i,j) \qquad (4)$$

or

$$\sum_k r_k(i,j)c_k(i,j) \le v(i,j) \qquad (5)$$

We are given $m(i,s)$, the location function, $q_k(s,r)$, the rate function, and $v(i,j)$, the line constraint function. The *filter design problem* is to select f_k and g_k so that the inequalities (5) are satisfied.

For nonnegative $v(i,j)$'s, it is clear that a solution always exists. The optimization objective, then, is to select f_k and g_k to maximize *W*, where

$$W = \sum_k \sum_{\substack{i \\ i \ne j}} \sum_j w_k(i,j) \qquad (6)$$

and

$$w_k(i,j) = \begin{cases} 0 & \text{if } c_k(i,j) = 0 \\ 1 & \text{if } c_k(i,j) = L_0 \\ 2 & \text{if } c_k(i,j) = L_k \end{cases} \tag{7}$$

In other words, we want to design filters so that more classes of messages are transmitted in full. Therefore, in (7), we increase W by 2 if a certain message class can be transmitted in full, and we increase W by 1 if only alert messages are transmitted. These weights are admittedly arbitrary. However, both algorithms to be described in Section 4 can be adjusted for different weights.

4. Filter Design Algorithms

From (5), we need only consider the line traffic between two stations i and j to determine $f_k(i,j)$ and $g_k(i,j)$. After some notational simplification, the problem can be restated as follows:

Given: n message classes with rates $r_1, r_2, r_3 \ldots , r_n$ and lengths $L_1, L_2, L_3, \ldots , L_n$; alerter message length L_0.

Constraints:

$$\sum_{k=1}^{n} f_k r_k L_k + g_k r_k L_0 \leq v$$

f_k in $\{0,1\}$, g_k in $\{0,1\}$, $f_k + g_k \leq 1$.

Determine: f_k and g_k.

Maximize:

$$W = \sum_{k=1}^{n} (2f_k + g_k)$$

Algorithm I. This algorithm essentially orders the message classes by their relative merits to determine f_k and g_k.

STEP 1: Set m_k to greater of $\{1/r_k L_0, 2/r_k L_k\}$.
All m_k are unmarked in the beginning.
Set sum to 0. Set f_k and g_k to 0.

STEP 2: Select k having largest unmarked m_k.
Mark m_k.

If $(1/r_kL_0) \leq (2/r_kL_k)$, then set f_k to 1, else set g_k to 1.

Set sum to sum $+ f_kr_kL_k + g_kr_kL_0$.

STEP 3: If sum $\leq v$, then go to STEP 2, else reset the last modified f_k or g_k to 0.

Algorithm II. This algorithm uses a simple sorting and indexing scheme to select the filter function and alert function.

STEP 1: Sort r_kL_0 in increasing order to build an index table TABLE1, whose first row contains r_kL_0 and second row contains k, which is the index to keep track of the relative position of entries in TABLE1.

STEP 2. Sort r_kL_k in increasing order to build an index table TABLE2, whose first row contains r_kL_k and second row contains k, which is the index to keep track of the relative position of entries in TABLE2.

STEP 3: Find the largest n_1 such that sum1, the sum of first n_1 terms of TABLE1[1], is no greater than the constraint v.

Find the largest n_2 such that sum2, the sum of first n_2 terms of TABLE2 [1], is no greater than the constraint v.

Set $w_1 = n_1$, and $w_2 = 2n_2$.

If $w_1 > w_2$

then set $W = w_1$, opt$i = n_1$, opt$j = 0$

else set $W = w_2$, opt$i = 0$, opt$j = n_2$

STEP 4: For all indices i,j, such that $i + j \leq n_1$, $j \leq n_2$, and $i + 2j > w$, compute the sum of first j terms of TABLE2[1] and first i distinct terms of TABLE1[1], (i.e., in TABLE1, select only terms with indices different from those already selected from TABLE2).

If sum $\leq v$ and $W < i + 2j$ then set $W = i + 2j$, opt$i = i$, opt$j = j$.

STEP 5: The optimum pair (opt$i$,opt$j$) determines the functions f_k and g_k as follows. Set $f_k = 1$ for $1 \leq k \leq$ opt$j$, and 0 otherwise, where k is index in TABLE2[2]. Set $g_{h_k} = 1$ for $1 \leq k \leq$ opt$i$, and 0 otherwise, where h_k is selected index in TABLE1[2].

5. *Experimental Results*

In this section, we present the experimental results obtained in simulation experiments. All programs were written in the C language and run under the Unix operating system on a PDP11/70 minicomputer system. In the following example, we compare the two algorithms with the true optimum.

Example 1. We assume five different message classes, with rates 50, 30, 40, 10, and 20 and message lengths 25, 10, 7, 20, 15, respectively. The alert message length L_0 is assumed to be 5.

In Algorithm I, we first compute the merit figures m_k. The following table is constructed:

$2/r_k L_k$	$1/r_k L_0$	m_k
2/1250	1/250	1/250
2/300	1/150	1/150
2/280	1/200	2/280
2/200	1/50	1/50
2/300	1/100	1/100

Given the above table and traffic constraint v, f_k, and g_k can be determined. In Algorithm II, we first build TABLE1 and TABLE2:

			TABLE1		
$r_k L_0$	50	100	150	200	250
Index	4	5	2	3	1

			TABLE2		
$r_k L_k$	200	280	300	300	1250
Index	4	3	2	5	1

Suppose the traffic constraint v is 1000. We can sum up four terms in TABLE1 without violating the constraint. Therefore $n_1 = 4$. We can sum up three terms in TABLE2 without violating the constraint. Therefore $n_2 = 3$. We have $w_1 = 1 \times 4 = 4$, $w_2 = 2 \times 3 = 6$, and $W = 6$. The different index pairs (i,j) to be tried in STEP 4 of Algorithm II, and their corresponding W values, are listed in the following table:

i \ j	0	1	2	3
0	0	2	4	6
1	1	3	5	7
2	2	4	6	8
3	3	5	7	9
4	4	6	8	10
5	5	7	9	11

For example, if $i = 3$ and $j = 1$, then $W = 5$. We select index 4 from TABLE2, and indices 4, 5, and 2 from TABLE1. Since sum $= 200 + (50 + 100 + 150) = 500 < 1000$ and it has the largest W, it is selected as the best solution.

In Figures 2 and 3, the performance of Algorithms I and II is illustrated. It is seen that Algorithm II performs better than Algorithm I in the sense that for the same constraint v, it designs filter with larger W. In this example, it actually achieves the true optimum. We have performed other simulation experiments, which all indicate the superiority of Algorithm II over Algorithm I. On the other hand, both algorithms may not achieve the true optimum. The computation time of Algorithm I is approximately 1.5 sec, and that of Algorithm II is approximately 2.0 sec.

If we are given the traffic constraint, the filter characteristics can be

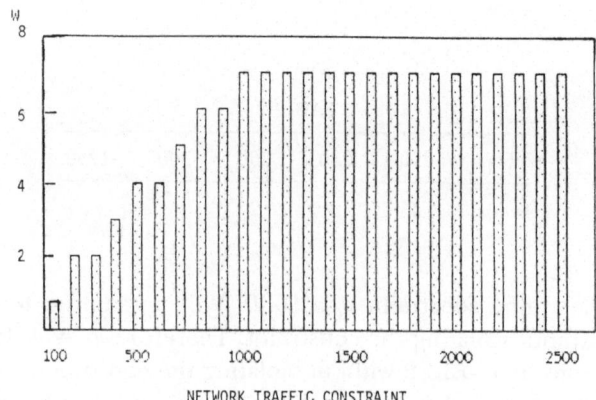

FIGURE 2. Performance of algorithm I.

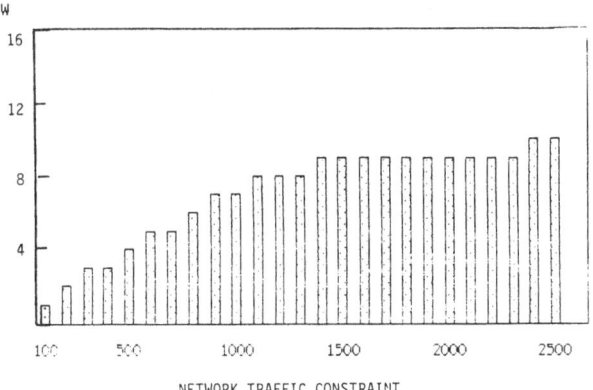

FIGURE 3. Performance of algorithm II.

determined from Figure 2 for Algorithm I, or Figure 3 for Algorithm II. In other words, we can design filters by a simple table look-up method.

6. Linguistic Filter Design Algorithms

The message filter described in Section 4 does not take into account user profiles and message profiles. A simple model for user profile could be a set of keywords with associated degree of relevancy. A simple model for message profile could also be a set of keywords with frequency of occurrences. We can then determine the degree of relevancy of a message with respect to a given user profile or a collection of user profiles. In designing message filters, we will consider network traffic, message rate, message length, and message relevancy. These considerations lead to the following formulation of a linguistic filter.

We assume there are p keywords. There are m users. Each user s is characterized by a *user profile vector* defined as follows:

$$x^s = (x_1^s, x_2^s, \ldots, x_p^s) \tag{8}$$

where $0 \leq x_i^s \leq 1$ is the relevancy of the ith keyword to user s.

There are n message classes. Each message class k is characterized by a *message profile vector* defined as follows:

$$y^k = (y_1^k, y_2^k, \ldots, y_p^k) \tag{9}$$

where y_i^k is 0 or 1, denoting the presence or absence of the ith keyword. The relevancy of message class k to user s is defined as

$$e_s^k = \frac{1}{p} \sum_{i=1}^{p} x_i^s y_i^k \tag{10}$$

If the y_i^k are numbers between 0 and 1, denoting frequency of occurrences of keywords, then the above equation (10) can be replaced by

$$e_s^k = \sum_{i=1}^{p} x_i^s y_i^k \tag{11}$$

The relevancy of message class k is defined as

$$e_k = \frac{1}{m} \sum_{s=1}^{m} e_s^k \tag{12}$$

The objective then is to give higher priority to more relevant messages. Therefore, we can replace (7) of Section 3 by

$$w_k(i,j) = \begin{cases} 0 & \text{if } c_k(i,j) = 0 \\ e_k & \text{if } c_k(i,j) = L_0 \\ 2e_k & \text{if } c_k(i,j) = L_k \end{cases} \tag{13}$$

The linguistic filter design problem can now be stated as follows:
    Given: n message classes with rates $r_1, r_2, r_3, \ldots, r_n$ and lengths $L_1, L_2, L_3, \ldots, L_n$ and relevancies $e_1, e_2, e_3, \ldots, e_n$; alerter message length L_0.
    Constraints:

$$\sum_{k=1}^{n} f_k r_k L_k + g_k r_k L_0 \leq v$$

f_k in $\{0,1\}$, g_k in $\{0,1\}$, $f_k + g_k \leq 1$.
    Determine: f_k and g_k.
    Maximize:

$$W = \sum_{k=1}^{n} (2f_k + g_k)e_k$$

Two heuristic algorithms for linguistic filter design are presented below.

Algorithm III. This algorithm first orders the message classes into a list by decreasing figure of merit, which is defined to be message relevancy over the product of message rate and average message length. Then f_k and g_k are determined, by going from top to bottom according to their positions in the list.

STEP 1: Sort $e_k/(r_k L_k)$ in decreasing order and place indexes into LIST.

STEP 2: Set t to 0. Let v be the traffic constraint.

STEP 3: For each i from 1 to n, do the following:
   Set k to LIST(i).
   If $v \geq t + r_k L_k$, set f_k to 1, g_k to 0, and t to $t + r_k L_k$;
   else if $v \geq t + r_k L_0$, set f_k to 0, g_k to 1, and t to $t + r_k L_0$.

The above algorithm favors the transmission of full messages over alert messages.

Example 2. The message rate, length, and message profile vector for each message class are given below:

Message-class	Rate	Length	Profile-vector
1	50	25	(1,0,0,0)
2	30	10	(0,1,0,0)
3	40	7	(0,0,1,0)
4	10	20	(0,0,0,1)
5	20	15	(0,1,0,1)

Suppose there is only one user ($m = 1$), and the user profile vector is (0.164, 0.06, 0.36, 0.04). Using equations (10) and (12), we can find the relevancy of each message class. We can also regard the single user profile vector as the average of m user profile vectors. Combining (10) and (12), the relevancy is computed as follows:

$$e_k = \frac{1}{p} \sum_{i=1}^{p} y_i^k \left(\frac{1}{m} \sum_{s=1}^{m} x_i^s \right) \tag{14}$$

Message-class	Rate	Length	Relevancy
1	50	25	0.042
2	30	10	0.015
3	40	7	0.090
4	10	20	0.010
5	20	15	0.025

After sorting, LIST contains: 3, 5, 4, 2, 1. Let L_0 be 5. We can compute the traffic:

Class	Full-message-traffic	Alert-message-traffic
3	280	200
5	300	100
4	200	50
2	300	150
1	1250	250

Let the traffic constraint v be 500. We have $f_3 = 1$, $g_5 = g_4 = 1$, and all remaining f_k and g_k are 0. The resultant W is $2 \times 0.09 + 0.025 + 0.01 = 0.215$. The resultant traffic is 430.

Algorithm IV. This algorithm applies the filter design algorithm, Algorithm II, iteratively. Each time we consider only message classes with relevancy above a threshold. We then reduce the threshold to consider more message classes, and so on.

STEP 1: Set relevancy threshold t to 1.

STEP 2: Select only those message classes with relevancy $\geq t$. Apply filter design algorithm to determine f_k and g_k.

STEP 3: Delete those message classes with f_k or g_k equal to 1. Reduce constraint v by the amount of traffic for these message classes, as specified by f_k and g_k.

STEP 4: Reduce relevancy t by dt. If $t \geq 0$, go to STEP 2.

Example 3. We apply Algorithm IV to the data given in Example 2. We choose dt equal to 0.1. Therefore, Algorithm IV will go through eleven iterations, with $t = 1.0, 0.9, \ldots, 0.1, 0$. When $t = 0.9$, we apply Algorithm II to class 3, and set $f_3 = 1$, $g_3 = 0$, and v is reduced to $500 - 280 = 220$. When $t = 0.2$, we apply Algorithm II again to set $f_5 = 0$, $g_5 = 1$, and v is reduced to 120. When $t = 0.1$, we set $f_4 = 0$, $g_4 = 1$, and v is reduced to 70. The resultant W is $2 \times 0.09 + 0.025 + 0.01 = 0.215$. The resultant traffic is 430.

The performances of Algorithms III and IV are illustrated in Figures 4 and 5, respectively. The computation time of Algorithm III is approximately 1.6 sec, and that of Algorithm IV is approximately 2.2 sec.

7. A Message Classifier for the Linguistic Filter

The heuristic algorithms presented above have computation time proportional to $n \log n$, where n is the number of message classes. With a large

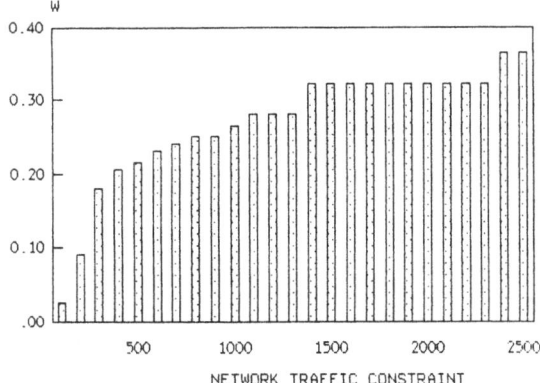

FIGURE 4. Performance of algorithm III.

number of keywords, the algorithms could become time consuming, because there may be as many as $n = 2^p$ different message classes, where p is the number of keywords.

Another reason that we prefer to have a smaller number of message classes is because in the message management system, each message class corresponds to a separate relation or separate file. A large number of message classes will increase the system overhead of the message management system.

The linguistic filter can be augmented by a *message classifier*, which

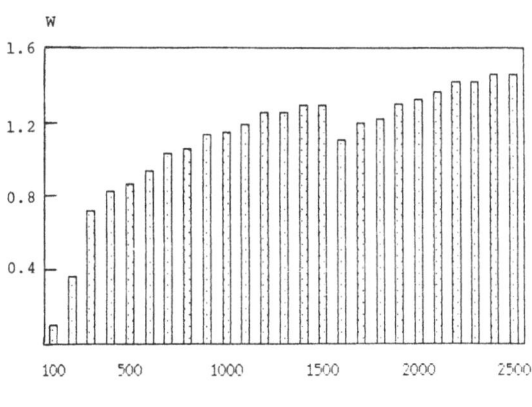

FIGURE 5. Performance of algorithm IV.

classifies messages into a smaller number of classes. The problem can be formally stated as follows:

Given n message classes, each characterized by a message profile vector $y_k = (y_{k1}, y_{k2}, \ldots, y_{kp})$, which is assumed to be a p-place binary vector. In other words, each message class k is characterized by a set of keywords. As before, each message class k has a rate r_k and an average length L_k. Applying the linguistic filter (based upon Algorithm III) will give a figure of merit W_n, defined as W over n.

A message classifier T is a mapping from the p-dimensional binary vector space into itself. T partitions the original message space into T-equivalent message classes as follows:

Message classes i and j are T-equivalent, if $T(y_i) = T(y_j)$.

The T-equivalent classes are considered new message classes. For each new message class k, $1 \leq k \leq n'$, the new rate r'_k is

$$r'_k = \sum_{\substack{\text{message class} \\ i \text{ is in this} \\ T\text{-equivalent} \\ \text{class}}} r_i \qquad (15)$$

the new average length L'_k is

$$L'_k = \frac{1}{r'_k} \sum_{\substack{\text{message class} \\ i \text{ is in this} \\ T\text{-equivalent} \\ \text{class}}} L_i r_i \qquad (16)$$

and the new message relevancy e'_k can be computed using equation (14) of Section 6.

The linguistic filter can now be applied to the new message classes. Our objective is to make n' smaller than n, and keep the new figure of merit $W_{n'}$, defined as W' over n', close to the original figure of merit W_n, *under the same traffic constraint v.* In other words, we want to find a mapping T, such that

1. the number of new message classes n' is small, and
2. the change in figure of merit $|W_n - W_{n'}| = |W/n - W'/n'|$ is less than a given tolerance t.

We will consider only those mappings T such that

$$y \leq T(y) \qquad \text{for all } y \text{ in message space}$$

where $y \leq z$ means $y_i \leq z_i$ for corresponding components y_i in y and z_i in z. In other words, the keyword set of the new message class contains the keyword sets of its constituent message classes.

We will first present an example, and then explain the algorithm.

Example 4. The message rate, length, relevancy, and message profile vector for each message class are the same as those given in Example 2 of Section 6.

Class	Rate	Length	Relevancy	Profile
1	50	25	0.042	(1,0,0,0)
2	30	10	0.015	(0,1,0,0)
3	40	7	0.090	(0,0,1,0)
4	10	20	0.010	(0,0,0,1)
5	20	15	0.025	(0,1,0,1)

The message class relevancy is computed based upon a combined user profile vector (0.168, 0.06, 0.36, 0.04). As in Example 2 of Section 6, traffic constraint is 500 and alert message length is 5. Applying Algorithm III, we have $f_3 = g_5 = g_4 = 1$, and all remaining f_k and g_k are 0. W is 0.215; therefore W_n is $0.215/5 = 0.043$. The resultant traffic is 430.

Suppose the mapping T is $T(1,0,0,0) = T(0,1,0,0) = (1,1,0,0)$ and $T(0,0,1,0) = T(0,0,0,1) = T(0,1,0,1) = (0,1,1,1)$.

The mapping T partitions the message space into two new message classes. The rate, length, relevancy, and profile vector for each new message class are given below:

Class	Rate	Length	Relevancy	Profile
1	80	19.375	0.057	(1,1,0,0)
2	70	11.142	0.115	(0,1,1,1)

Applying Algorithm III, we have $g_2 = 1$, and $g_1 = f_1 = f_2 = 0$. The resultant W' is 0.115. Therefore $W_{n'}$ is 0.0575. The resultant traffic is $70 \times 5 = 350$.

Suppose the mapping T is $T(0,1,0,0) = T(0,0,0,1) = T(0,1,0,1) = (0,1,0,1)$, and $T(1,0,0,0) = T(0,0,1,0) = (1,0,1,0)$. The rate, length, relevancy, and profile vector for the two new message classes are given below:

Class	Rate	Length	Relevancy	Profile
1	60	13.34	0.025	(0,1,0,1)
2	90	17	0.132	(1,0,1,0)

Again, applying Algorithm III, we have $g_2 = 1$, and $g_1 = f_1 = f_2 = 0$. The resultant W' is 0.132. Therefore $W_{n'}$ is 0.066. The resultant traffic is $90 \times 5 = 450$. This is the optimal solution.

Algorithm V. This algorithm attempts to detect common patterns in the message profiles.

STEP 1: Let $Y = \{y_1, \ldots, y_n\}$ be the set of message profiles. Let Z be the set of all p-place binary vectors.

STEP 2: For each z_i in Z, let a_i be the number of zeros in the vector z_i, $0 \leq a_i \leq p$. For each z_i in Z, Let b_i be the number of times such that $z_i \geq y_k$ for y_k in Y.

STEP 3: Sort z_i by decreasing $a_i \times b_i$. In case of ties, sort by decreasing a_i. In case of further ties, sort by decreasing relevancy. Let the sorted sequence be $\{z_1, z_2, \ldots, z_n\}$.

STEP 4: Determine mapping T as follows: for each y_k in Y, find first z_i in the sequence $\{z_1, \ldots, z_n\}$ such that $y_k \leq z_i$, and let $T(y_k) = z_i$. Let $\{z_1, z_2, \ldots, z_{n'}\}$ be the new message classes, where for each z_i, there is some y_k in Y such that $T(y_k) = z_i$. Let n' be the number of new message classes thus established.

In Example 4, $(0,1,0,1)$ and $(1,0,1,0)$ are the first two z-vectors in the ordered z-sequence. They are therefore selected as the new message classes.

8. Adaptive Protocol Design

After the message filter has determined to send a short alert message or a long message (or no message at all), the adaptive protocol handler is invoked to perform actual message transmission. The details of the protocol handler, and the analysis of its performance characteristics, are described in Ref. 5. We will summarize the main points to illustrate our design approach.

First, we consider the transmission of short messages such as the alert message or the acknowledgment message, which is assumed to have uniform length.

The usual protocol will send a message, wait t time units for the acknowledgment message, and then retransmit. This procedure is repeated m_1 times, before it is given up. This is referred to as Protocol #1.

In the second protocol, Protocol #2, we will transmit k_2 messages all at once. The receiving protocol handler will also send back k_2 acknowledgment messages all at once, if it receives at least one message.

Intuitively, Protocol #1 may need more time, before a message is successfully transmitted, and Protocol #2 may send out more messages.

In the analysis of the two protocols, the following assumptions are made:

a. The probability of successful transmission of a single message through the noisy channel is p.
b. The transmission time through the channel is a constant $t/2$, so that the round-trip transmission delay is a constant t.
c. There is no transmission delay between two messages.
d. Each message has a uniform length of 1.

Based upon these assumptions, we can design protocols #1 and #2 to achieve a given reliability r ($1 > r > p > 0$), defined as the overall probability of successful message transmission, by choosing the appropriate m_1 (number of retransmissions) or k_2 (number of multiple transmissions). Once m_1 and k_2 have been determined, we can calculate N (the expected number of messages exchanged), and T (the expected time delay before a message is successfully transmitted). The overall figure of merit for a protocol is defined as $C = N \times T$.

In Figure 6, we plot $C(p)$ by varying p (the probability of successful transmission of a single message), under the assumed reliability r equal to 0.98. It can be seen that Protocol #2 is better when p is small (i.e., when the channel is very noisy). However, when p is large (i.e., when the channel is less noisy), Protocol #1 is better.

Intuitively, we can hope to gain better performance by combining the two protocols. The result is Protocol #3. In Protocol #3, we will transmit k_3 copies of a message all at once (as in Protocol #2), and retransmit m_3 times (as in Protocol #1). Now we need to determine both k_3 and m_3 to achieve a given reliability r. The result for r equal to 0.98 is also plotted in Fig. 6. It can be seen that Protocol #3 is uniformly better than either Protocol #1 or Protocol #2.

If the noisy channel condition changes (p varies), then we need to adjust m_3 and k_3 accordingly. An adaptive protocol to dynamically adjust k_3 is presented in Ref. 5. The basic idea is to increase k_3 when the channel gets noisy, and decrease k_3 when the reverse is true.

The above-described Protocol #3 (and its adaptive version) is for short messages. For long messages, it is intuitively clear that we do not want to transmit multiple copies of a long message all at once. Therefore, another protocol, Protocol #4, is devised. Basically, it uses Protocol #3 to send the short alert message to the receiving protocol handler. Then, the receiving handler will send negative acknowledgment messages (i.e. retransmission requests), if the long message is not received. This procedure is repeated m_4 times, before it is given up. Protocol #4 also has good performance characteristics, and can be modified to become an adaptive

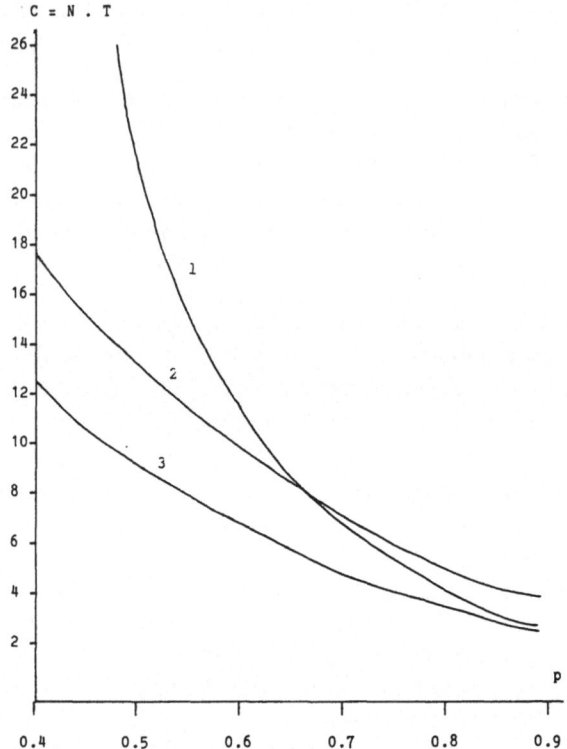

FIGURE 6. Comparisons of Protocols #1, #2, and #3 for $r = 0.98$.

protocol. In the adaptive version, the number of retransmission requests will be increased or decreased based upon variations in noisy channel conditions.

9. Discussion

The function of the linguistic filter can now be summarized: each message has a message profile characterized by a keyword set. Based upon the keyword set, the message is classified by the classifier as belonging to a certain message class. The filter then determines whether to transmit the full message or a short alert message, or to store this message in the message database, waiting for later transmission. The message classifier and filter are designed based upon user profiles, message profiles, message rates, and average message lengths.

The effects of message filters with changing traffic conditions can be summarized as follows. When we reach peak traffic hours, maximum allowable traffic will decrease. The message filters will first prevent the transmission of long-frequent messages. If traffic constraint continues to decrease, either average-length-average-frequency messages or long-infrequent messages or short-frequent messages will be replaced by alert messages. Further reduction of traffic constraint will lead the message filters to replace even short messages by alert messages. When network gets really congested, the message filter will temporarily suspend the transmission of all types of messages.

The four algorithms described in Sections 4 and 6 can be invoked whenever the traffic constraint changes, or whenever the message length or message rate changes. If message length and rate are known and relatively stable, we can use a simple table look-up method to determine filter characteristics (i.e., the f_k and g_k functions), thus saving computation time.

The message filter proposed in this paper has been incorporated into an experimental message management system implemented in C running under the Unix operating system on a PDP11/70 minicomputer. In this experimental system, the filter is dynamically designed based upon simulated traffic conditions.

We plan to use this experimental message management system in IIT's remote teaching environment, where students from local industries attend classes by listening to live TV broadcasts, and submit homework exercises using the experimental message management system. For this application, each message file corresponds to a homework exercise. A message file is owned by an instructor (the receiver). The student (the sender) may deposit his exercise in the message file. Using the DEFINE command, the instructor can specify whether the student is allowed to inspect his exercise or make changes, whether the student can submit an exercise twice, etc. The instructor can then use the SEND and RECEIVE commands to collect homework submitted by students.

10. Toward an Intelligent Filter

Our current effort is to expand the linguistic filter to an intelligent filter by incorporating a small rule-based system interfaced with a database system, so that certain classes of messages can be processed by the filter itself. For example, student inquiries about grades can be answered immediately by querying a grades database. Such an intelligent filter can replace part of the functions of the secretary, thereby increasing office productivity.

To study the "filtering function" provided by a human secretary, we monitored the interruptions received by the departmental secretary, and compiled the following statistics:

Total number of interruptions : 137
          Phone calls : 75
          Face visits : 62

Phone calls from:		Face visits from:	
professors	1	professors	23
students	26	students	24
clerks	3	clerks	10
others	45	others	5

Phone calls to:		Face visits to:	
professors	51	professors	5
clerks	6	clerks	1
information	15	information	56
wrong call	3		

Actions taken for face visits:

direct info	46.8%
retrieval	17.8%
action	24.1%
leave note	3.2%
none	8.0%

Actions taken for phone calls:

direct info	33.3%
retrieval	13.3%
transfer	25.3%
action	2.6%
leave note	16.0%
none	9.3%

The above statistics are for one typical working day for one departmental secretary. We have monitored the departmental office for a period of a week, and the above statistics can be considered quite typical. It is interesting to note that for both phone calls and face visits, the departmental secretary is able to handle more than 30% of the interruptions directly. Less than 20% of the interruptions are database retrieval requests, for which the secretary must do file searching to answer the request.

The above discussion motivates the design of an intelligent filter. The intelligent filter utilizes a linguistic filter to classify incoming messages into three *superclasses:* (1) Superclass I for *immediate action,* (2) Superclass G for *general processing,* and (3) Superclass J for *junk mail.* The linguistic filter will filter out Superclass J messages. An expert system consisting of a collection of alerter rules will then process those messages in Superclass I and Superclass G

The linguistic filter described in Section 6 is designed based upon message rates, message lengths, message relevancies (which are calculated from the message profile and the user profile), and the traffic constraint.

We can use the same general approach to design the intelligent filter, provided that the various parameters are interpreted as follows:

(a) Message rate r_k is interpreted as the estimated number of message arrivals per working period.
(b) Message length L_k is interpreted as the estimated message processing time to process a class k message.
(c) Alerter message length L_o is interpreted as the background process scheduling time.
(d) Traffic constraint v is interpreted as total time available for message processing per working period.

The reason for making these changes in interpretation is because the intelligent filter does not send messages to other stations. It will either process a message, or dispose of it as junk mail. Therefore, in designing the intelligent filter, we should consider message processing time instead of message transmission time.

The result of applying the linguistic filter design algorithm should also be interpreted differently. If fk is 1, this class of messages will be processed immediately, i.e. it belongs to Superclass I. If gk is 1, this class of messages will be processed generally as background processes, i.e. it belongs to Superclass G. If both fk and gk are 0, this class of messages will be regarded as junk mail, i.e. it belongs to Superclass J.

After the incoming messages have been classified into their respective superclasses and their relevancy calculated, the expert system processes these messages. The expert system consists of database alerter rules. As described in [2, 3], the alterer rules monitor message database updates. Since each message class corresponds to a separate database relation, the arrival of a new message will trigger the alerters associated with this message class. If several alerters are triggered simultaneously, they will be handled according to the relevancy of the messages they monitor.

An alerter rule has the following form:

If (CONDITION)
then

(ACTION)

The condition part is a function of the following: (1) receiver profile, x^r, which identifies who the receiver is, (2) receiver state, z^r, which dynamically changes with respect to time, and (3) message profile y^k. The action part of the rule is a function of the following: (1) sender profile, x^s, which identifies who the sender is and (2) message profile y^k.

As an example, the condition part of the alerter rule determines whether the receiver can receive a message. The action part of the alerter rule activates the voice synthesizer. The alterter rule is:

If (receiver.profile.job is 'professor')

and

(receiver.state.status is 'not in')

then

VOICE_SYNTHESIZE("Professor $receiver.profile.name is not in.
Please leave your name and message")

With this approach, Superclass J messages (junk mail) can be filtered out automatically, thus reducing the number of messages which must be handled by an office information system.

Acknowledgment

This research was supported in part by the National Science Foundation under grant no. MSC-8306282.

References

1. S. K. CHANG, Protocol analysis for information exchange, *Int. J. Policy Anal. Inf. Syst.* **6**, 1–23 (1982).
2. J. M. CHANG and S. K. CHANG, Database alerting techniques for office activities management, *IEEE Trans. Commun.* **COM-30**(1), 74–81 (1982).
3. S. K. CHANG, Office information system design, in *Management and Office Information Systems*, S. K. Chang (Ed.), Plenum Press, New York, 1984.
4. S. K. CHANG and W. L. CHAN, Modeling office communication and office procedures, Proceedings of National Electronics Conference, Chicago, October 4–6, 1982, pp. 250–257.
5. T. P. DONNADIEU, An adaptive protocol for the transmission of two message types, M.S. thesis, Department of Electrical Engineering, Illinois Institute of Technology, December, 1983.
6. C. A. ELLIS and M. BERNAL, Officetalk-D: An experimental office information system, Technical Report, Xerox Palo Alto Research Center, 1982.
7. R. A. MYERS, Trends in office automation technology, Technical Report, RC-9321, IBM Watson Research Center, Yorktown Heights, New York, April, 1982.
8. N. C. SHU, F. C. TUNG, and C. L. CHANG, Specification of forms processing and business procedures for office automation, *IEEE Trans. Software Eng.* **SE-8**(5), 499–512 (1982).
9. D. TSICHRITZIS, Form management, *Commun. ACM* **25**,(7), 453–477 (1982).
10. TSICHRITZIS et al., A system for managing structured messages, *IEEE Trans. Connum. COM-30*(1), 66–73 (1982).

15

AUDIO-DOCUMENT TELECONFERENCING IN THE AUTOMATED OFFICE

KAZUNARI NAKANE, HIROSHI KOTERA, MASAO
TOGAWA, AND YOSHINORI SAKAI

1. Introduction

Recently, demand for teleconferencing seems to have increased rapidly, particularly in the office environment, in order to save time and reduce the traveling expenses involved in participating in business meetings. This is one of the reasons that teleconferencing services and systems such as audio teleconferencing, audiographics teleconferencing, video teleconferencing, and computer teleconferencing have already been put into commercial or trial use.[1-7]

In most business meetings, handwritten or printed paper materials are frequently used in support of oral presentations to make understanding easier and clearer. Thus, teleconferencing would be more useful if such documents could be easily and efficiently distributed, presented, and explained among participants in distant locations. However, there are few teleconferencing systems[8-11] that have these document-oriented facilities.

From this viewpoint, the Yokosuka Electrical Communication Laboratory (Yokosuka ECL) of Nippon Telegraph and Telephone Public Corporation (NTT, Japan) has been working on a document-based teleconferencing system called an Audio-Document teleConferencing System (ADCS).

Presented in this paper are the ADCS concept, a system organization outline, and comments on the evolution of such system in an automated office environment.

KAZUNARI NAKANE, HIROSHI KOTERA, MASAO TOGAWA, and YOSHINORI SAKAI • Visual Communication Development Division, Yokosuka Electrical Communication Laboratory, NTT., Yokosuka, Kanagawa, 238 Japan

2. Automated Office Environment

There has been a great metamorphosis in the office environment in the past ten years. Many kinds of office automation aids, such as plain paper copiers (PPC), word-processors (WP), facsimile transceivers (FAX), small-size office computers, and the like, have emerged and pervaded offices during this period. Advances in microelectronics represented by microprocessor chip realization can be considerd to be one of the main inducements of this metamorphosis. During the last several years, other remarkable changes have also occurred in the office environment. New types of OA equipment and systems, such as Teletex(TTX), Videotex(VTX), personal computers, digital PBX, electronic telemailing systems, teleconferencing systems, and the like, have been occupying places in offices.

Trends in these changes in the office environment may be summarized as follows:

1. electronizing to telecommunicating;
2. centralized on-line system to distributed on-line system;
3. individual system to multipurpose integrated system;
4. single communication media to multimedia and new media utilization.

Very large scale integrated circuit (VLSI), artificial telecommunications satellite (CS), integrated service digital network (ISDN), and the open telecommunications policies can be thought to be principal inducements for these changes.

The above-mentioned advances in office environment seem to have made great contributions to office worker productivity and efficiency but have not had much of an effect on management and knowledge workers.

Therefore, development of OA aids for improving management and knowledge worker productivity is a considerable objective for office automation in the near future.

On the other hand, it is said that almost half of management and knowledge workers' office hours are spent in meetings, conferences, or negotiations related to decision making. It is therefore important to make business meetings more efficient in the sense of time and cost savings and of increasing business productivity.

Accordingly, teleconferencing systems and services are expected to play a significant role in the automated office environment as a substitute for conventional face-to-face meetings.

3. Teleconferencing Hierarchy

Teleconferencing systems can be classified into the four types shown in Table 1 according to communications media used. These teleconferencing systems have the following characteristics and advantages.

3.1. Audio Teleconferencing

This type of teleconferencing is an extension of ordinary telephone communications, extending person-to-person conversations to a group discussion. It is convenient for making confirmations or for obtaining mutual consent among participants on relatively simple matters. However, it is not suitable for explaining or discussing complicated subjects. Nevertheless, its great advantage lies in its low communications cost.

Audio teleconferencing services or facilities are widely provided as part of a public telephone network or as an addition to PBXs.

3.2. Audiographics Teleconferencing

This type of teleconferencing provides conferees with supplemental visual information. Facsimile transceivers, electronic blackboards, and

TABLE 1
Teleconferencing Hierarchy

Type	Communication media	Terminal equipment	Conference cost	Application
Audio teleconferencing	voice	Telephone set, Speaker-phone set	Low	Confirmation, Discussion on simple matter
Audiographics teleconferencing	Voice, Telewriting signals, Still pictures	Electronic blackboard, Sketchphone, Slow-scan TV	Relatively low	Explanation, Discussion on cc mplicated subjects
Video teleconferencing	Voice, Motion picture etc.	Video camera, Video monitor etc.	High	Debates, Executive meetings, etc.
Computer teleconferencing	Typed text, etc.	Data terminal	Low	Notification, Discussion, Voting, etc.

telewriting terminals can be used to present visual information. Audiographics teleconferencing seems useful in explaining or discussing complicated matters, because it provides both audio and visual communications capabilities. This service also provides relatively low communications cost teleconferencing.

Some types of audiographics teleconferencing terminal instruments such as the GEMINI 100 electronic blackboard[12] and the Sketchphone[13,14] have been already released for commercial use.

3.3. Video Teleconferencing

Teleconferencing has a disadvantage in that it lacks a feeling of presence, closeness, or contact. To overcome this problem, motion picture capability has been added to video teleconferencing. Motion pictures in video teleconferencing transmit conferees' live action and thus conferees can see the expressions and actions of conferees at other sites. Although video teleconferencing is superior to other types of teleconferencing in this respect, it is expensive since motion picture transmission requires broadband signal transmission. Thus, it seems suitable for debates, executive meetings, and for meetings in which several strangers participate.

Many video teleconferencing systems and services, for instance the Picturephone Meeting Service,[6] the New Video Conference System,[7] and the BIGFON,[15] have been put into commercial use.

3.4. Computer Teleconferencing

This type of teleconferencing differs slightly from the above-mentioned real-time based teleconferencing in its communicating facility. In typical computer teleconferencing, conferees communicate with each other via an on-line computer system or an electronic telemailing system by way of a store-and-forward type multiple addressing message communications network. In other words, this type of teleconferencing provides non-real-time based teleconferencing. Each conferee can participate in or access the conference and can exchange messages with other conferees any time at his convenience. It seems that computer teleconferencing is a suitable way for busy people to discuss with each other without scheduling troubles.

Various types of distributed computer network provide telemailing services, or computer teleconferencing facilities. Some store-and-forward type facsimile networks such as the Facsimile Intelligent Communication System[16–18] provide facsimile-based telemailing as a public network service.

Of these four types of teleconferencing, the second offers the most promise. It will see widespread use in the automated office environment, despite a lack of feeling of presence, because it is economical and it offers real-time based communications support on complicated matters.

4. ADCS Concept

4.1. Design Concept

As mentioned before, many kinds of printed or handwritten paper materials are used in business meetings. Thus, the ability to efficiently distribute, present, and explain documentary materials in teleconferencing is significantly valuable. ADCS is an improved version of audiographics teleconferencing that has been developed on the basis of the following concept.

4.1.1. Document Distribution. Documentary materials are usually sent by mail or by facsimile before teleconferencing starts. It takes time and involves several people to distribute documents in this way. To overcome this problem, it is necessary to integrate electronic filing and distribution of documentary materials.

4.1.2. Document Retrieval and Presentation. The following capabilities are important for effective documentary material use in teleconferencing:

a. Quick retrieval of a particular document page.
b. Simultaneous presentation of the same document page to all conferees.
c. Unrestricted arbitrary document page retrieval by each individual conferee at any time during the conference.

These capabilities are realized by employing an image reader, electronic storage, and high-resolution CRT display to present document pages in soft copy form instead of hard copy. Document name and page number are sent during the conference to enable page retrieval and display simultaneously at each conferee's terminal.

4.1.3. Document Explanation and Discussion. For effective discussion and explanation of documents and complicated subjects, the following functions are desirable in addition to voice communications capabilities:

a. Ability to indicate a particular spot on a page with simultaneous display of this pointing work to other conferees.
b. Memo-writing capability similar to an electronic blackboard to enable conferee to add notes to a displayed document page or to a blank page.

These capabilities can be realized by utilizing a high-resolution display and an electronic writing-pad and by sending pointing data and telewriting signals instantly whenever a confereee points or writes a memo.

The above-mentioned communications capabilities provide a common visual space[19] among participants in addition to a common acoustic space. Common visual space combined with common acoustic space enables participants to deal with complicated subjects in teleconferencing.

4.2. Service Overview

Basic service images for audio-document teleconferencing are described as follows.

4.2.1. Preparation Stage or Document Distribution Phase. Documents prepared for conference are read in using an image reader and stored in the conference terminal's electronic storage. Then, document data are sent to other terminals and stored in each terminal's storage before teleconferencing begins.

4.2.2. Conferencing Stage. When a conferee explains something along with documents, a particular document page is displayed on the terminal screen at his direction. The document name and page number are sent to other terminals at the moment the page is turned. Thus, the same page can be presented to all conferees simultaneously and automatically without transmission speed problem. When a participant points or writes a memo, telewriting signals consisting of cursor position data and memo-writing data are sent to and displayed on each terminal screen.

Basic ADCS service images are shown in Figure 1.

4.3. Multimedia Communication Protocols

In ADCS, multiple forms of media such as voice, documents, telewriting signals, and control data are transmitted concurrently and alternatively during a teleconferencing session. Furthermore, the communication control functions required for each type of communications media differ as shown in Table 2.

It is, therefore, important to design and provide multimedia communication protocols which enable the efficient and concurrent transmission of these various types of data.

Multimedia communication protocols[20] featuring concurrent multiple-session dialogue control as well as single-session connection control functions have been designed as follows.

4.3.1. Basic Protocols Architecture. The basic protocol architecture conforms to the OSI 7 layer reference model[21] recommended by ISO for the purpose of interworking communication worldwide.

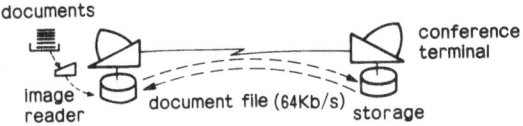

documents

image reader

document file (64Kb/s)

conference terminal

storage

(a) Preparation stage (Document Distribution)

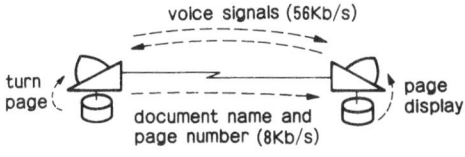

voice signals (56Kb/s)

turn page

document name and page number (8Kb/s)

page display

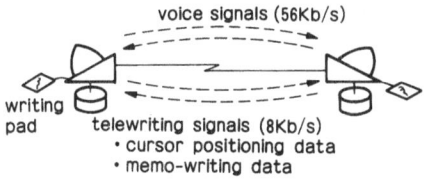

voice signals (56Kb/s)

writing pad

telewriting signals (8Kb/s)
• cursor positioning data
• memo-writing data

(b) Conferencing stage

FIGURE 1. Basic ADCS service images.

4.3.2. Multimedia Control Functions. To provide multimedia control functions, a two-level hierarchy architecture has been introduced to the session layer protocols. The session upper layer, named super-session, provides functions of token control for Two Way Alternative (TWA) communications and checkpoint control for large amounts of data transmission as session dialogue control functions. The session lower sublayer, named basic-session, provides the function of session connection control. Furthermore, concurrent multiple super-sessions can be set up on a single

TABLE 2
Required Communication Control Functions

Media type	Control function			
	Token control	Checkpoint control	Error control	Signals mixing
Voice	No (TWS[a])	No	No	Yes
Document	Yes (TWA[b])	Yes	Yes	No
Telewriting signals	Yes (TWA)	No	Yes	No
Speaker identifier	No (TWS)	No	Yes	No

[a]TWS: Two-way simultaneous communication.
[b]TWA: Two-way alternative communication.

basic-session which facilitates negotiations required for session connection by virtue of its single-session characteristics, thus supporting concurrent multimedia communication control.

5. System Organization

5.1. Data Transmission

In ADCS, transmitted data can be classified into (a) voice signals, (b) documents, (c) telewriting signals, and (d) control data such as the document name and page number to display common images on each screen. In the case of documents prepared in advance, large amounts of document data are sent speedily on a batch basis before the meeting starts over the INS[22,23] or ISDN digital network at the speed of 64-Kb/s. Other nonvoice data are smaller in data quantity than documents but have to be sent frequently and concurrently with voice signals during teleconferencing. For such concurrent voice and nonvoice data transmission, a 64-Kb/s channel is separated into 56-Kb/s and 8-Kb/s subchannels. These subchannels are assigned to voice and nonvoice transmission, respectively. If document transmission is required during the conference, a 64-Kb/s channel is fully utilized for this purpose. This is allowable because of 64-Kb/s high-speed transmission requiring interruption of voice and other nonvoice data transmission only a few seconds per page.

5.2. Multipoint Connection

In multipoint teleconferencing, conference terminals can be efficiently connected to each other via a multipoint connecting node (MPCN) located in the network as shown in Figure 2. The MPCN bridging facility is required to have various communication control capabilities related to transmission data.

1. Voice data must be mixed, and distributed or broadcast to all terminals except the originating one.
2. Nonvoice data cannot be mixed because the aforementioned protocols handshake and token priority management are required in MPCN to avoid race conditions among conference terminals.
3. Demultiplexing and multiplexing of voice and nonvoice data are also required to perform the above-mentioned media-dependent communications control, respectively, in MPCN.

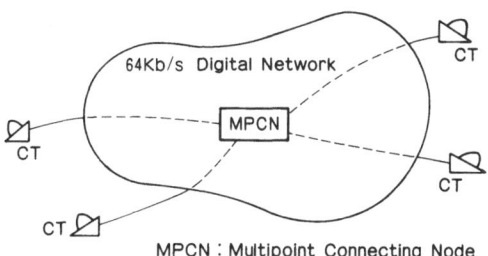

FIGURE 2. Multipoint connection via MPCN.

MPCN : Multipoint Connecting Node
CT : Conference Terminal

5.3. System Configuration

Conference terminals are connected to each other via MPCN over a 64-Kb/s digital network. The MPCN consists mainly of an audio and visual signal multiplexer (AVMUX), an audio-bridge (AB) and a graphics-bridge (GB). The AVMUX controls voice and nonvoice data multiplexing and demultiplexing. The AB mixes and distributes voice data and cancels echos. The GB controls protocols handshake and arbitrates token requests coming from conference terminals. ADCS conference terminal tentative specifications are shown in Table 3. Picture signals input from the image reader are sent to electronic storage directly through a parallel bus interface at a few seconds per page. An electronic writing-pad is employed for pointing and memo writing. A high-resolution CRT display monitor is also

TABLE 3
ADCS Terminal Specifications

Subsystem	Item	Specification
Image reader/printer	Resolution	8 lines/mm
	Speed	less than 5 sec/page
	Paper size	ISO A4
Writing-pad	Resolution	8 lines/mm
	Sensor type	Electromagnetic
	Size	ISO A4
Display monitor	Resolution	8 line/mm
	Addressable point	1728 × 2400
	Monitor size	monochrome CRT
Data transmission	Audio-document mode	56 Kb/s (voice)
		8 Kb/s (nonvoice)
	Document mode	64 Kb/s (nonvoice)

employed to enable ISO A4 full-size document page display with the same resolution as facsimile equipment.

6. Evolution of ADCS in the Automated Office Environment

As mentioned above, ADCS distributing, retrieving, presenting, and explaining capabilities facilitate teleconferencing dealing with rather complicated subjects in actual business. In addition, concurrent voice and nonvoice data transmission over a 64-Kb/s channel is effective in reducing communications cost. Moreover, document handling capabilities including storing, retrieving, and managing documents are integrated in ADCS. In other words, ADCS has an affinity with electronic filing systems. In the automated office environment, an enormous amount of data traditionally stored in paper form can be filed in the electronic filing system, resulting in savings in floor space and retrieval time. Other aids to office work such as word processing and data processing can also be integrated in ADCS. Therefore, ADCS will be augmented into an integrated workstation system, involving multiple communications media such as voice and various kinds of nonvoice data.

7. Conclusion

Teleconferencing is becoming increasingly widespread in the business world by virtue of its time- and money-saving capabilities. With actual business practice in mind where many kinds of paper materials are used to explain and discuss complicated subjects, a new type of teleconferencing system has been proposed. The outline of this system and its evolution in the automated office environment has been discussed.

References

1. M. MIDORIKAWA, K. YAMAGISHI, K. YADA, and K. MIWA, TV conference system, *Rev. ECL* (published by NTT, Japan) **23**(5-6), 453–468 (1975).
2. K. HIRATSUKA and H. KAKIHARA, Video conference system, *Jpn Telecommun. Rev.* (published by NTT), **18**(3), 145–151, (1976).
3. R. W. HOUGH and P. R. PANKO, Teleconferencing systems: A state-of-the-art survey and preliminary analysis (published by SRI), April, 1977.
4. M. KIKUCHI and M. YOSHIKAWA, New video conference system, *Jpn. Telecommun. Rev.* **22**(2), 112–119, (1980).

5. C. Ungaro, Teleconferencing focuses on a new era, *Data Commun.* 85–97, September (1982).
6. D. B. Menist and B. A. Wright, Picture phone meeting service: The system, *Teleconferencing and Electronic Communications*, (L. A. Parker and C. H. Orgren, eds.,) University of Wisconsin-Extension, Madison, Wisconsin, 1982.
7. T. Tanaka, Y. Nakatani, and T. Doi, NTT new video conference system, Proceedings of the International Teleconference Symposium, 241–247, April, 1984.
8. T. Omachi, Y. Takashima, T. Usubuchi, T. Sakabe, and H. Iijima, High resolution image system for teleconferencing, Proceedings of the International Conference on Communications, 2G.3.1, June, 1982.
9. J. O. Boese, R. J. Jaeger, Jr., and R. C. Thanawala, Audiographics teleconferencing, Proceedings of the ICC, 4E.1.1, June, 1982.
10. H. Yamaguchi, K. Yamada, K. Maya, and I. Hidaka, A graphic system for international teleconference, Proceedings of the ITS, 221–226, April, 1984.
11. T. Kamae, S. Ohtsuka, and Y. Sato, Sketchfax—A terminal having telewriting and facsimile capabilities, Proceedings of the ICC, D3.4.1, June, 1983.
12. D. R. Fishell and C. Stockbridge, The quorum teleconferencing system: Going the distance for business customers, *BLR* December, 278–283 (1982).
13. T. Kishimoto and Y. Sato, Simultaneous transmission of voice and handwriting signals: Sketchphone system, *IEEE Trans. Commun.* **COM-29**(12), 1982–1986, December (1982).
14. Telewriting terminal: Sketchphone, ECL Technical Publication (published by NTT, Japan), No. 282, 1983.
15. K. D. Shenkel and E. Braun, BIGFON—An all-service subscriber communication system goes into operation, Telecom Forum '83, p. 3.11.7.1-6, 1983.
16. T. Kamae, Development of a public facsimile communication system using storage and conversion techniques, Conference Record of NTC '80, p. 19.4.1, 1980.
17. T. Kamae, T. Endo, and S. Nakabayashi, Facsimile intelligent communication system, *Rev. ECL* (NTT, Jpn) **29**(7-8), 649–662 (1981).
18. K. Yuki, F. Kanaya, and H. Ishikawa, Facsimile intelligent communication system (FICS-2) outline, *Rev. ECL* (NTT, Jpn) **31**(4), 467–475, (1983).
19. G. B. Thompson, Shared space—A vital concept in successful teleconferencing, *Telephony* **203**(4), 27–29 (1982).
20. M. Matsumoto, K. Nakane, and H. Tanigawa, A proposal on interactive multiple super-session protocol applied for telematic conferencing system, *Tech. Group Inf. Network*, (IECE, Jpn) (IN84) **34**, 25–30 (1984).
21. H. Zimmermann, OSI reference model—The ISO model of architecture for open system interconnection, *IEEE Trans. Commun.* **COM-28**(4), 425–432, April (1980).
22. Y. Kitahara, *Information Network System*, Telecommunication Association, Tokyo, 1982.
23. Y. Kitahara, Telecommunications for the advanced information society—INS (information network system), Telecom Forum '83, Part 1, Vol. 2, 1983.

16

MODULE LEVEL COMMUNICATION PROTOCOL SPECIFICATIONS OF APPLICATIONS SOFTWARE FOR DISTRIBUTED COMPUTING SYSTEMS

Sol M. Shatz

1. Introduction

While the advantages of distributed computer systems are well established and documented,[1] the technology for developing distributed software is still very immature. A distributed software system consists of a number of distributed processing components (DPCs), which are each executing on a different processing node of the distributed computer system. Each DPC, in turn, consists of a number of functional modules; modules which reside in different DPCs must communicate via message passing. Naturally, the most important aspects of distributed software systems are the communication mechanisms which are required for both message exchange based communication and synchronization. These communication mechanisms are referred to as communication protocols. The ISO Reference Model[2] defines these communication protocols into seven different layers. The "lowest" layer is called the physical layer and the "high-

SOL M. SHATZ • Department of Electrical Engineering and Computer Science, University of Illinois, Chicago, Illinois 60680

est" layer is called the applications layer. In this paper, we are only concerned with the communication protocols of the applications layer. Specifically, we present guidelines for deriving the applications layer communication protocols for modules which comprise a distributed software system. The techniques which we are presenting are intended for use during the design phase of the software lifecycle, as opposed to the implementation phase. In order to represent (document) the protocols used, we use a communications-based design representation scheme which is based on the Petri net model.[3]

2. An Overview of the Representation Scheme

In this section, we would like to give an overview of the representation scheme which we are using. A more detailed presentation of this scheme can be found in Refs. 4 and 5.

Our representation scheme is oriented towards representing the communication and synchronization aspects of a distributed software system's design, i.e., the applications level communication protocols. We refer to this level of representation as a "Level One DPC specification." "Level Two DPC specifications" are used to represent the sequential *local* computation operations of the modules under consideration. Because our interest in this paper is with the communication protocols, we will not discuss Level Two DPC specifications further. The specification itself is based on the Petri net model.[3]

Definition 1. A *Level One DPC Specification* is a control flow specification which is defined as a 4-tuple

$$L1 = (LCT, LCP, CM, A)$$

where LCT is a finite set of local computation transitions, LCP is a finite set of local control flow places, CM is a finite set of communication modules, and A is a finite set of directed arcs, with $A \subseteq (LCT \times (LCP \cup CM)) \cup (LCP \times LCT) \cup (CM \times (LCP \cup CM))$.

In Level One DPC specifications, all aspects of the software which do not involve any interaction with other components (i.e., the local computation operations) are specified using a single Petri net transition, a Local Computation Transition (LCT). The fact that we compress all the details into that form does not reflect on the complexity of that local computation. Thus, a level one local computation transition in the Petri net specification may represent an operation such as adding two numbers together, or it may represent a complex statistical computation. The important point

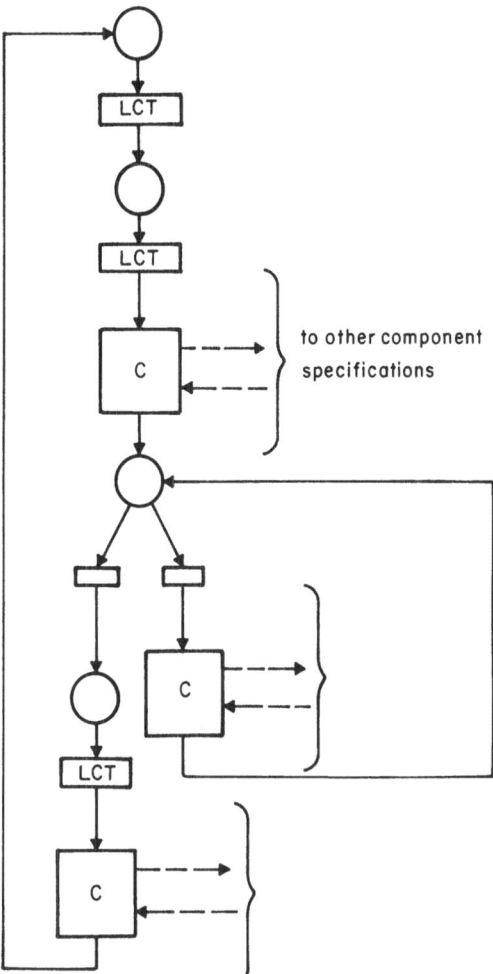

FIGURE 1. A general component's
level one representation.

is that the operation does not require any interaction with other components of the system.

The local control flow places are used to direct the "execution" of the computations. They provide for the specification of sequencing, selection, and iteration. For instance, a decision point is represented by using a local control flow place to produce a nondeterministic "branch" in the Level One DPC specification.

Those aspects of the software which relate to interactions with other components are represented in the specification by communication modules. These modules have fixed interpretations which define their behavior. As we would expect, the arcs are simply used to connect together selected elements of a specification as defined in Definition 1. Figure 1 is a sketch of a general distributed software component specification. Local computation transitions are labeled "LCT" and communcations modules are represented by square boxes labeled with a "C".

2.1. Communication Modules

Definition 2. A *Communication Module* is defined as a 5-tuple

$$CM = (IM, OM, IT, OT, B)$$

where IM is a finite set of input message ports, OM is a finite set of output message ports, IT is a finite set of input test ports, OT is a finite set of output test ports, and B is some "well-defined" communications-oriented behavior.

The messages of a CM are of two types: *Local* (or internal) messages flow within a component's specification and can represent the control flow of that component. *Nonlocal* (or external) messages flow between components and represent communication or synchronization messages.

The input message ports allow messages to flow into a communication module. These input messages may be local or nonlocal (depending upon the location of their source) with respect to the given communication module. Output message ports allow messages to flow out of a given communication module. Again the messages may be local or nonlocal (depending upon the location of their destination).

Test ports are required when the communication mechanism must allow for the explicit specification of a bounded buffer. Input test ports are used by message sending modules in order to "test" a destination buffer to determine if it is currently full. Output test ports are used by message receiving modules in order to provide the buffer status information to message sending modules which use test ports. In this paper, we will be restricting our attention to communication modules which do not use test ports.

The behavior (B) of a CM defines the communication mechanism which is specified by that communication module. For example, a CM's behavior may define the module as being a selective, asynchronous receiver communication module. In general, a CM's behavior is used as the name of the CM.

As mentioned previously, all distributed communication operations are specified by the use of communication modules. In general, this is carried out by selecting a suitable communication module from a set of standard modules. This issue is explicitly addressed by the guidelines which we will present in the next section of this paper. At this point, we would like to present some specific communication modules which are used by our guidelines for deriving module level communication protocol specifications.

2.2. Some Specific Communication Modules

For each CM, we will present a "module view" which hides that communication module's internal details, and an "internal view" which shows (using the Petri net formalism) how that communication module is constructed and thus behaves. Typically, local (nonlocal) input message ports are illustrated as directed arcs which enter the CM from the top (side), and local (nonlocal) output message ports are illustrated as directed arcs which exit the CM from the bottom (side).

The *Asynchronous Broadcast* (AB) communication module is used by a source DPC when it requires to transmit (broadcast) a common message to a set of destination DPCs, and it does not wish to wait for any acknowledgments. Figure 2 illustrates the AB communication module specification. The *Asynchronous Send* (AS) communication module is a degenerate case of the AB communication module. By this we mean that the AB can be used as an AS if we only use one nonlocal output message port. Thus, the AS communication module is used to asynchronously send a message to a specific destination DPC. This is similar to the communication in PLITS[6] and extended CLU.[7] Figure 3 illustrates the AS communication module specification.

The *Selective Asynchronous Receive* (SAR) communication module is used by a destination DPC when it wishes to receive a single message from any of a set of source DPCs—it is selective. Since the SAR does not reply

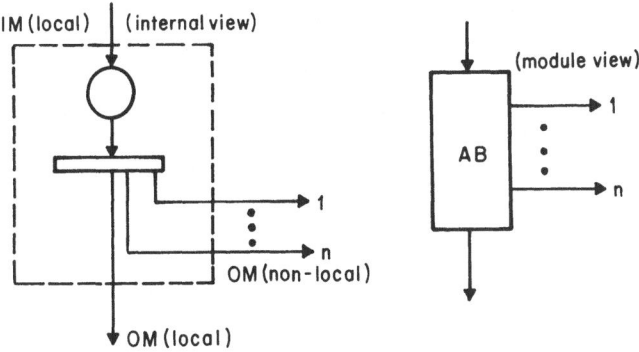

FIGURE 2. An asynchronous broadcast module, © 1983 IEEE.

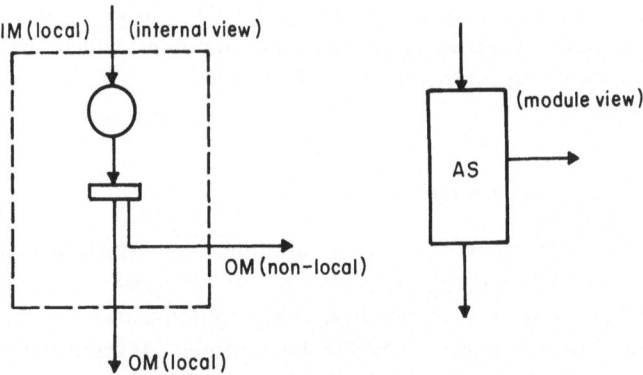

FIGURE 3. An asynchronous send module, © 1983 IEEE.

with an acknowledgment when it actually receives the message, it is asynchronous. Figure 4 illustrates the SAR communication module specification. A *Nonselective Asynchronous Receive* (NAR) communication module is a degenerate case of the SAR communication module. An NAR communication module is used to asynchronously receive a message from a specific source DPC. Figure 5 illustrates the NAR communication module specification.

The *Synchronous Send* (SS) communication module is used by a source DPC when it wishes to send a message to a specific destination DPC in a synchronous manner. Thus, after the message is sent, the source DPC waits for an acknowledgment message. This communication is similar to that of CSP[8] in that the exchange of the message is totally synchronous.

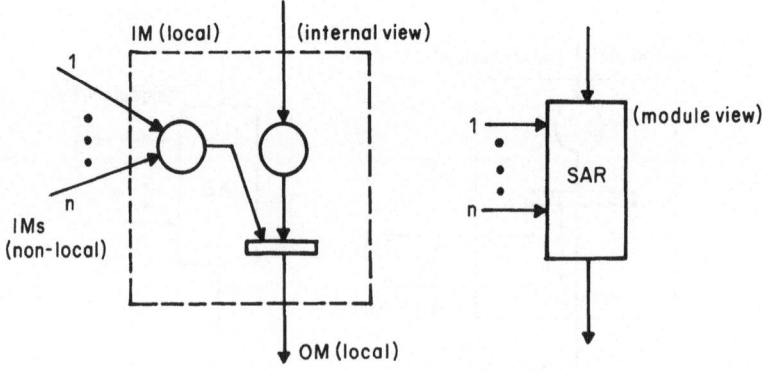

FIGURE 4. A selective asynchronous receive module, © 1983 IEEE.

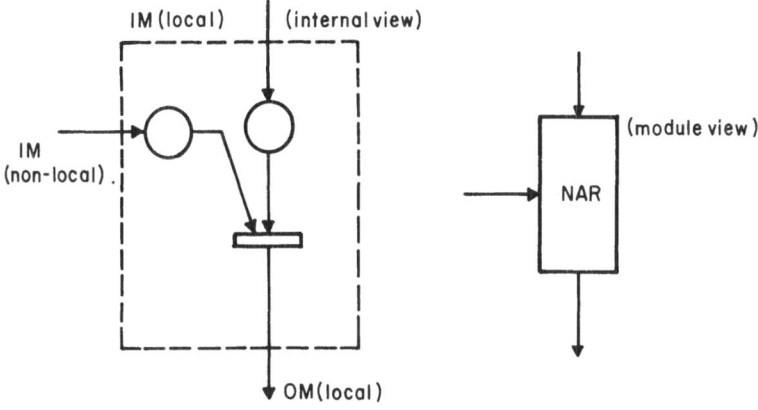

FIGURE 5. A nonselective asynchronous receive module, © 1983 IEEE.

One interesting property that this communication module possesses is that its specification is produced by combining other communication modules. Figure 6 illustrates the SS communication module specification.

The *Selective Synchronous Receive* (SSR) communication module is used by a destination DPC when it wishes to use the SAR in a synchronous manner, i.e., it does send an acknowledgment message to the source DPC which sent the message. Note that if multiple messages are available (from multiple source DPCs), the selection of which message to actually receive is done in a nondeterministic fashion. Figure 7 illustrates the SSR module

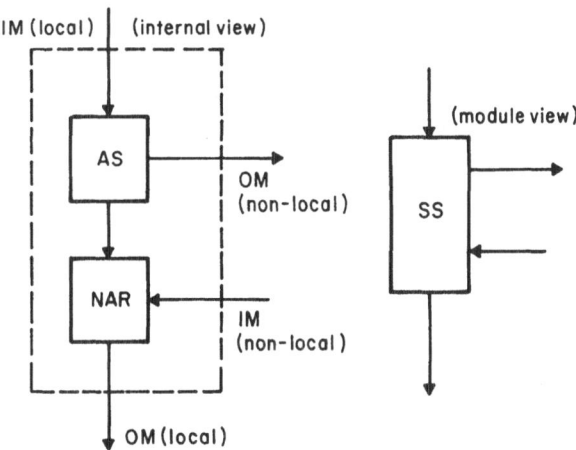

FIGURE 6. A synchronous send module, © 1983 IEEE.

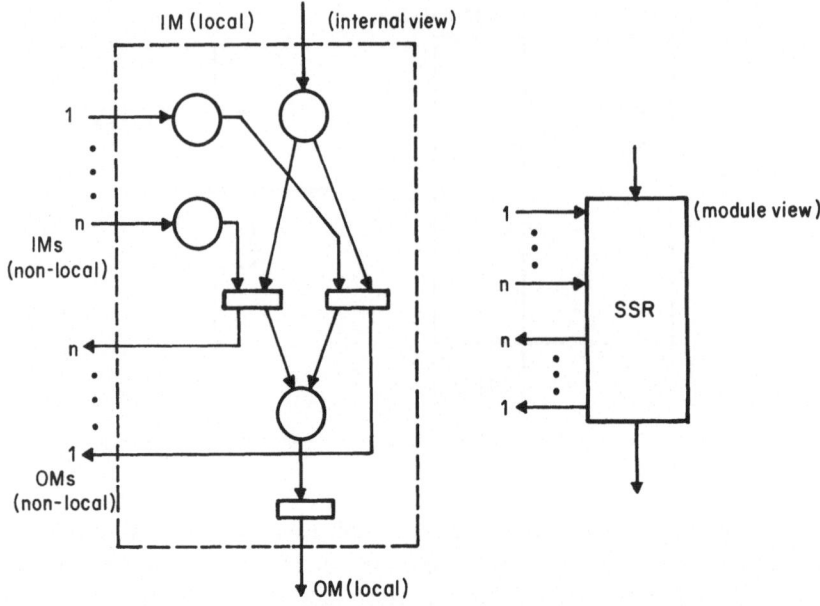

FIGURE 7. A selective synchronous receive module, © 1983 IEEE.

specification. The *Nonselective Synchronous Receive* (NSR) communication module is used by a destination DPC when it wishes to receive a message from a specific source DPC in a synchronous manner, i.e., it does send an acknowledgment. The NSR can be considered a degenerate case of the SSR, although it can be specified more simply by combining an NAR communication module and an AS communication module. Figure 8 illustrates the NSR communication module specification.

The *Remote Invocation Receive* (RIR) communication module is used by a destination DPC when it wishes to provide a function or service that can be invoked by other source DPCs. This communication module supports the remote invocation form of communication (from the receiver's perspective). The destination DPC relies on the RIR communication module to control the acceptance of "remote invocation" from source DPCs and to control the acknowledgments which must be sent *after* the service has been provided. This mechanism is similar to that of Ada[9] and other remote invocation based languages. Figure 9 illustrates the RIR communication module specification.

Because resource sharing is a very common reason for communication in a distributed environment, we now want to illustrate the use of communication modules in specifying a more complex module—a

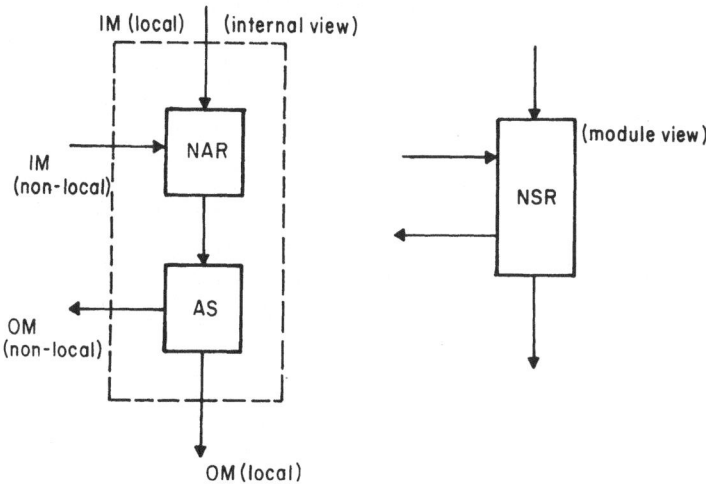

FIGURE 8. A nonselective synchronous receive module, © 1983 IEEE.

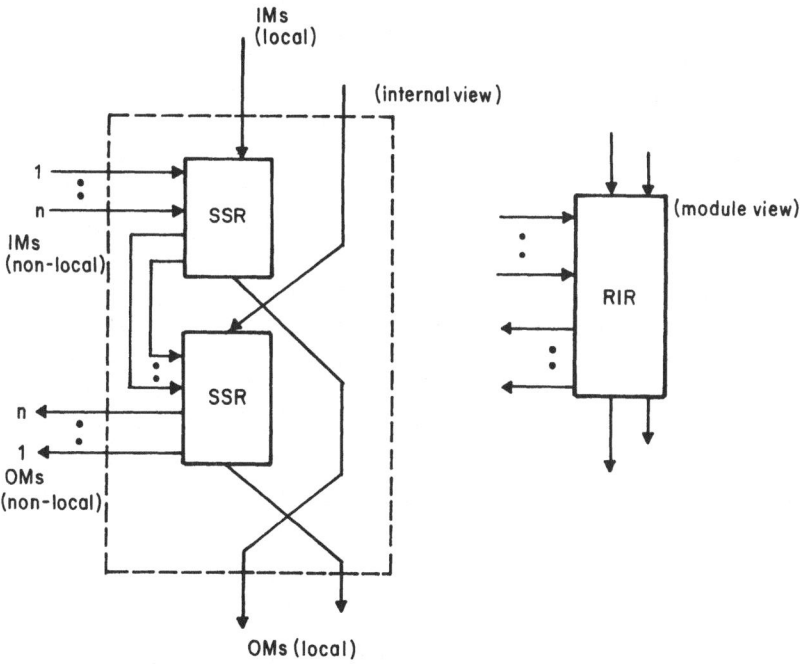

FIGURE 9. A remote invocation receive module, © 1983 IEEE.

Resource Guardian (RG) communication module. The RG communication module is a very basic synchronization module and is similar to the guardian construct of the extended CLU language.[7,10] Its purpose is to coordinate access to some system resource which is local to a DPC. The RG accepts requests for access to the resource and grants access according to some rule (such as a mutual exclusion rule). Once a component completes its operations on the resource, it notifies the RG via a release message.

The RG specification is derived from other communication modules. The initialization is accomplished by marking an internal Petri net place with some number of tokens; this number establishes an upper bound on concurrency. Figure 10 illustrates the specification of a two-way RG, i.e., one that supports two components. The more general n-way RG is constructed similarly.

2.3. Properties of DPC Specifications

The following important properties of Level One DPC specifications, which have been shown in Ref. 5, ensure that the protocol specifications derived using the representation scheme do indeed form a legal Petri net when the internal views of the communication modules are considered. This is important from the perspective of performing analysis on the resulting protocols, although we will not discuss that subject in this paper.

Theorem 1. Given a Level One DPC specification L1 = (LCT,LCP,CM,A), replacing each LCP_i of LCP with a node P_i, each LCT_j of LCT with a node T_j, and each CM_k of CM with its internal view minus any nonlocal message ports or test ports, produces the Petri net $G = (P', T',A')$, where $P' = \{P_i | i=1,2, \ldots ,|LCP|\} \cup \{P(k)|k=1,2, \ldots ,|CM|\}$, $T' = \{T_j|j=1,2, \ldots ,|LCT|\} \cup \{T(k)|k=1,2, \ldots ,|CM|\}$, $A' = A \cup \{A(k)|k=1,2, \ldots ,|CM|\}$), and P(k) is the set of Petri net places in the internal view of CM_k (the kth communication module). $T(k)$ is the set of Petri net transitions in the internal view of CM_k, and $A(k)$ is the set of Petri net arcs in the internal view of CM_k, except those arcs which act as nonlocal message ports or test ports.

Theorem 2. A set of Level One DPC specifications, which collectively comprise all components of a distributed software system, can be represented by a Petri net graph which is equivalent in structure and behavior to the Level One DPC specification.

2.4. An Example

Consider the classical dining philosophers' problem.[11] The distributed solution consists of five active processes or DPCs (one for each phi-

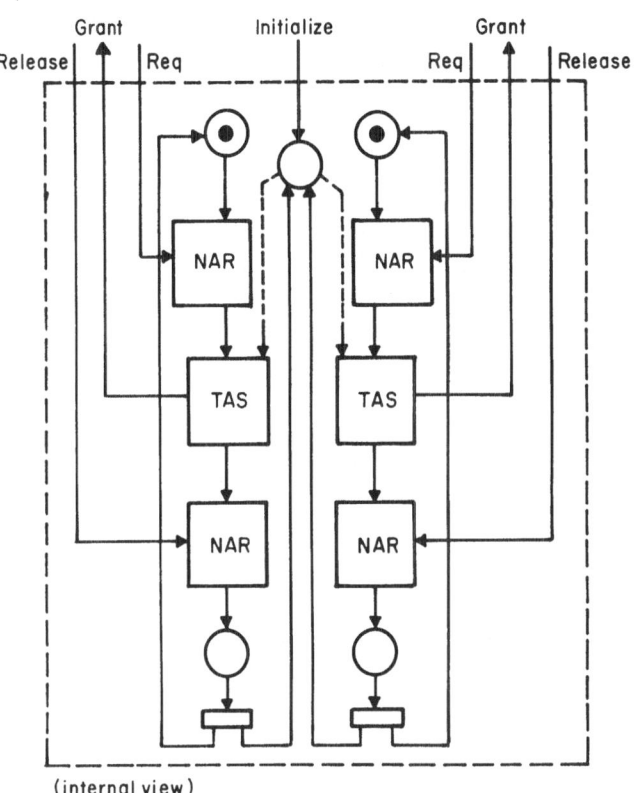

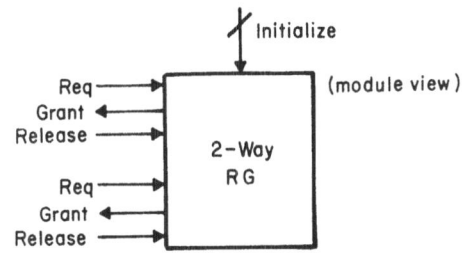

FIGURE 10. A two-way resource guardian module, © 1983 IEEE.

losopher), and five system resources (one for each fork). In order to avoid deadlock, we can introduce one more system resource which will model a doorway into the dining room. This strategy is to allow a maximum of four philosophers in the dining room at any given time, thus ensuring that no deadlock can occur (solution due to Dijkstra[8]). A design representation, which uses the scheme previously described, for this problem is illustrated and discussed in Refs. 4 and 5. As one would expect, the representation uses Resource Guardian communication modules to represent each fork, and the doorway to the room. Each philosopher is represented by a Level One DPC specification which uses SS communication modules to "get access to" the room and the particular forks needed. After eating (represented by a LCT), the philosopher must then release the resources it is holding; this is represented by the use of AS communication modules.

We now turn our attention to the guidelines for deriving module level design specifications for distributed software. Of course, the resulting specifications are represented using the representation scheme just discussed.

3. Deriving the Protocol Specifications

Because we are discussing module level communication protocol specifications for distributed software, we are necessarily assuming that the software system has already been decomposed into functional modules[12] and that these modules have been partitioned into one or more "clusters." Each cluster contains a set of functional modules which are to be assigned to a processing node of the distributed computer system. Thus, note that the existence of only one cluster implies that the software is to be implemented in a centralized manner as opposed to a distributed manner. In this paper, we refer to these clusters as DPC Clusters. A partitioning algorithm which forms the DPC Clusters has been developed,[13] but will not be discussed here.

What we are interested in discussing is how to derive the design representation of each of the DPCs in such a way as to emphasize the communication protocol requirements. To do this, each functional module of the software system is considered separately and a design representation for the functional module is produced.

For many applications, it is possible to end up with a number of functional modules which are in fact functionally identical. This is common in applications which possess multiple occurrences of a given functional entity, such as in Brinch Hansen's sorting array problem[14] and the classical dining philosophers problem mentioned earlier. As illustrated in Refs. 4 and 5, we can allow these functional modules to share a common specification and use an index labeling in order to distinguish sources and destinations of nonlocal messages. In order to give guidelines for deriving the design specifications for functional modules, we must define a means of classifying each type of module. Our classification is based on the method of invocation that can potentially be used to invoke the given module. The four types of modules that we will consider are: modules which are locally invocable only (we will refer to these as LI modules), modules which are remotely invocable only (we will refer to these as RI modules), modules which are both locally invocable and remotely invocable (we will refer to these as LI/RI modules), and modules which are potentially concurrently invocable (we will refer to these as PCI modules).

In centralized software systems, all modules are of the LI type. An LI module is only capable of being invoked by other modules which reside in the same DPC as the given LI module (i.e., they are all members of the same DPC Cluster). The invocation of an LI module is performed using only local messages (messages which "flow" within a DPC); thus, this form of invocation is quite efficient.

Because LI modules are invoked by local messages, their representation is straightforward and simple. The representation of an LI module consists of an LCT (Local Computation Transition) used to symbolize the initialization operations of the module, followed by a series of Level One DPC specification elements used to symbolize any local computation operations (represented by LCTs) and distributed communication operations which define that module's actual behavior. These elements are then followed by a "termination" LCT which is labeled RETURN. This LCT obviously symbolized the natural type of a return associated with subroutines and functions. Figure 11 illustrates the representation of a general LI module.

The actual invocation of an LI module is represented by an LCT labeled with a call to the LI module. For example, consider the invocation of an LI module, named VERIFY, by another module named ACCEPT PASSWORD. The invocation would be represented by an LCT labeled CALL VERIFY in the representation for the module ACCEPT PASSWORD. Note that due to the way that the LI module is represented, we can actually replace the "calling LCT" with the representation for the called module. Thus, the individual module specifications can be connected together to give a connected representation of the system's design.

An RI module is only capable of being invoked by modules which reside in DPCs different from that DPC in which the given RI module resides. The invocation of RI modules is performed using only nonlocal messages; thus, this form of invocation is more "expensive" than the LI form of invocation, in terms of longer message propagation delay and decreased reliability of message transfer. The advantage of using this "expensive" invocation is that it allows the designer to exploit potential concurrency that may exist between the RI module and other modules that reside in the same DPC as that which contains the module which is making the remote invocation.

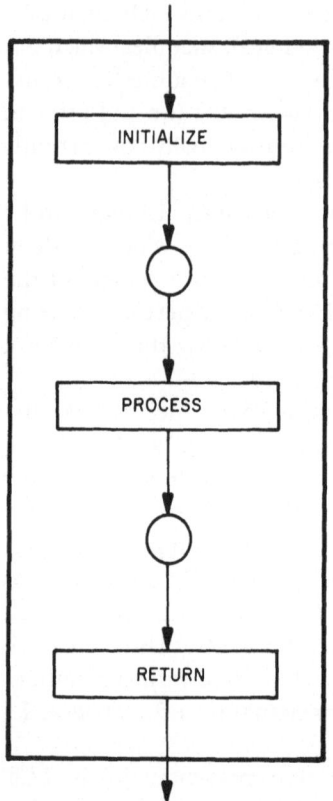

FIGURE 11. The representation of a general LI module.

The representation of an RI module is based on the Remote Invocation Receive (RIR) communication module previously described. The RI module performs an initialization (if necessary) and then waits for the presence of a remote invocation via a nonlocal message. Upon acceptance of the invocation, the module then performs its actual function. This function would, of course, be represented by a Level One DPC specification. When the module has completed its task, it then replies to the invoking module and transfers control to another module (possibly itself) in the same DPC. Figure 12 illustrates the representation of a general RI module.

As we would expect, RI modules are invoked via nonlocal messages that are sent from Asynchronous Send (AS) communication modules. The RI module's "return" message (sent by the RI module after it has completed its task) is accepted by a receiving communication module in the invoking module's representation. This receiving communication module would typically be a Nonselective Asynchronous Receive (NAR) communication module.

An LI/RI module is one which is capable of being invoked both locally (i.e., by modules that belong to the same DPC as it does) and remotely (i.e., by modules that belong to a DPC which is different from the one to which it belongs). As an example of this condition, consider the following system of five modules: module S (the start module) invokes modules A, B, and C, while modules B and C invoke module D. The ordering is such that module S invokes modules A and B (which can execute concurrently), and when both have completed, module A then invokes module C. One possible partitioning of these functional modules is to create the two DPC Clusters { S,A,C} and {B,D}. Note that module D is an LI/RI module as it is invoked by module B (local invocation) and by module C (remote invocation).

Because a remote invocation requires the ability to accept and transmit nonlocal messages, we must force the representation of an LI/RI module to explicitly deal with these messages. Thus, in order to make the representation uniform, we will choose to represent LI/RI modules as if they were RI modules. This implies that the local invocations to these modules will in fact be treated as if they were remote invocations. Of course, these

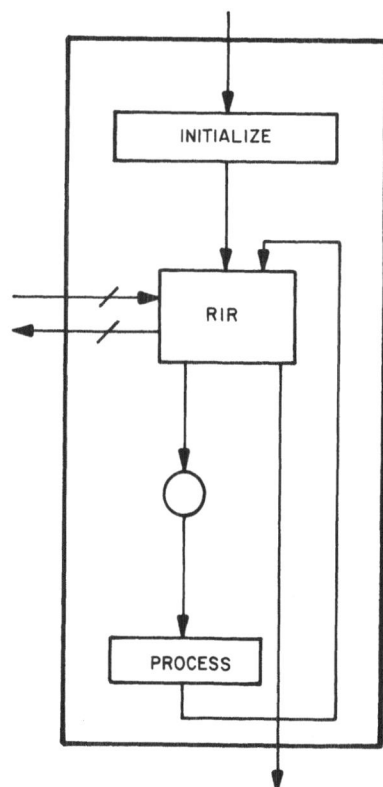

FIGURE 12. The representation of a general RI module.

"remote invocations" will result in "nonlocal" messages being routed from the source DPC to the destination DPC—in this case, these DPCs are not distinct.

LI/RI modules are invoked in the same way that RI modules are invoked. But, because the modules which "locally" invoke the LI/RI module cannot continue execution until the LI/RI module completes its task, we will use synchronous send (SS) communication modules to represent those invocations.

A PCI module is one which may be invoked concurrently, i.e., it can be invoked *during* its execution. We claim that PCI modules should always be invoked remotely. We now argue this claim informally. First, assume that a module C is a PCI module. Now, for C to be a PCI module, it must be true that there exist two modules A and B (not necessarily distinct) such that both A and B directly invoke C, and A and B may execute concurrently (i.e., they are elements of different DPC Clusters or are themselves PCI modules). Since A may execute concurrently with B, and since C is "a descendant" (in terms of invocation) of B, C can potentially execute concurrently with A. Thus, in order to allow the exploitation of this potential concurrency, we conclude that C should not be placed in the same DPC cluster which holds A. Module A must, therefore, invoke module C remotely. A similar argument leads us to conclude that module B must also invoke module C remotely.

We choose to further classify the PCI modules into two groups: modules which represent physical system resources which other modules may access after being granted permission (e.g., a printer), and modules which represent pure software resources which provide a service which is generally of use to other modules (e.g., a sort routine). In general, we would represent the physical system resources using Resource Guardian (RG) communication modules. Of course, we must determine the degree of concurrent access that the resource can support so that the RG communication module can be properly initialized. The invoking (requesting) functional module can represent its interactions with the RG in one of two basic ways: using a Synchronous Send (SS) communication module or using an Asynchronous Send (AS) communication module. The SS communication module is used if the invoking functional module does not have any useful computations which it can perform during the time that it is waiting to be granted access to the resource. Thus, the invoking functional module uses the SS communication module to handle both the sending of its request for access and its receipt of the grant message. In order to release the resource, the requesting functional module would use an AS communication module to send a release message to the RG.

If the invoking functional module can perform some useful computation during the time that it is waiting to be granted access to the resource, then it can use an AS communication module to handle the request for access and then "later" (after the computation is completed) use a Nonselective Asynchronous Receive (NSAR) communication module to handle the receipt of the grant message. Again the release message would be handled by an AS communication module.

3.1. Representing the Invocation of Concurrently Executable Modules

In order to explain how we represent the invocation of concurrently executable modules, we will demonstrate the procedure using an example. In this example system, functional module A invokes modules B, C, and D, and modules B, C, and D can all execute concurrently. As a specific example, consider a simple patient-monitoring system which assigns the following interpretations to the modules A, B, C, and D:

A: Monitor patient
B: Check patient's temperature
C: Check patient's pulse
D: Check patient's blood pressure

Assume that the system is partitioned into the following DPC Clusters: {A,B}, {C}, {D}. Note that since B, C, and D can all execute concurrently, they appear in distinct DPC Clusters.

It is obvious that module A must invoke module B locally (they appear in the same DPC cluster) and it must invoke modules C and D remotely. When module A (locally) invokes module B, module A's execution will be suspended until module B "returns." Thus, in order to exploit the concurrency between modules B, C, and D we must design module A such that it (remotely) invokes modules C and D *before* it invokes module B. Because we assume that module A cannot continue processing until modules B, C, and D are finished, it does not matter in which order the remote invocations are performed. We can now describe the representation of module A. Module A's preliminary local computation (if any) is represented by an LCT. Following this, we would use an AS communication module to represent the sending of the invocation message to functional module C (or D). Following this, we would use another AS communication module to represent the sending of the invocation message to functional module D (or C). Now, we represent the local invocation of module B using an LCT labeled with the call to module B (as described previously). Following the local invocation LCT, we use a series of NAR communication modules to accept the "return" messages from functional modules C and D (or D and C), respectively.

3.2. An Example

The example system is similar to the one used in Ref. 15. We are interested in designing a distributed software system which is part of a ballistic missile defense (BMD) system.[16] In simple terms, the requirements of the software system are that it gather and analyze radar signals and that it track and intercept any threatening objects. Of course, the actual requirements (and design) of such a system would be much more complex and detailed;

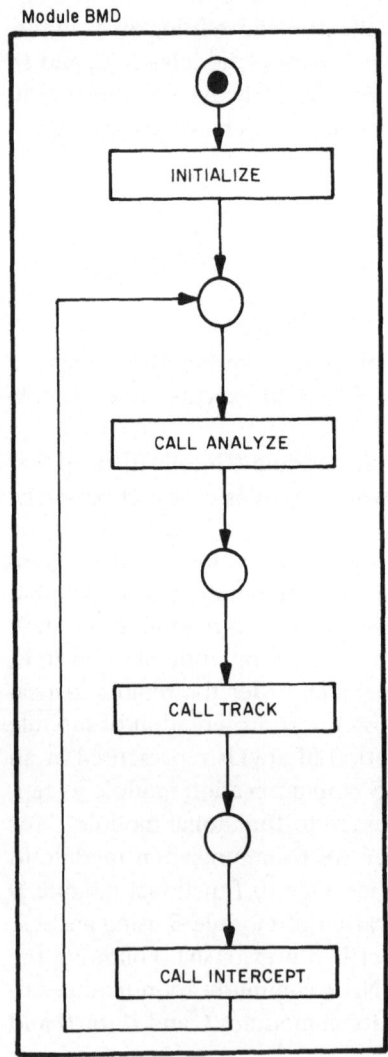

FIGURE 13. The design representation of module BMD.

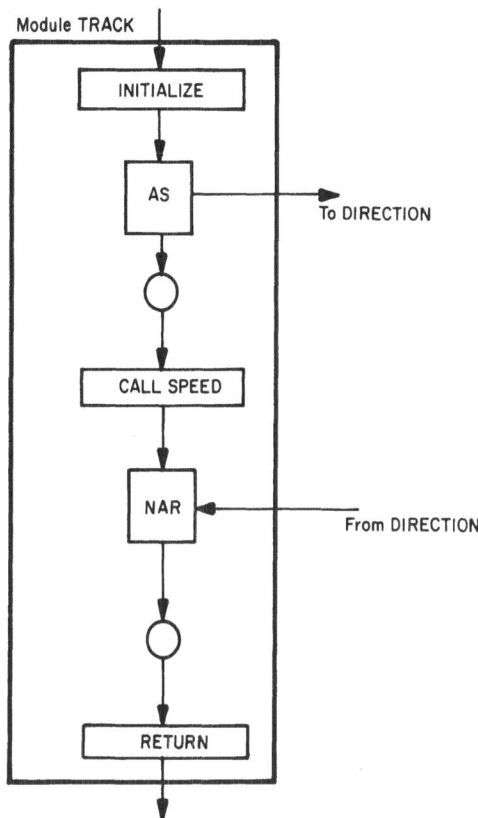

FIGURE 14. The design representation of module TRACK.

our example is intended to illustrate the design representation guidelines being proposed here.

Assume that the software system is decomposed into the following functional modules:

1. BMD: the main module for the system;
2. ANALYZE: to gather and analyze radar signals;
3. GET SIGNALS: to gather and edit the radar signals;
4. TRACK: to track threatening objects within the radar space;
5. SPEED: to calculate the threatening object's speed of flight;
6. DIRECTION: to calculate the threatening object's direction of flight;
7. INTERCEPT: to intercept threatening objects.

Also, assume that these functional modules are partitioned into two DPC clusters (note that the degree of potential concurrency is two; in tracking an object, we can calculate the object's speed and direction con-

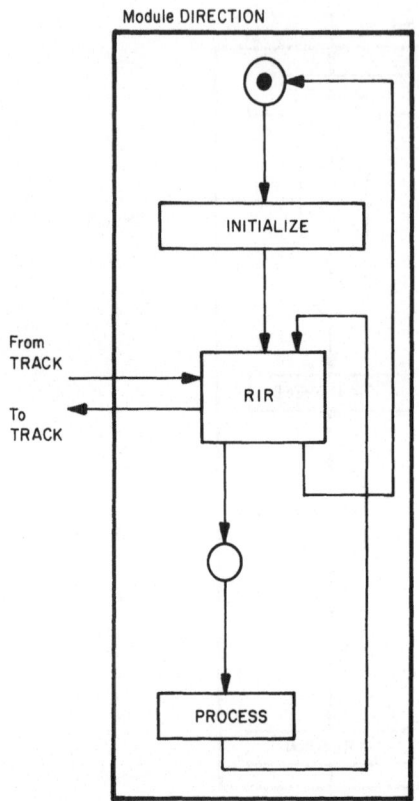

Module DIRECTION

INITIALIZE

From
TRACK

To
TRACK

RIR

PROCESS

FIGURE 15. The design representation of module DIRECTION.

currently): {BMD, ANALYZE, TRACK, INTERCEPT, GET SIGNALS, SPEED} and {DIRECTION}.

At this point, we classify each module and then represent the design of each module using the representation scheme. Based on the DPC clusters, we can see that each of the modules, except DIRECTION, is an LI module. DIRECTION is, of course, an RI module. Figure 13 illustrates the design representation of the BMD module. Because the design representations of modules ANALYZE, GET SIGNALS, SPEED, and INTERCEPT are similar to the design represented in Fig. 12, we do not illustrate their designs here. The design representations of modules TRACK and DIRECTION are given in Figures 14 and 15, respectively.

4. Conclusion

Development of distributed software is a relatively unexplored discipline. Much research is needed in all phases of the software lifecycle for distributed software. In this paper, we have focused attention on the design phase of the lifecycle. We have presented some general guidelines for deriving communication protocol specifications for distributed software modules. These specifications are represented using a Petri net based representation scheme which explicitly addresses the communication and synchronization activities. It is expected that these protocol specifications will greatly facilitate the implementation and even testing of the resulting software product.

In order to help the software designer in developing design specifications, some sort of automated design aid would be useful. Because the specifications themselves are graphical in nature, a tool which supports graphic input and display would be most convenient, especially if it is coupled to some predefined communication module library. Finally, since the specifications use the Petri net model, automated design analysis can be performed by using techniques which have evolved from Petri net analysis research.

Acknowledgment

The author wishes to thank S. S. Yau for his support during the term of this research.

References

1. J. A. STANOVIC and A. VAN DAM, Research directions in (cooperative) distributed processing, in *Research Directions in Software Technology*, P. Wegner, Ed., MIT Press, Cambridge, Massachusetts, 1979, pp. 611–638.
2. H. ZIMMERMANN, OSI reference model—The ISO model of architecture for open systems interconnection, *IEEE Trans. Commun.* **28**, 425–435 (1980).
3. J. L. PETERSON, *Petri Net Theory and the Modeling of Systems*, Prentice-Hall, Englewood Cliffs, New Jersey, 1981.
4. S. S. YAU and S. M. SHATZ, On communication in the design of software components of distributed computer systems, *Proceedings of the Third International Conference on Distributed Computing Systems*, October, 1982, pp. 280–287.
5. S. M. SHATZ and S. S. YAU, The application of Petri nets to the representation of communication in distributed software systems, *Proceedings of the IEEE Workshop on Languages for Automation*, November, 1983, Chicago, Illinois, pp. 163–173.
6. J. A. FELDMAN, High level programming for distributed computing, *Commun. ACM* **22**(6), 353–368 (1979).
7. B. LISKOV, Primitives for distributed computing, *Proceedings of the Seventh ACM Symposium on Operating Systems*, 1979, pp. 33–42.
8. C. A. R. HOARE, Communicating sequential processes, *Commun. ACM* **21**(8), 666–677 (1978).
9. J. D. ICHBIAH *et al.* Preliminary Ada reference manual, *SIGPLAN Not.* **14**(6) June (1979).
10. B. LISKOV, On linguistic support for distributed programs, *IEEE Trans. on Software Eng.* **8**(3), 203–210 (1982).
11. R. C. HOLT, G. S. GRAHAM, E. D. LAZOWSKA, and M. A. SCOTT, *Structured Concurrent Pro-*

gramming with Operating Systems Applications, Addison–Wesley, Reading, Massachusetts, 1978, 84.

12. E. YOURDON and L. L. CONSTANTINE, *Structured Design: Fundamentals of a Discipline of Computer Program and Systems Design,* Prentice-Hall, Englewood Cliffs, New Jersey, 1979.

13. S. M. SHATZ and S. S. YAU, A partitioning algorithm for the design of distributed software systems, (to appear) in *Inf. Sci.*

14. P. BRINCH HANSEN, Distributed processes: A concurrent programming concept, *Commun. ACM* **21**(11), Nov. 934–941 (1978).

15. S. S. YAU, C. C. YANG, and S. M. SHATZ, An approach to distributed computing system software design, *IEEE Trans. Software Eng.* **SE-7**(4), 427–436.

16. C. G. DAVIS and R. L. COUCH, Ballistic missile defense: A supercomputer challenge, *Computer* November 37–46 (1980).

17

THE GCP LANGUAGE AND ITS IMPLEMENTATION

G. CASTELLI, F. DE CINDIO, G. DE MICHELIS, AND C. SIMONE

1. Introduction

Guarded Communicating Processes is a complete language for distributed applications based on Hoare's CSP.[1] CSP are an "ambitious attempt to find a single simple solution" to the communication and synchronization problems of concurrent processes, that is, to represent in a single frame semaphores, critical regions, monitors, queues, and so on. They are characterized by some choices:

- synchronous communication;
- no shared data among processes;
- nondeterminism.

These features deal with concurrent systems through the notion of communication state machine components. This approach has been enriched by many contributions, especially by Milner's CCS[2] and related works. It is becoming more relevant not because of optimization purposes (optimization is for Peter Wegner[3] an issue of smart compilers and it is related to the ratio between logical concurrency and physical concurrency), but because it is a simple and conceptually powerful kind of process abstraction. The technology development pressing towards fully distributed applications and, consequently, requiring highly distributed languages, shows clearly the great impact of the CSP and related languages on the design and the implementation of such systems.

Furthermore, the trend shows that the area of application is very interesting for the use of such systems and languages, even if the problems to be solved are very different from the original ones, i.e., communication

G. CASTELLI, F. DE CINDIO, G. DE MICHELIS, and C. SIMONE. • Istituto di Cibernetica, Università di Milano, 20133 Milano, Italy

and synchronization between processes in a multiprocessor machine. Just
to recall a relevant example: the LAN technology is supporting an increasing number of distributed applications in the office automation and process control areas.

A number of experimental implementations[4-6] and communication-
algorithms have been presented in literature since 1978, but very rarely
they can actually be used for distributed applications.

In fact, the proposed languages concentrate mainly on the communicating mechanism, ignoring all the other necessary features that make a
language usable.

Furthermore, very often the implementations were simulations on a
single processor of a distributed application: that means that little consideration was given to the performance issues within a network. CSP inspired
also the development of different concurrent languages, one of the most
recent being OCCAM.[7]

The GCP language (Guarded Communicating Processes language, the
name is still provisional) has been defined, deriving with some modifications its control mechanism from Hoare's CSP,[8] embedding them in a
fully defined concurrent programming language. The GCP language, in
fact, even if it is still in an experimental phase, is completely defined except
for two very critical areas still under investigation: exception handling and
real time features. Of course even the consolidated features could be subject to minor changes based on the judgment of a small community of
"guinea pigs."

The GCP implementation consists of an easily retargetable compiler,
a configurator/distributor to configurate the GCP program allocating processes to processors (nodes), to define the mapping of logical communication channels on physical links, and to build the run time tables. The
development of a distributed symbolic debugger is planned. The current
implementation uses Olivetti M40[9] machines both as implementation and
application nodes. This is due to the fact that currently just one compiler
exists running on the M40 and generating code for the M40 under MOS.

2. A Short Introduction to CSP

Tony Hoare proposed in 1978 a prototype of a concurrent programming language "Communicating Sequential Processes" (CSP[1]), based on
the idea that input and output are basic primitives of programming and
that parallel composition of communicating sequential processes is a fundamental program structuring method.

CSP combine therefore two basic communication primitives, the input command, whose syntax is the following:

$$\langle source \rangle \ ? \ \langle target \ variable \rangle$$

and the output command, whose syntax is the following:

$$\langle destination \rangle \ ! \ \langle expression \rangle$$

with Dijkstra's guarded commands.[14] Input commands (and also output commands in more recent versions of CSP) may occur also as guards in a guarded command. If the processes with which a given process is waiting to communicate are terminated, then the given process terminates too.

CSP have been widely discussed in the computer science community from the semantics and implementation points of view,[4,5,7,10-13] and have inspired other programming language as ADA (with respect to the extended rendezvous mechanism[15])

3. Communication and Termination

The purpose of this chapter is to describe in some level of detail the main differences between GCP and Hoare's CSP, first of all the interprocess communication primitives.

The way GCP processes communicate between each other is through a guarded communication primitive. A guarded communication statement is a coupling of a communication condition (indicating the logical channel previously linked to the process with which the communication must occur) and of the messages that have to be exchanged, e.g.,

$$COMM \ (delta) \rightarrow (IN \ x,y; \ OUT \ i + j * k);$$

where "delta" is the name of the channel to be used for the communication. After the keyword IN the names of the local variables are indicated where the received messages have to be stored; after the keyword OUT the expressions are listed whose values have to be sent. The communication primitive is therefore an exchange statement.

Hoare's input and output are represented omitting, respectively, the OUT or IN lists. Such a schema allows a clear distinction between the condition enabling the communication and the action performing the communication. If compared with the original Hoare schema, GCP programs are more homogeneous because of the absence of any kind of asymmetry between Boolean conditions and input/output guards (the asymmetry between input and output guards, as proposed in Ref. 1, has already been eliminated by many authors and implementors.[10,11]

Processes communicate through channels dynamically created to bind two processes. The creation of a channel is a synchronization action between two processes. The actual communication may occur only after the channel has been created.

While a communication action (message exchange) has a system defined time out, so that other statements can be executed instead of blocking the process, the creation of a channel is always suspensive, until the synchronization occurs. The reason for this choice is that the opening/ closing of a channel defines the frame where communication may occur.

Secondarily, the problem of distributed termination has been deeply analyzed by a number of authors.[10,12,13] As already stated in a previous paper,[6] we agree with Francez' ideas in classifying processes into two categories: endoprocesses and exoprocesses, and specifying the semantic differences also through a different syntax.

As stated in Ref. 13, a process can terminate because of one of two conditions: the reaching of its final state (endotermination) or the impossibility of communicating with other processes because of their termination (exotermination). The "distributed termination convention" applies to exoprocesses waiting for opening a channel or wishing to communicate with already terminated processes.

4. Some Language Features

The following is a brief description of GCP. GCP has been influenced by a number of different languages, but the basic criterion we followed was to reach a difficult compromise: keep the language as small as possible, quite safe, flexible, and orthogonal.

4.1. Data Types

GCP has the usual predefined data types INTEGER, REAL, CHAR, and BOOLEAN, but it is possible to have some control on their representation. The programmer has the capability of controlling the representation in terms of addressable units. The boundaries of this choice are fixed by the implementation.

Enumerated types can be introduced by the user as in Pascal, but unlike in Pascal, it is possible to force a value from which start (or restart) the enumeration.

Type constructors are: ARRAY, RECORD, STRING, SET, and POINTER. The FILE constructor is omitted as in C, Modula 2, and Ada with the same motivation.

The user has explicit control over type compatibility. Type compatibility is based on name equivalence, except for the exchanged messages. This choice is motivated by the fact that the language allows separate and independent (maybe even distributed) compilations. No intercompilation unit check is done because the only intercompilation unit communication is through exchanged messages and processes have no parameters nor shared data.

To simplify the configurator/distributor task, checks on message types are structural based on a compact coding of the type that does not require the propagation of the symbol table from the compiler front end to it.

4.2. Processes

A compilation unit is a collection of processes. A process has a declaration part and a body. The former includes TYPE, CONST, and PROCESS declarations; the latter contains the executable statements.

All the processes declared within a compilation unit must reside on the same node; more than one compilation unit may reside on one node.

Processes communicate in a synchronous way through message exchange.

Process declarations can be nested; processes enclosed in a process (children) are activated explicitly in the body of the enclosing process (parent), inside a parallel control structure.

The parent process is suspended until all the activated children processes terminate. It is therefore impossible to have any message exchange between a parent and its children; for this reason a controlled sharing of data between a parent and its children is allowed. Children processes can INHERIT constants, types, and variables from the parent; inherited variables are copied during children activation into their own data area and become local variables. It is possible for children to change their parent's variables; however, to guarantee data integrity, the intersection of the so-called YIELD variables must be empty. Modified values are returned to the parent upon termination of all children processes.

A declaration of ARRAY PROCESS generates a fixed number of identical copies of the same process, each copy retaining its identity. Because the array processes are identical copies of the same object, it is necessary to specify which process returns a value through the YIELD statement. This is done by a selection associated to the variable. For example,

$$\text{YIELD } x, z, y(3)$$

means that the third element of the array process (i.e., the third instance of the process described in the array processess declaration) will return to the parent the value of the variable *y*.

4.3. Channels and Communication

Channels are logical links between two processes. Channels are not explicitly declared in process declaration area, but are created and destroyed in the process body by means of the TIE and UNTIE statements. A process Pi declaring a channel named ALFA as the communication link with a process named Pn contains the following:

$$\text{TIE (ALFA: Pn)}$$

This is a synchronizing operation and the process Pi is suspended until the operation succeeds, that is until the partner process Pn executes a corresponding statement:

$$\text{TIE (BETA: Pi)}$$

Communication is always point to point, but it is possible to select randomly a partner specifying ANY or ANY of ⟨partner list⟩ instead of a specific name. After a TIE operation has been successfully executed, communications may occur, until an UNTIE statement is executed, removing the link between the processes.

Communication is performed by the guarded communication statements COMM (see above). The communication statements must always be guarded by a communication expression specifying the channel and the type of the message. Because GCP calls for synchronous communication and nondeterministic control, a communication failure within a fixed amount of time makes possible the execution of another statement, and the communication, if possible, is retried later.

4.4. Control Structures

GCP control structures are directly derived from Dijkstra[14] guarded commands allowing just three structures to specify sequence (;), selection *(if-fi)*, iteration *(do-od)* and extended with the parallel control structure (p1 // p2 // ... // pn).

4.5. Distributed Termination

As mentioned above endoprocesses and exoprocesses are distinguished by the syntax. In fact, the two keywords ENDOPROCESS and EXOPROCESS, respectively, precede the declaration of the two types of processes. The distributed termination convention applies to exoprocesses, bringing them to termination when they are waiting to communicate (or to open a channel) with already terminated processes.

5. Current Implementation

The description of a current implementation is subdivided into three parts: a description of the programming environment, a description of the abstract communication machine, and some measures of critical times on the implementation machine.

5.1. The GCP Programming Environment

The GCP programming environment (see Figure 1) is a very simple one; except for the tools usually available on a development system it currently consists of a GCP compiler, of a configurator/distributor, and of an arbiter managing nondeterminism and communication. The development of a distributed symbolic debugger is planned in late 1985.

The compiler is a classic two-pass compiler, the two passes communicating through an intermediate representation. It is written in Pascal+,[16] the M40 system implementation language.

The configurator/distributor is written in Pascal+ too. It operates on the object files and on a set of compiler generated files of information (one for each module) describing all the message structures and the requested logical link.

Network-independent description tables are built by the first phase of the configurator/distributor to define all the possible communications among processes. Checks are done to verify that for each TIE/UNTIE and COMM, at least one correspondent communication statement exists.

If the first phase completes correctly, the second phase, interacting with the user, distributes the processes on the network and modifies the tables generated by the first phase with information about the network structure and the physical links. The logical links are mapped on the phys-

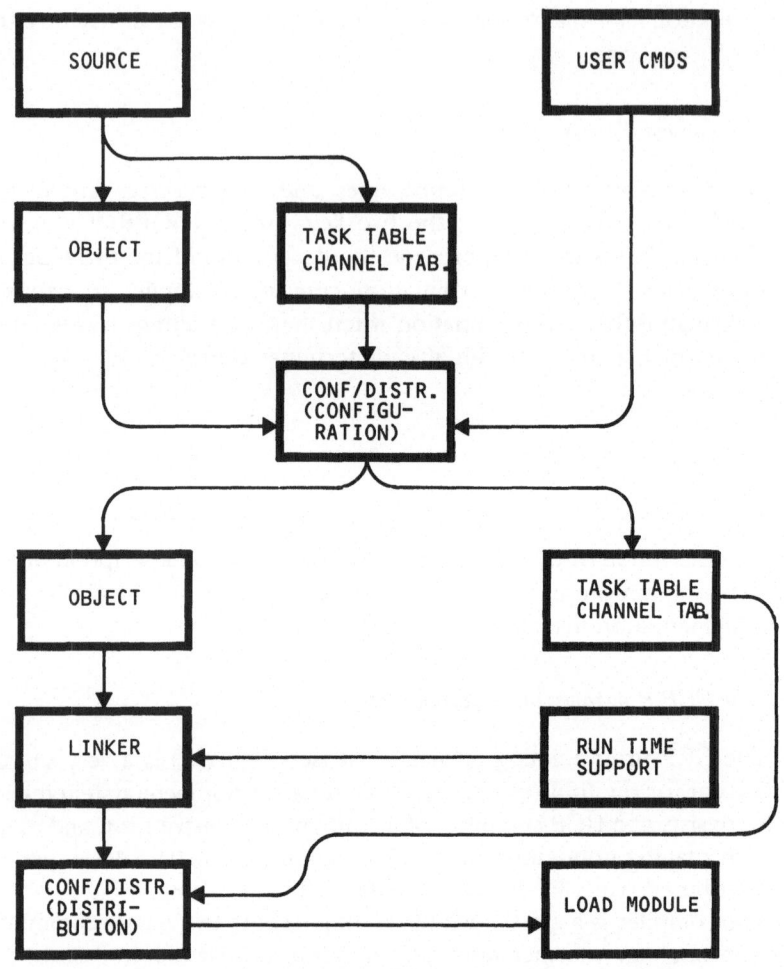

FIGURE 1. The GCP environment.

ical ones. The run time support is driven by these tables, guaranteeing that if a process is moved from one node to another, then only the configuration/distribution process has to be redone, recompiling and relinking being not necessary. This is the only centralized phase in the development and execution of an application written in GCP.

After the configuration/distribution activity is terminated the processes can be started locally on each node or remotely from the machine where the configuration/distribution activity took place.

5.2. The Abstract Communication Machine

Machines can be connected on the same network through physical communication media that can be very different: point to point or multi dropped lines, Ethernet, X.25, and so on.

The effort we made in this area is to define an Abstract Communication Machine to hide all the details of the physical media and regardless of whether communication is performed locally, between processes running on the same node, or remotely.

Another important issue is that the intersystem communication had to be as light as possible. These requirements were very well supported by the Cosmos to Cosmos Communication Subsystem (COSCOMS)[17] running on the M40 under MOS. The main difference between our needs and the functionalities offered by COSCOMS was the latter's asynchronous way of communicating. Therefore we integrated the set of primitives into the run time support where the Buckley–Silberschatz[11] algorithm was also implemented to support the communication schema.

The set of primitives introduced to support communication are the following ones:

OPENCHANNEL (ChannelId, LinkedChannel, TransRec, Bufsize)

This primitive opens a channel in transmit or receive mode and allocates the space for the buffers containing the messages to be exchanged. It is suspensive without timeout. The link between the channels is performed. It is invoked when a TIE statement is executed.

The second one is

CLOSECHANNEL(ChannelId)

This primitive closes a channel opened by OPENCHANNEL. It is suspensive without timeout. The unlink between the channel and its partner is performed. It is invoked when an UNTIE statement is executed.

Furthermore:

READCHANNEL(ChannelId, Where, TimeOut)
WRITECHANNEL(ChannelId, What, TimeOut)

These primitives read/write a message from/to a channel. The read/written data are put in the Where/What parameter, driven by their length. The read/write is suspensive, but if the waiting time specified by TimeOut expires, the run time selects, if possible, another statement to be executed and the process restarts.

6. Some Timing Measurements

The timings described here are mainly related to the process activation, the process switching, and the send/receive operations. These timings are very important to evaluate the overall system performance because GCP have no procedure/function concept so that, even locally, the structuring is obtained through processes.

The timings are those of a M40 under MOS. The CPU is a segmented version of the Zilog Z8000 processor running at 4 MHz.[18]

Our net is based on Omninet, with a speed of about 1 Mbit per second. The performances described below are those at the abstract machine. For the MOS operating system there is just one process on each node for a given GCP application.

All the process management is done by the local run time support, acting as process scheduler, communication handler, and as the unique interface towards the actual operating system.:

Process activation:	800 μsec
Process switching:	600 μsec
Send/receive in memory	
(with no message):	30 μsec
Remote send/receive	
(with no message, using Omninet):	70 μsec

7. Conclusions

As Hoare has pointed out,[1] the synchronous communicating primitives underlie many familiar synchronization methods and therefore the communicating nondeterministic sequential processes can constitute a synthesis of a large number of distributed programming concepts.

In particular, communicating processes allow high modularity in distributed system design, because it is possible to create a library of exoterminating processes that accomplish particular tasks.

GCP will be very well suited for the design of distributed applications both in the areas of the office networking, and of the distributed control.

The nondeterminism of the iterative (do-od) and alternative (if-fi) command does not create implementation problems. In fact, since the nondeterminacy of the processes is bounded by their postcondition, any order in the evaluation of the guards can be followed. It is, anyway, open to question whether a run time support can be modeled in such a way that it can use the nondeterminism in order to optimize its performance.

The language is completely defined[19] (the report is available upon request to the authors) and the formal semantic specification is currently under development, based on Petri nets and denotational methods.

Some possible extensions with respect to fault tolerance and real-time systems are also planned.

The revised report, together with the definitive name of the language, will follow the above-mentioned activities.

The architecture of a distributed symbolic debugger will be defined during the first half of 1985. An engineered version of GCP programming environment for the Olivetti M40/M60 will be accomplished in time depending also on the availability of the required hardware. The same planning governs the development of tools for GCP implementation on a nonhomogeneous network.

Acknowledgments

The authors wish to thank U. Passalacqua and S. Scola for supporting the development of the language and the architecture of the configurator/distributor, and A. Osnaghi, F. Lacapra, and G. Tamburelli of Olivetti for providing useful information for the implementation.

This work is a result of cooperation between the Università di Milano and Honeywell-ISI (CP Project).

References

1. C. A. R. HOARE, Communicating sequential processes, *Commun. ACM* **21**,(8), 666–677 (1978).
2. R. MILNER, *A Calculus for Communicating Systems*, Lecture Notes in Computer Science 92, Springer-Verlag, New York, 1980.
3. P. WEGNER, Perspectives of capital intensive software technologies, presented at Capri '84 Advanced Personal Computer Technology, 1984.
4. L. SHRIRA and N. FRANCEZ, An experimental implementation of CSP, Proceedings of the IEEE Conference, 1981.
5. T. J. ROPER and C. J. BARTER, A communicating sequential language and implementation, *Software Pract. Exper.* **11**, 612–623, 1981.
6. F. DE CINDIO *et al.*, A prototype of a distributed application language for a real time microcomputer based system, Proceedings of the EUROMICRO Conference 82, North-Holland, Amsterdam, 1982.

7. INMOS, *OCCAM,* Prentice Hall, Englewood Cliffs, New Jersey, 1984.
8. F. DE CINDIO *et al.,* Guarded communicating processes, Proceedings of the fifth Honeywell International Software Conference, Minneapolis, 1981.
9. MOS—Modular operating system, Olivetti, 1981.
10. A. J. BERNSTEIN, Output guards and nondeterminism in communicating sequential processes, *ACM TOPLAS* **2,**(3) 316–323, (1980).
11. G. N. BUCKLEY and A. SILBERSCHATZ, An effective implementation for the generalized input output construct in CSP, *ACM TOPLAS* **5,**(2) 223–235, (1983).
12. R. B. KIEBURZ and A. SILBERSCHATZ, Comments on communicating sequential processes, *ACM TOPLAS* **1,**(2) 218–225, (1979).
13. N. FRANCEZ, Distributed termination, *ACM TOPLAS* **2,**(1) 42–55, (1980).
14. E. W. DIJKSTRA, Guarded commands, nondeterminacy and formal derivation of programs, *Commun. ACM* **18,**(8) 653–657, (1975).
15. *ADA: The Programming Language ADA Reference Manual,* Lecture Notes in Computer Science 155, Springer-Verlag, New York, 1983.
16. PASCAL+ Reference Manual, Olivetti, 1981.
17. MOS System Primitives, Reference Manual, Olivetti, 1981.
18. Z8000 CPU Technical Manual, Zilog Inc.
19. G. CASTELLI *et al.,* The GCP Report: first Draft Version, I.C. Internal Report, 1984.

A LANGUAGE FOR THE DESCRIPTION OF CONCURRENT SYSTEMS MODELED BY COLORED PETRI NETS: APPLICATION TO THE CONTROL OF FLEXIBLE MANUFACTURING SYSTEMS

JAVIER MARTINEZ AND MANUEL SILVA

1. Introduction

The writing of software for the control of medium- and large-scale concurrent systems is quite a complex task. To facilitate this it is convenient to have languages which permit a description of these control systems based on easily constructed models.

Petri nets (PN) are very suitable tools for the construction of these models because of their power of expression, their graphic nature, and the possibility of validating the constructed model. Their use may be limited by the size of the systems to be modeled since a large-scale system is, in general, complex and difficult to manipulate.

Colored Petri nets (CPN) are tools of the family of PN, with identical power of description,[1] but which are more concise from a graphic point

JAVIER MARTINEZ and MANUEL SILVA • Departamento de Automatica, E.T.S. Ingenieros Industriales, Universidad de Zaragoza, 50009 Zaragoza, Spain

of view. They provide a good solution for the modeling of concurrent systems of some size, as is shown in several papers where CPNs are used to model telephone systems,[2,3] distributed data bases,[3] and flexible manufacturing systems.[4] A decisive factor in choosing CPNs lies in the possibility of verifying the quality of the constructed model (validation) using a set of invariants of the marking of the CPN (the invariant method).[5,6]

The language we have developed has been designed so as to facilitate the description of the CPNs which model concurrent systems. It is of the declarative type (nonprocedural) and allows not only a simple syntactic analysis of a program, but also a certain semantic verification of the described model (validation).

The work of programming an application must necessarily pass first through a stage of construction of a model (CPN) of the control system to be designed. It is clear that this stage is of capital importance as it is essential to have a good method for constructing models.

In Section 2 we outline the problems in controlling a flexible manufacturing system. In Section 3 modeling with CPNs is discussed. Firstly, we present an industrial system for flexible manufacturing whose control system we propose to design (Section 3.1); then we outline a method for modeling with CPNs and we detail the interpretation associated with the net which constitutes the model (Section 3.2), and, finally, (Section 3.3) we describe the process of modeling the system described in Section 3.1. Section 4 considers the process of validating a model. Finally a language for describing systems modeled by CPNs is presented (Section 5).

This chapter has been written with the assumption that readers are familiar with CPNs and their associated terminology. For this reason we have omitted the formal presentation of the tool, which can be found in Refs. 5 and 6.

2. The Hierarchical Structure of the Control System of a Flexible Manufacturing System

In the control system of a flexible manufacturing system (FMS) we can distinguish three levels[7]:

(a) *Low level* or *local control* of the elements which make up the system. Among these we can mention the units of the transport subsystem (e.g., trolleys, roller tables, etc.), the robots and the units of production (e.g., machine tools).

(b) The *level of coordination* between controllers of low level which can evolve concurrently and maintain relationships of *cooperation* (e.g., the transport subsystem must provide the production units with tools and raw materials, and take away their finished products) and of *competition* (e.g., competition between various production units for the resources they share, such as shared tools or the transport subsystem).

(c) The *level of scheduling* which must make the highest level decisions of the system, such as what to produce at any point in time, how to produce it and with what means, and the management of the preventive maintenance policy. Other possible functions could be those relating to control during breakdowns or unforeseen incidents, information on the current state of the system, and the production of periodic reports.

There will be various types of low-level controls which will carry out local functions such as sequential control of logic automatisms, control of robots, or numerical control of machine tools. The techniques for designing and programming these systems are classics in their respective fields and they will not be considered here.

The design of the coordinator is, in principle, a more complex task owing to the high level of concurrency which is common among elements of a FMS. In Section 3 the basic concepts of its modeling by CPN will be presented, and applied to an industrial problem.

The design of the scheduler is difficult to systematize. Production and maintenance policies, event management, the handling of exceptions, communication with the manager of the system, etc. must all be considered.

The coordinator will be related with the lower and upper levels. Thus, it must send orders to the low-level controls to carry out certain local tasks and receive information on their completion. On the other hand, its evolution will be conditioned, in certain cases, by the scheduler's decisions, or simply by events produced in the system and managed by the scheduler.

3. Modeling of Complex Concurrent Systems with CPNs

In this section we present the key ideas of a methodology for modeling concurrent systems using Petri nets, and we use it to model the coordination of a flexible production system. Since we propose to give an intuitive presentation of ideas, we will first outline the control problem to be resolved (Section 3.1), then discuss methodological aspects of the modeling (Section 3.2), to conclude with the modeling of the coordinator of the FMS (Section 3.3).

3.1. Statement of the Problem of Control of a FMS

The system to be controlled is, in fact, part of a flexible workshop of the company Renault, which has previously been the subject of various studies.[8,4] The layout of the flexible workshop is shown in Figure 1. The

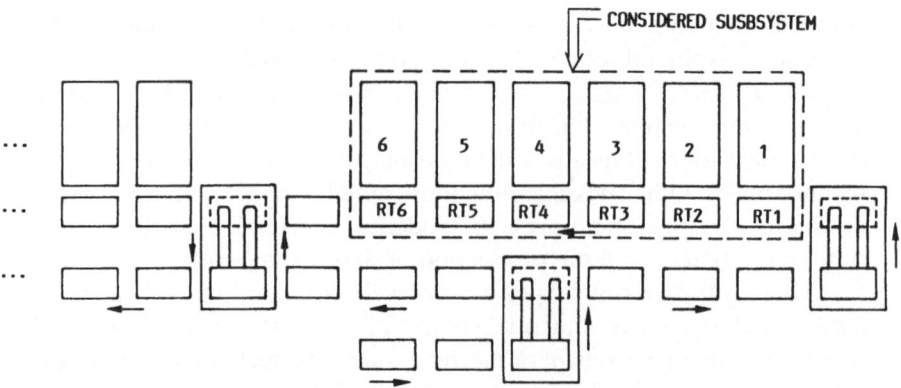

FIGURE 1. Layout of the flexible workshop.

system we will consider is that enclosed by the broken line, and we will leave aside the rest of the workshop.

The purpose of the system is to mastic seal the body work of automobiles. For this purpose it has six workstations (or simply stations), each consisting of one work place P, sufficient for one car body, and also one transfer bench with two roller tables, one designed to load the station (TL) and the other to unload it (TU). The bench, which is mobile, can adopt two positions: left, to load, or right, to unload the car bodies.

A system of transport, made up of six roller tables (RT1, ... , RT6), has the task of transporting all the car bodies which enter the system (via table RT1) to the sealing station designated by the scheduler, and of later unloading and removing them to the exit via RT6. Figure 2 shows the path a car body must follow within the subsystem under consideration.

It is worth noting that the tables RTi can pivot to permit loading and unloading of the corresponding station.

Each table (TL,TU,RTi) can only carry one car body and at any station there can be a maximum of two (P-TU, P-TL, or TU-TL) at any given time.

3.2. The Construction of Models Using CPNs: General Ideas

The approximation we propose combines a set of ideas aleady applied to modeling with PN and adapted to this new tool, with others which are peculiar to CPNs. Basically it consists of the modular design of an autonomous colored net, and the association of an interpretation with its evolution.

In *modeling the autonomous CPN,* we can distinguish three stages:

(1) *Independent modeling* of different subsystems which form part of

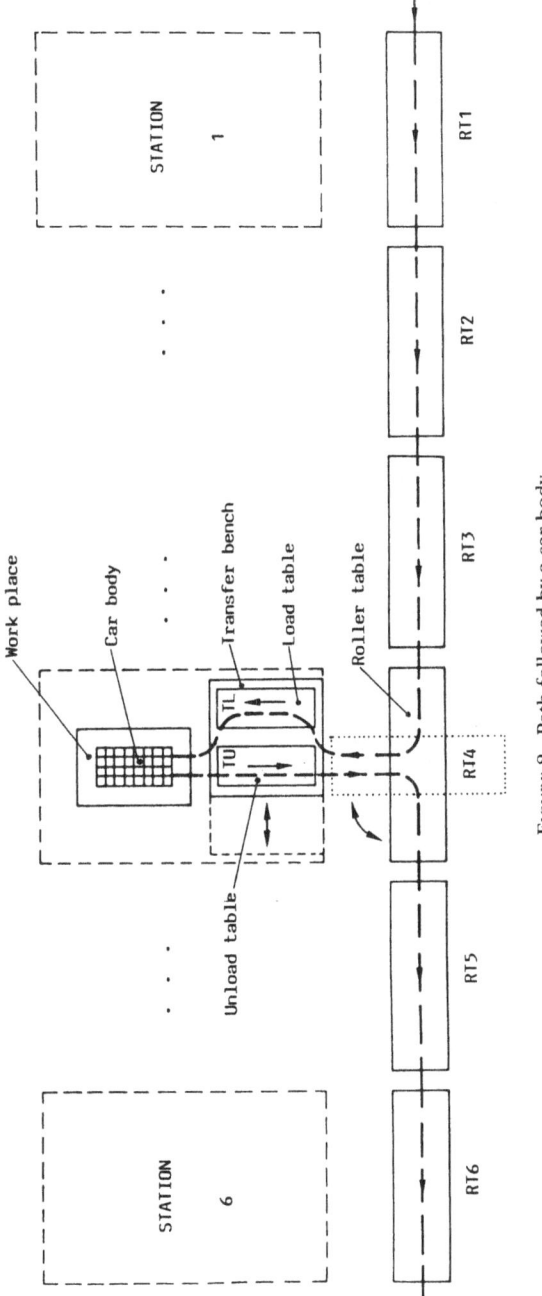

FIGURE 2. Path followed by a car body.

the system under consideration, using CPN modules. Thus, in the system described in Section 3.1 we can distinguish two subsystems: transport (roller tables) and stations.

(2) *Fusion* of the modules by fusing the transitions which represent synchronizations between subsystems. Thus, for example, the loading and unloading of the stations requires synchronization between the transport and station subsystems.

(3) The eventual *simplification* of the CPN obtained by eliminating *implicit places*. These are places whose marking can be expressed as a positive linear combination of a set of places of the net and whose elimination preserves the liveness and boundedness of the net.[9]

A *model* is a CPN whose places represent the states of tasks and resources of a subsystem (e.g., stations, roller tables or car bodies) and whose transitions define the different possible evolutions of the state of the subsystem. To each place we associate a set of colors such that we can distinguish all the tasks and resources associated with it. The colors can be *simple* or *compound* (n-tuples) depending on whether the associated tasks or resources are defined by one parameter or whether they need several. Thus, for example, to each roller table it is sufficient to associate the color "table," $\langle t(i) \rangle$, where the index i refers to its position in the transport system. The state of a car body, on the other hand, will be characterized by two colors: the mastic station to which it has been assigned, and the roller table which is carrying it ($\langle s(i), t(j) \rangle$).

To each transition we associate a set of colors such that its cardinal is equal to the number of different possible evolutions of the state of the system, which are defined by their possible firings. The nature of the colors will be such that it perfectly describes the evolution which each firing implies. In general, to a transition we associate compound colors. The functions associated with the arcs of a net define the movement of tokens in a net on the firing of a transition. Our experience in modeling with CPNs has shown that there exists a set of very frequently used functions and that a model, in general, can be constructed using functions from a *catalog* or *library* (simple functions or their combinations).

The interface between the control system and the station will consist of (1) a set of *events* which, when produced in the controlled systems, can set off an evolution in the state of the control system and (2) a set of *tasks* which this must actuate under certain circumstances.

The proposed interpretation is based on associating *events* and *tasks* with transitions of the CPN, which can eventually be individualized by colors. The firing of a color-enabled transition will take place when the event associated with that transition and color is verified. At that moment the tokens which correspond to the input places of the transition will be removed and the tasks associated with the transition and color considered will be activated. The marking of the output places of the transition will be made once the activated tasks have concluded.

3.3. Application to the Modeling of the Coordinator of a FMS

3.3.1. Construction of the Autonomous Model. To do this we will follow the steps outlined in the previous section, but first we define the colors which will appear in the model.

(a) Colors. To those elements of the system which have similar behavior we will associate a set of colors with the same name. They will be distinguished by subindices. In this way we define two sets of colors:

$\langle s(i) \rangle$, where $1 \leqslant i \leqslant 6$, which will be the set of station colors
$\langle t(i) \rangle$, where $1 \leqslant i \leqslant 6$, which will be the set of roller-table colors

The index of a color is useful for modeling the functioning of a system whose elements present a certain order relationship (e.g., the roller tables of the transport system).

A single color is not always enough to identify an element of the system. This is the case of a car body. Its characterization requires double the information: its mastic station and its position in the transport system. In these cases, *compound* or *n-tuple colors* will be used:

$\langle s(i),t(j) \rangle$,   where $1 \leqslant i \leqslant 6$ and $1 \leqslant j \leqslant 6$ represent the set of car body doublets

(b) Modules. The object of this stage of the design is to construct the two CPNs of Figs. 3 and 4 which model the station and transport subsystems, respectively.

The modeling of the station subsystem is relatively simple since all the stations have identical functioning, from the coordinator's point of view. The five places of the net correspond to the five possible permanent states of each station ("empty," "P occupied," "TU occupied," "P and TL occupied," "P and TU occupied"). The states "TL occupied" and "TU and TL occupied" are transitory since they occur during the firing of the transitions of loading and unloading the stations. The transitions represent possible evolutions in the state of the system ("loading" and "unloading of a station" and "end of sealing of a car body").

In Figure 3, contrary to classical procedure with CPNs,[5] the arcs of the net have not been labeled with their corresponding functions. Instead, we have chosen a labeling similar to that of predicate-transition nets,[10]

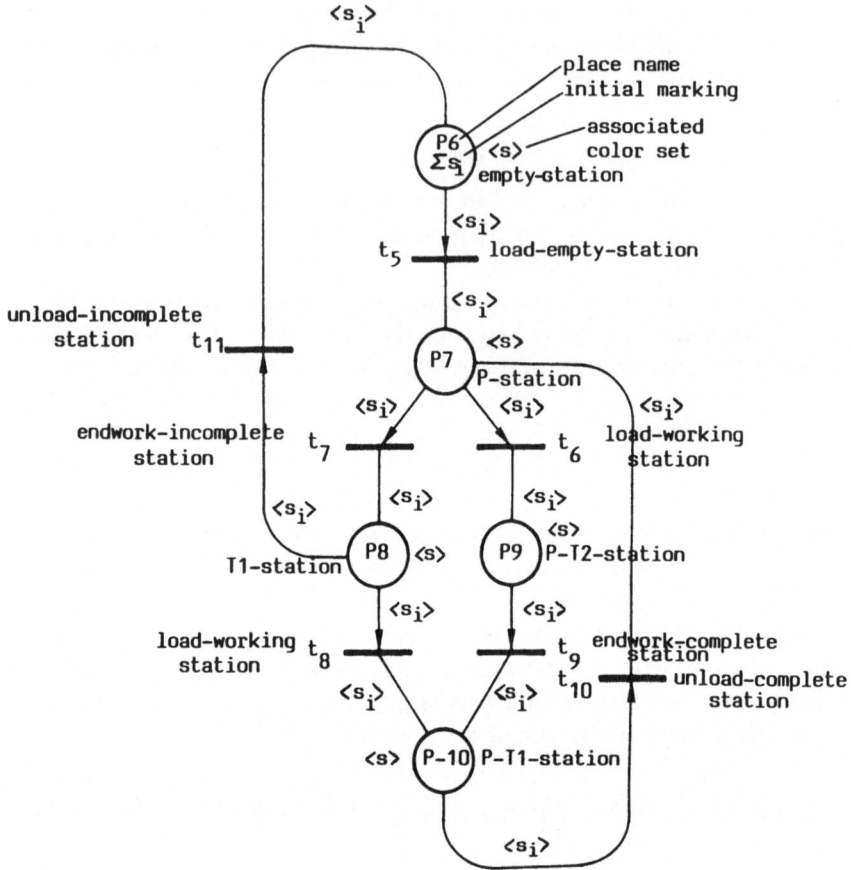

FIGURE 3. CPN modeling the station subsystem.

which consists of associating, with each arc, the colors which must be added or removed from the places of entry or exit of a transition when this is fired. This notation, which in our case is equivalent to the use of functions, has the occasional advantage that it improves the legibility of the graphic model.

In the model of the transport subsystem (Figure 4), the places represent both the states of the car bodies and resources (roller-tables) and the transitions represent evolutions between states.

It is worth noting that this subsystem behaves, in some sense, like a FIFO queue. The transitions t1 and t2 represent input and output from the queue while t3 and t4 represent the progression of a car body towards the exit. The special feature of this progression is that it allows, under certain conditions, the loading or unloading of any one of its elements (tl and tu, respectively).

(c) *Fusion of Modules.* The two subsystems considered cooperate. This relationship can be modeled by fusing the transitions of the two models which correspond to a synchronization between them. Thus the operations of loading and unloading a station will require the synchronization of subsystems, and the transitions to be fused are

- tl of the transport CPN with t5, t6, and t8 of the station CPNS;
- tu of the transport CPN with t10 and t11 of the station CPNs.

(d) *Simplification.* The fused net has an implicit place, p4. Its marking can be expressed as a positive linear combination of the markings of other places:

$$m(p4) = m(p7) + m(p8) + 2m(p9) + 2m(p10)$$

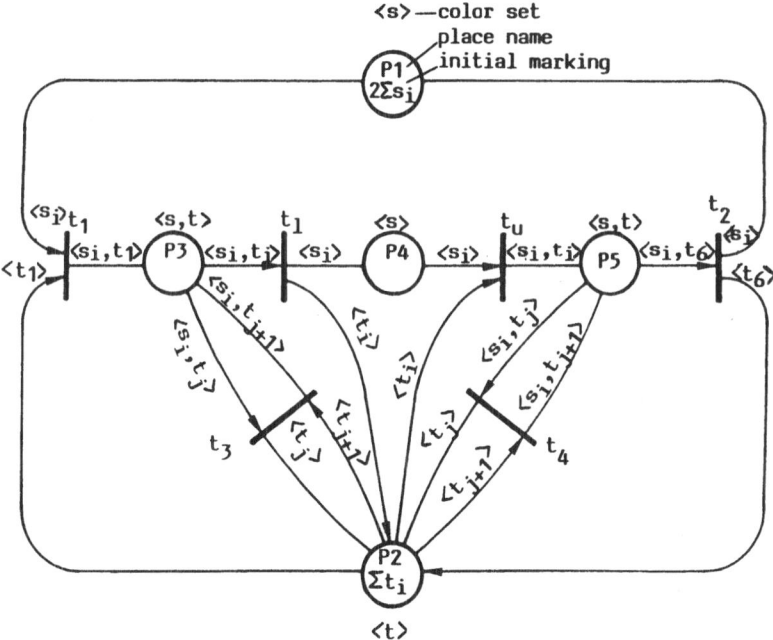

FIGURE 4. CPN modeling the transport subsystem. t_1, input body car; t_2, output body car; t_3, next table input; t_4, next table output; t_1, load station; t_u, unload station; P_1, remaining body cars; P_2, empty tables; P3, loading body cars; P4, working body cars; P5; unloading body cars.

and its elimination does not affect the transition firing sequences of the net. Thus we can eliminate this place, and we obtain the net shown in Figure 5.

The calculation of implicit places can be carried out during the process of validating the model (Section 4).

(e) Functions. All that remains to complete the autonomous model is to define the set of colors associated with each transition and the functions which must label each arc. The former are defined following criteria given in Section 3.2. Thus, for example, to the transition t1 we associate a set of colors of the type $\langle s(i),t(1)\rangle$ $(1 \le i \le 6)$ which indicate that the roller table RT1 (color $\langle t(1)\rangle$) and the mastic station represented by the color $(\langle s(i)\rangle)$ are involved in the firing of the transition.

Having defined the domain of each transition, the functions associated with the arcs of the net are perfectly specified. Their identification with *library functions* (identity, projection, successor, etc.) on the basis of the labelling of Figs. 3 and 4, is relatively simple and can even be automated. To make the model more legible, a specific name (with a clear meaning for the modeler) is associated with each of the functions used.

Specific name	Standard name	Examples
id	identity (colorset)	$id(\langle s(i)\rangle) = \langle s(i)\rangle$
station	projection (1,colorset)	$station(\langle s(i),t(j)\rangle) = \langle s(j)\rangle$
table	projection (2,colorset)	$table(\langle s(i),t(j)\rangle) = \langle t(j)\rangle$
nextable	sucessor (2,colorset)	$nextable(\langle s(i),t(j)\rangle) = \langle s(i),t(j+1)\rangle$

In the net in Fig. 5, the arcs have been labeled with their corresponding functions.

3.3.2. The Associated Interpretation. The interfaces between the coordinator and the scheduler of the system and between the coordinator and the low-level controllers are shown in Table 1.

Events managed by the scheduler (input-carbody and end-mastics) or tasks which must be carried out under the control of the low-level controllers (loading and unloading of tables or the transfer of a car body to the next table of the transport system) are associated to each transition. Events and tasks can have an associated color, st or tbl, which refers to the station or roller table in question, respectively.

The CPN of Fig. 5, together with its associated interpretation, constitute a complete model of the coordinator of a control system for a flexible workshop. For its implementation in computer (Section 5) a language for the description of the model (net and interpretation) is proposed. A program, which describes a model, will be analyzed syntactically and validated

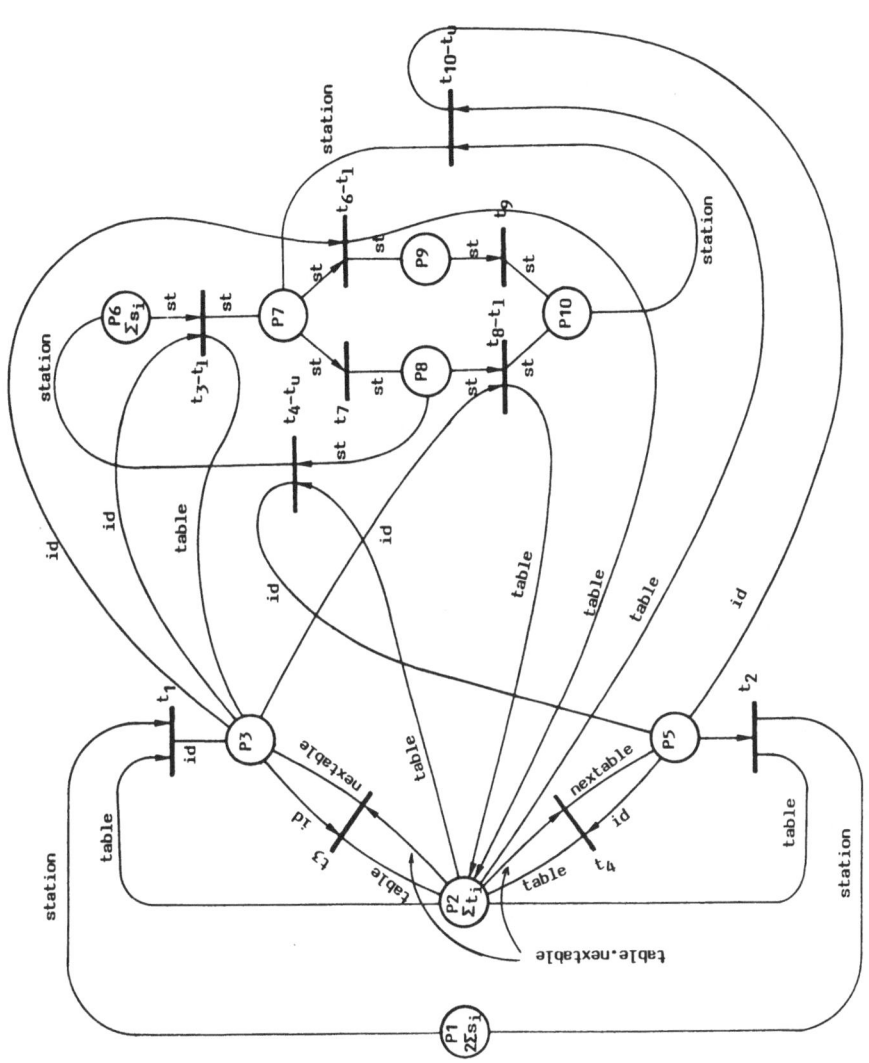

FIGURE 5. CPN obtained by fusing CPNs of Fig. 3 and 4. (Remark: st = station.)

TABLE 1
Interpretation Associated with the Fig. 5 CPN

Transition	Event	Tasks
t1	arrival_carbody(st)	load_rt1
t2		unload_rt6
t3		next_rt(tbl)
t4		next_rt(tbl)
t5		load_t1(st),load_p(st)
t6		load_t1(st)
t7	end_mastics(st)	load_tu(st)
t8		load_t1(st),load_p(st)
t9	end_mastics(st)	load_tu(st),load_p(st)
t10		unload_tu(st)
t11		unload_tu(st)

(Section 4) to detect errors both in the program-writing and in the model itself. The result of this analysis phase will, normally, be a data structure interpretable according to the rules of evolution of the CPN and of the interpretation. The interpreting system can be seen as the kernel of the control system being designed. This kernel can be designed specifically or can be constructed on the basis of the kernel of a commercialized operating system. An interpreter of CPNs implemented under the kernel of the system RMS68K is presented in Ref. 7.

4. Validation of the Model

As has been commented before, one of the principal attractions of CPNs is the possibility of checking the quality of the constructed model. This validation is restricted to the autonomous model. In spite of this, the results obtained are of great value to the designer.

The structural analysis, carried out on the basis of a set of invariants of the marking, constitutes today the most viable technique for validating a CPN[3,6,9] and allows two sets of properties of the model to be characterized:

   a. *General properties* of the model such as: conservativeness, boundedness, or nonblocking.
   b. *Specific properties* of the considered model (mutual exclusions, etc.).

Returning to the model of the coordinator constructed in Section 3.3, and applying the method presented in Ref. 6 for calculating marking invariants, we obtain the four invariants shown in Table 2. Their associated left

TABLE 2
A Set of Marking Invariants of the Fig. 5 CPN

i1	$m(p1) + \text{station}[m(p3) + m(p5)] + m(p7) + m(p8) + 2.m(p9) + 2.m(p10) = 2\Sigma s_i$
i2	$m(p2) + \text{table}[m(p3) + m(p5)] = \Sigma t_i$
i3	$m(p6) + m(p7) + m(p8) + m(p9) + m(p10) = \Sigma s_i$
i4	$m(p4) + 2.m(p6) + m(p7) + m(p8) = 2\Sigma s_i$

null vectors constitute a basis of left null vectors of the CPN incidence matrix. By considering this set, the following general properties can be automatically deduced:

1. The net is conservative;
2. The net is bounded (it is possible, in the same way, to characterize the limit of each place);
3. The (autonomous) net is without blocking.

From the same set of invariants, a set of specific properties can be inferred. For this purpose, each invariant must be interpreted in terms of the model. This allows us to make a series of statements which should agree with the specifications of the system. Many of them will actually coincide with the specifications, which will be a symptom of the "goodness" of the model. If any invariant contradicts the specifications the model should be revised.

The following could be a direct interpretation of the invariants calculated:

i1. At any given moment, there can be a maximum of two car bodies assigned to each station $(2\Sigma s_i)$. A car body which has been assigned to the station s_i can be located in the transport system either on the way to the station $[m(p3)]$, or on the way to the exit $[m(p5)]$, or at the mastic station, s_i, to which it has been assigned $[m(p7) + \cdots + m(p10)]$.

i2. A roller table can be free $[m(p2)]$ or occupied by a car body on its way to the mastic station $[m(p3)]$ or the output $[m(p5)]$.

i3. A station must be in one of the five permanent states represented by the places $(p6, \ldots, p10)$.

i4. Represents the balance of occupation of stations by car bodies. The maximum occuption is of two car bodies per station $(2\Sigma s_i)$. The marking $m(p4)$ represents the effective occupation of stations, $m(p6)$ means that there are free stations (with two free sites), and $m(p7)$ and $m(p8)$ represent stations occupied by just one car body and which have one free site as a result.

Finally, it is worth noting that implicit places of the net can be detected by structural analysis. Thus, in the present case, the place p4 is implicit. Its marking can be expressed as the linear combination of the markings of the places p7, p8, p9, and p10:

$$2(i3)\text{-}(i4) \rightarrow m(p4) = m(p7) + m(p8) + 2m(p9) + 2m(p10)$$

As a result, the suppression of the place p4 in the CPN does not alter its functioning.

5. Description Language

The description language of concurrent systems which is given below has been defined in terms of the chosen modeling tool, CPN, and of the proposed methodology for constructing models.

A description program of a concurrent system consists of a series of declarations which together shape the designed model. These declarations refer to an autonomous colored net (colors, functions, modules, and the operations for module fusion), to its initial marking, and to its associated interpretation.

The syntax of the language is not presented formally here; rather its most notable aspects are described. By way of illustration, Table 3 contains the description program of the coordinator of the FMS modeled in Section 3.

A program written in the above-mentioned language contains the following declarations:

(a) *Name* of the program or of the model described.

(b) *Constants.*

(c) *Types.* In this section scalars and simple and compound colors are declared. An identifier can be associated to one single color (e.g., green = COLOR) or to various single colors (e.g., s = COLOR(index)). In this case each color, $\langle s(i) \rangle$, will be distinguished by the value which the index i takes within the scalar type "index."

A set of colors is declared by specifying the types of its elements and optionally by a definition of these. The elements of a set are, in general, compound colors (n-tuples of single colors). If the definition of elements is omitted, all colors of the type specified will be assumed to belong to it. By means of the definition it is possible to state which colors (n-tuples) belong to the set, using a compact algorithmic notation.

Thus, in Table 3, the declaration of the car body t4 specifies that it is

TABLE 3

Description Program of the Fig. 5 CPN

CPN_flexible_manufacturing_system ;
CONST n = 6 ;

TYPE
  index = (1..n) ; s,t = COLOR(index) ;
  stationset = SET OF ⟨s⟩ ; tableset = SET OF ⟨t⟩ ;
  carbody = SET OF ⟨s,t⟩ ;
  carbody_t1 = SET OF ⟨s,t⟩ DEFINED BY
                ⟨s(i),t(1)⟩ FOR i:index
        END :
  carbody_t2 = SET OF ⟨s,t⟩ DEFINED BY
                ⟨s(i),t(6)⟩ FOR i:index
        END ;
  carbody_t3 = SET OF ⟨s,t⟩ DEFINED BY
                ⟨s(i),t(j)⟩ FOR i,j:index AND i>j
        END ;
  carbody_t4 = SET OF ⟨s,t⟩ DEFINED BY
                ⟨s(i),t(j)⟩ FOR i,j:index AND i<n AND i<=j
        END ;
  carbody_st = SET OF ⟨s,t⟩ DEFINED BY
                ⟨s(i),t(i)⟩ FOR i:index
        END ;

FUNCTION station = proj (1,⟨s,t⟩) ;
FUNCTION table = proj (2,⟨s,t⟩) ;
FUNCTION nextable = succ (2,⟨s,t⟩) ;

MODULE stations ;
  {Places}
    empty_station , P_station , TU_station , P_TL_station ,
    P_TU_station : PLACE OF stationset ;
  {Transitions}
    load_empty_station : TRANSITION OF carbody_st ; {⟨s(i),t(i)⟩}
      PRE   station → empty_station ;
      POST   station → P_station ;
    load_working_station : TRANSITION OF carbody_st ; {⟨s(i),t(i)⟩}
      PRE   station → P_station ;
      POST   station → P_TL_station ;
    load_waiting_station : TRANSITION OF carbody_st ; {⟨s(i),t(i)⟩}
      PRE   station → TU_station ;
      POST   station → P_TU_station ;
    endwork_incomplete_station: TRANSITION OF carbody_st ; {⟨s(i),t(i)⟩}
      PRE station → P_station ;
      POST   station → TU_station ;
    endwork_complete_station : TRANSITION OF carbody_st ; {⟨s(i),t(i)⟩}
      PRE   station → P_TL_station ;
      POST   station → P_TU_station ;

TABLE 3
Description Program of the Fig. 5 CPN (continued)

```
    unload complete_station : TRANSITION OF carbody_st ; {⟨s(i),t(i)⟩}
      PRE    station → P_TU_station ;
      POST   station → P_station ;
    unload_incomplete_station : TRANSITION OF carbody_st ; {⟨s(i),t(i)⟩}
      PRE    station → TU_station ;
      POST   station → empty_station ;
END MODULE stations ;
MODULE transport ;
  {Places}
    remaining_carbodies, working_carbodies : PLACE OF stationset ;
    empty_tables : PLACE OF tableset ;
    loading_carbodies, unloading_carbodies : PLACE OF carbody ;
  {Transitions}
    input_carbody : TRANSITION OF carbody_t1 ; {⟨s(i),t(1)⟩}
      PRE    station → remaing_carbodies , table → empty_tables ;
      POST   id → loading_carbodies ;
    output carbody : TRANSITION OF carbody_t2 ; {⟨s(i),t(6)⟩}
      PRE    id → unloading_carbodies ;
      POST   station → remaining_carbodies , table → empty_tables ;
    next_table_input : TRANSITION OF carbody_t3 ; {⟨s(i),t(j)⟩,i>j}
      PRE    table*nextable → empty_tables , id → loading_carbodies ;
      POST   table → empty_table , nextable → loading_carbodies ;
    next_table_output : TRANSITION OF carbody_t4 ; {⟨s(i),t(j)⟩,i<=j<6}
      PRE    table*nextable → empty_tables , id → unloading_carbodies;
      POST   table → empty_tables , nextable → unloading_carbodies ;
    load_station : TRANSITION OF carbody_st ; {⟨s(i),t(i)⟩}
      PRE    id → loading_carbodies ;
      POST   station → working_carbodies , table → empty_tables ;
    unload_station : TRANSITION OF carbody_st ; {⟨s(i),t(i)⟩}
      PRE    station → working_carbodies , table → empty_tables ;
      POST   id → unloading_carbodies ;
END MODULE transport ;

FUSION stations(load_empty_station,load_working_station,
  load_waiting_station) WITH transport(load_station) ;

FUSION stations(unload_complete_station,unload_incomplete_station)
  WITH transport(unload_station) ;

INITIAL MARKING
  remaining_carbodies := 2*all(stationset) ;
  empty_tables := all(tableset) ;
  empty_station := all(stationset)
END MARKING ;

INTERPRETATION
  VAR
    st:⟨s⟩ ; tbl:⟨t⟩ ;
```

TABLE 3
Description Program of the Fig. 5 CPN (continued)

WHEN EVENT arrival_carbody(st) AND input_carbody(st,t(1))
   THEN load_rt1;
WHEN next_table_input(st,tbl) THEN next_rt(tbl) ;
WHEN next_table_output(st,tbl) THEN next_rt(tbl) ;
WHEN output_carbody(st,t(6)) THEN unload_rt6 ;
WHEN load_empty_station(st,tbl) THEN load_tl(st),load_p(st) ;
WHEN load_working_station(st,tbl) THEN load_tl(st) ;
WHEN load_waiting_station(st,tbl) THEN load_tl(st),load_p(st) ;
WHEN EVENT end_mastics(st) AND endwork_incomplete_station(st)
   THEN load_tu(st),load_p(st) ;
WHEN unload_complete_station(st) THEN unload_tu(st) ;
WHEN unload_incomplete_station(st) THEN unload_tu(st)
END INTERPRETATION ;

---

a subset of the set car body and that all doublets $\langle s(i),t(j)\rangle$ belong to it, if the indices i and j simultaneously satisfy

1. i,j:index $(1 \leqslant i,j \leqslant n)$;
2. $i < 6$;
3. $i \leqslant j$.

(d) *Functions.* In principle, only the use of library functions will be considered. The declaration of a function will take the following form:

specific_name = standard_name (list of arguments)

Thus the functions "station" and "table" are defined using the library function "projection," parametrized by two arguments: the component to be projected (first and second, respectively) and the type of color on which to apply the function ($\langle s(i),t(j)\rangle$ in both cases). The function "nextable" is the library function "successor" applied to the second component (first argument) of the elements of type $\langle s(i),t(j)\rangle$ (second argument).

Functions which cannot be expressed in terms of library functions can be defined using a syntax which is not dealt with here.

(e) *Modules.* A module describes a colored subnet. It can contain declarations of types (colors) and of local functions. The specific declarations of a module are those which correspond to the places and transitions of the subnet to be described.

The declaration of a place contains the specification of its name and of the set of colors associated with it. The declaration of a transition contains, as well as that of the name and the set of associated colors, the lists of its input places or *preconditions* (PRE) and of its output places or *post-*

conditions (POST). A function is associated to each place of these lists and it labels the arc which joins it to the transition in question, according to the following scheme:

$$\text{Function} \rightarrow \text{name of place}$$

A series of comments, written in brackets, helps to greatly improve the legibility of the description of the module (see Table 3).

(f) Fusion of transitions from different modules. Each declaration contains the lists of transitions of each module which must be fused with the corresponding transitions of the other modules:

$$\text{FUSION module1 (list of transition) WITH}$$
$$\text{module2 (list of transition) WITH}$$
$$\cdots$$
$$\text{moduleN (list of transition);}$$

(g) Initial marking consists of declaring the marking initially stored at each place in the net. By default, those places whose initial marking has not been declared are considered to be initially unmarked. The initial marking of a place should be *compatible* with the set of colors associated with it in the description of the model to which it belongs.

A marking is, in principle, expressed as weighted symbolic sums of colors, allowing the use of parentheses:

$$\text{remaining carbodies} = 2.[s(1)+s(2)+s(3)+s(4)+s(5)+s(6)]$$

In order to make the notation more compact, certain *standard functions* can be used. Thus, for example, the function *all(color set)* represents the symbolic sum of all the elements of the set of colors in question, "color set":

$$\text{remaining carbodies} = 2.\text{all(stationset)}$$

(h) Interpretation associated to the CPN. To each transition we associate a *guarded command* whose condition of execution is a Boolean expression where, as well as the condition of sensitization of the transition, events, which are Boolean functions of the scheduler of the control system, can intervene. The enable conditions and the events can be parametrized by one or more colors. The verification of the guard condition implies the execution (sequential, in principle) of a list of tasks, associated with the low-level control devices. This does not prevent some of these tasks from having internal concurrency.

The syntactic scheme of a guarded command, associated with a transition, is the following:

WHEN guard condition THEN list of tasks ;

6. Conclusion

A language for the description of complex concurrent systems modeled with CPNs has been presented. It is adapted to the modular description and allows the validation of the model described. In the future, it would be interesting to incorporate the possibility of a top-down description similar to those developed for PN.

To model the control system of a FMS, a CPN has been provided with an interpretaiton, which is defined in terms of the language by the association of a guarded command to each transition of the net.

The result of processing the description program will, generally, be a data structure. The nucleus of the control system will consequently be an interpreter of CPNs designed according to the interpretation associated with the net. It can be designed specifically, or on the basis of the nucleus of another already existing operating system.

To describe other types of concurrent systems (e.g., distributed data bases, or communication protocols) it will, in any case, be sufficient to modify the net's associated interpretation.

Finally, we point out that the description program of a system can also be used to simulate the described system or to evaluate its performances. For this, an extension of timed Petri net techniques may be realized by considering the special role of colors.

References

1. J. L. PETERSON, A note on colored Petri nets, *Inf. Process. Lett.* **11**(1) 40–43 (1980).
2. J. DORTH and T. MURATA, Use of colored Petri nets for modeling PBX systems, IEEE Workshop on Languages for Automation, Chicago, November, 1983, pp. 196–201.
3. K. JENSEN, How to find invariants for coloured Petri nets, *Mathematical Foundations of Computer Science.* Lecture Notes in Computer Science 118. Springer Verlag, Berlin, 1981.
4. H. ALLA, P. LADET, J. MARTINEZ, and M. SILVA, Modeling and validation of complex systems by coloured Petri nets: Application to a flexible manufacturing system, Fifth European Workshop on Petri net applications and theory, Aarhus (Denmark), June, 1984, pp. 122–141, paper selected to be published by G. Rosemberg in *Advances in Net Theory*, Springer Verlag.
5. K. JENSEN, Coloured Petri nets and the invariant method, *Theor. Comput. Sci.* **14**, 317–336 (1981).

6. M. SILVA, J. MARTINEZ, P. LADET, and H. ALLA, Generalized inverses and the calculation of symbolic invariants for coloured Petri nets, *Tech. Sci. Inf.*, 4,(1), 1985, 113–126.
7. E. THURIOT, R. VALETTE, and M. COURVOISIER, Implementation of a centralized syncronization concept for production systems, IEEE Real-Time Symposium. Arlington, Virginia, December, 1983.
8. R. VALETTE, R. COURVOISIER, and E. MAYEUX, Control of flexible production systems and Petri nets, Third European Workshop on Applications and Theory of Petri nets. Varenna, Italy, September, 1982, pp. 426–439.
9. M. SILVA, *Las redes de Petri en la Automática y en la Informática.* Ed. AC, Madrid, 1985.
10. K. LAUTENBACH and A. PAGNONI, On the various high-level Petri nets and their invariants. Petri nets and related system models, Newsletter 16. Gesellschaft fur Informatck, February, 1984, pp. 20–36.

ROBOTICS AND THE CAD/CAM LANGUAGES

INTRODUCTION TO COMPUTER-AIDED DESIGN AND COMPUTER-AIDED MANUFACTURING

C. S. GEORGE LEE

1. Introduction

In recent years, intense international competition has forced companies to seek better manufacturing techniques and systems with a higher level of automation to increase productivity gain and improve quality end product.[1-4,13] One of the possible solutions is through the effective use of digital computers and their peripherals in the design and manufacture of products. Because of the advancement of hardware and software in digital computers, their use in the design and manufacture of products has greatly reduced the manufacturing cost and constitutes substantial productivity gains in various industries. This application of digital computers in manufacturing has evolved into a new technology commonly known as computer-aided design and computer-aided manufacturing (CAD/CAM). Computer-aided design can be described as any design activity that effectively utilizes digital computers to create, retrieve, modify, draft, and store an engineering design; while the Computer-Aided Manufacturing—International defines computer-aided manufacturing as the effective utilization of computer technology in the management, control, and operations of the manufacturing facility through either direct or indirect computer interface with the physical and human resources of the company to produce high-quality end products.

The use of CAD/CAM in the design and manufacture of products improves efficiency and flexibility because computers replace human labor

C. S. GEORGE LEE • School of Electrical Engineering, Purdue University, West Lafayette, Indiana 47907

for many routine and boring ativities. Computers optimize the design of the product and check all the tolerances before actual manufacturing takes place, then coordinate the various manufacturing processes, and control the movement of workpieces to optimize the utilization of machines.

Many view this trend of increasing use of computers in manufacturing as a second industrial revolution which will ultimately lead to the computer-integrated manufacturing (CIM) facility, often called "factory of the future." The factory of the future, as we know today, will exhibit two major changes in present manufacturing practice. First, most of the physical fabrication processes will be automated through direct or indirect use of computers. Second, most of the product design and information handling tasks will be done on computers with minimal human intervention. Factory management will exercise integrated control of all factory activities using a common integrated data base.

In this chapter, we shall address several important features and characteristics of computer-aided design (CAD) and computer-aided manufacturing (CAM) and their integration into an overall system for manufacturing. We shall start our discussion on the manufacturing system hierarchy and put CAD and CAM into perspective within this system hierarchy. Then we shall discuss CAD and CAM separately and elaborate more on their respective components and functions. Finally we shall discuss the integration of CAD and CAM.

2. Manufacturing System Hierarchy

Simply stated, CAD/CAM can be defined to be the application of computers to all activities of product design, production planning, production engineering, and tooling to produce a variety of high-quality end products. Within this context, a large-scale manufacturing system is hierarchically structured as shown in Figure 1. Three main divisions are managed by the centralized management: production and control, engineering analysis and design, and financial and marketing. The production and control division focuses on the control of various manufacturing processes and monitors the physical fabrication of the products. Its goal is to produce the products within the specifications done by the engineering analysis and design division. The engineering analysis and design division is part of the corporation research and development unit. It focuses on the design and production analysis of the products and specifies the specifications and tooling for producing the products. Most CAD/CAM systems focus on the utilization of computers for these two divisions while they

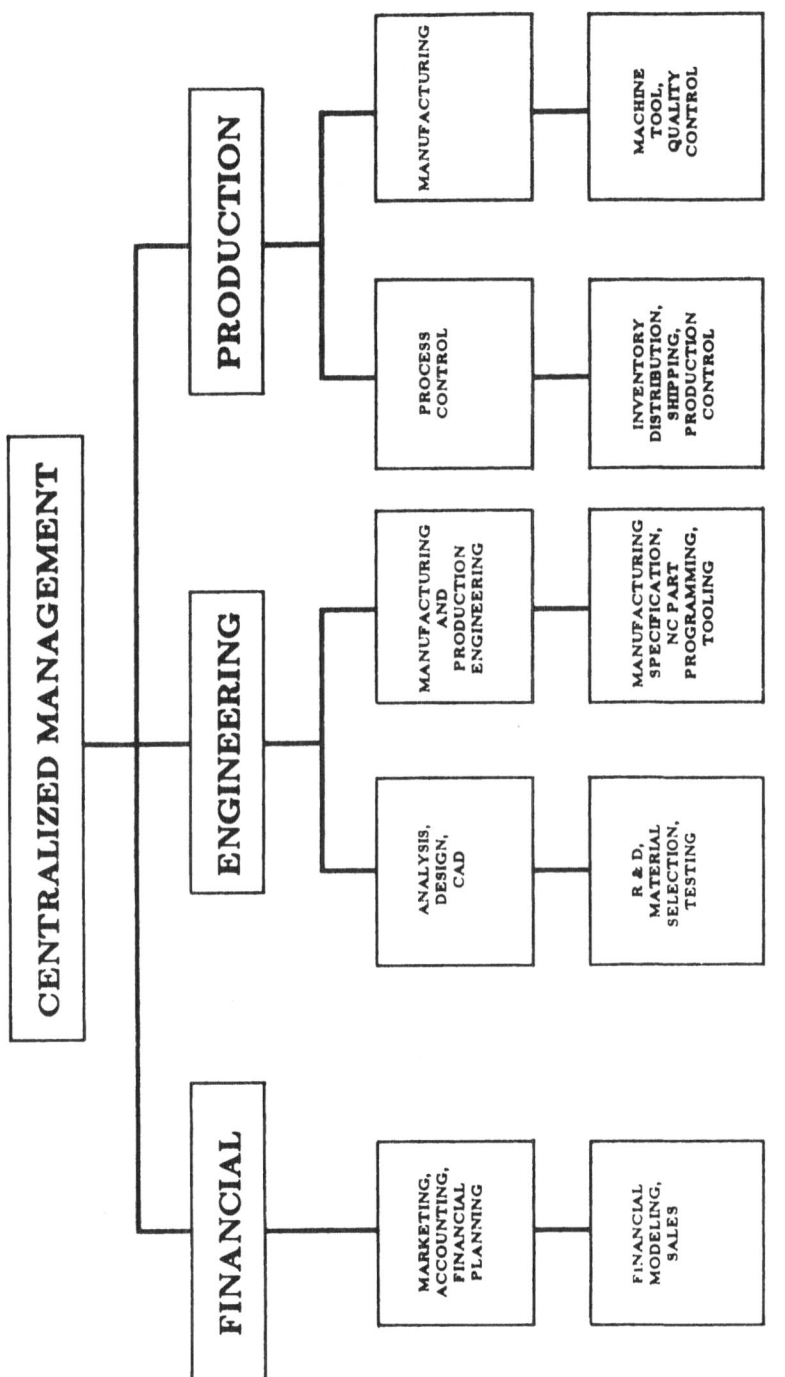

FIGURE 1.   Manufacturing system hierarchy.

loosely tie with the financial and marketing divison to form a distributed computer network for manufacturing.

Within the CAD/CAM system, there is again hierarchical structure in three levels as shown in Figure 2. The highest level represents the CAD in which a design engineer utilizes the interactive graphics tool to perform optimal design of the products for different options, display the design, and check for any inconsistency in the specification of the parts. This design process is done iteratively and the final result is the drafting of the product and its manufacturing specification of its constituent components. The next level down the CAD/CAM system hierarchy is production analysis and tooling in which computer technology is used to compute an optimized plan for manufacturing the product and select the appropriate tools and fixtures necessary to adapt the manufacturing facility to the production of the specific product. The lowest level represents the physical fabrication operations which utilize a distributed computer system to perform parts fabrication, material handling, assembly of parts, and finally quality inspection to assure the products are produced according to the specifications.

On this lowest level, large-scale general-purpose computers are used to control smaller computer systems (minicomputers). These minicomputers perform two distinct tasks: as control and monitor devices for various

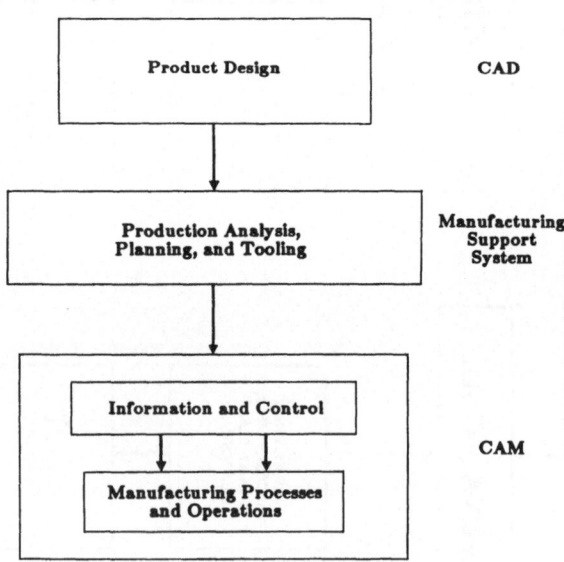

FIGURE 2. CAD/CAM hierarchical structure.

systems in real time, such as process monitoring and control, manufacturing quality control, and direct digital control; and as support systems for manufacturing systems, such as computer-aided process planning, computer-assisted numerical control part programming, computer-aided line balancing, production/inventory control, and material managing, etc. In the following sections we will discuss some of these processes on manufacturing operations.

3. Computer-Aided Design (CAD)

One of the major developments in computer technology is the computer-aided design which exhibits extensive graphics capability to assist design engineers to visualize the product and its component subassemblies and parts.[4-10,14] Computer-aided design can be described as an iterative design activity that effectively utilizes digital computers to create, retrieve, modify, and store an engineering design. The computer-aided design systems are usually coupled with a user-friendly interface device such as joystick, keyboard, light pens, menu commands, touch-sensitive screens, mechanical cursor, etc. to allow an engineer to interact with the graphics system and design a product in an iterative process. Thus, a design engineer can complete the analysis, synthesis, and documentation of a product design in a relatively short time.

Most of the CAD systems used in engineering design are based on interactive computer graphics technology. Its ability to create, transform (or modify), and display the product design on a CRT screen relies heavily on its geometric database to represent and describe the product's component subassemblies and parts.[6,8]

One of the most important features of CAD/CAM is the geometric model—the representation of the different elements in the design process, in size and shape in the computer memory. This is the starting point of many other designs and manufacturing functions used in CAD/CAM. These geometric models are usually created by the user through the use of interactive computer graphics. Depending on the capabilities of the CAD system and the user requirements, the geometric model may be: a 2D type (two dimensional); a 2½D model, which represents a constant section with no side-wall details; or a full 3D model.

The basic concept for creating a geometric model is basically the same regardless of the particular CAD system used. There are three basic ways of representing geometric models in a computer: wire-frame, surface models, and solid models. These representations of geometric models are discussed in the following sections.

3.1. Wire-Frame

Wire-frame is the simplest of the three possible models and is widely used for the drawing of parts and detail work in automatic drafting.[4] It represents part shapes with interconnected line elements in the computer and is often called edge-vertex representation or stick-figure models. Because of its simplicity in connecting the points and lines defining the parts, wire-frame models consume little computer time and memory.

Wire-frame models are created by specifying points and lines in space. The screen is divided into four major areas: the top, bottom, side, and isometric views. The designer builds the model by specifying points to the terminal and commands chosen from a menu. Automatic features accelerate the process and produce predetermined figures such as circles, conics, etc. Many other features such as automatic projection into other views, duplication of objects already defined, deletion of individual lines, temporary erasure of particular lines, enlargement or reduction of particular areas, and blank-out or dash-hidden lines give the view a solid appearance.

Today's 3D wire-frame systems exhibit serious deficiencies: wire-frame diagrams contain no information about surfaces, no differentiation between inside and outside objects, and when used to represent 3D objects may result in ambiguities in that they may represent different solid models at the same time. The user, rather than the system, is expected to detect such anomalies. Objects with curved surfaces are not well defined, and the user must supply a lot of information to describe some things as simple as a sculptured surface.

3.2. Surface Models

Many of the ambiguities of wire-frame models are overcome by using surface models. This system is precisely defined and may help produce numerically controlled (NC) machining instructions and other applications where part boundary definition is important. However, surface models only represent the external part of the object and do not contain information to describe the interior of the part. On the other hand, this system provides facilities such as automatic hidden lines, which makes the model appear as a solid model.

Surface models are created by connecting various types of surface elements to user-specified lines. The typical surfaces may be chosen from a menu. There are, for example, planes, tabulated cylinders, rule surfaces, and surfaces of revolution, along with sweep, filet, and sculptured surfaces.

The plane is built between two user-specified straight lines. A tabulated cylinder is the projection of a free curve into a three-dimensional space. A rule surface is produced between two different edge curves. A surface of revolution is created by revolving an arbitrary curve in a circle about an axis. Sweep surfaces are created by sweeping an arbitrary curve over another arbitrary curve instead of a circle. A filet surface is a cylindrical surface connecting two other surfaces in a smooth transaction. Sculptured surfaces represent the most general surfaces. They are also called curve-mesh surfaces, B-surfaces, or free-form surfaces. They are created from two different families of curves which are not necessarily orthogonals. This kind of surface cannot be described with the usual lines and curves of conventional models.

3.3. Solid Models

A solid model is a 3D model and represents the most sophisticated method in geometric modeling.[4−10,14] This technique promises to become the predominant method in geometric modeling and overcomes the drawbacks of wire-frames and surface models by defining parts mathematically as solid objects. More complicated solid models are then created from the existing primitive objects such as cube, cone, cylinder, etc. Unlike the other two techniques, solid modeling can determine if a specified point lies inside or outside of the solid, or even on the surface of a part. A cross section can be cut through the solid to expose its internal details. This technique allows the representation of a solid model in the computer, thus facilitating the calculation of parameters such as weight, moment of inertia, center of gravity; kinematic and dynamic studies for checking the action of motion of a moving part in three-dimensional space, as well as the position of parts in complex assembly lines; and many other parameters in the design process.

The most important characteristic of solid modeling is its ability to support any geometric applications, because it is based on unambiguous representation. In general, a geometric modeling system should provide structure and controls that guarantee the consistency of creating desired objects. A generic geometric modeling system consists of four major components: (1) data structures which represent solid objects; (2) processes which use such representations for answering geometric questions about the objects; (3) input facilities for creating and editing object representations and for evoking processes, and (4) output facilities and representations of results. The subsystem which provides facilities for entering, storing, and modifying object representation is called a *geometric modeling system*.[8] Figure 3 shows the internal structure of a general geometric modeling system. The system builds internal representation of objects from the

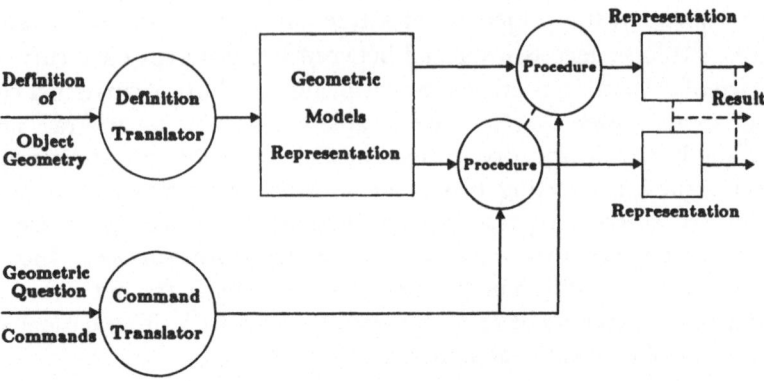

FIGURE 3. A geometric modeling system.

definition of data supplied by the user and it evokes application-dependent procedures to answer geometric questions about objects (weight, volume, moment of inertia, etc.).

The user interacts with the geometric modeling system through a definition language with which the user can enter information into the system.[6] The language is usually convenient and tailored to the needs of the designer. This language allows the designer to enter the shape description information in a way that can be interpreted by the computer. It incorporates many preprogrammed functions to draw predefined shapes and objects, thus reducing the input time and effort. It also has shape modifications as well as shape definition facilities. However, there is often little relationship between the definition language of the system and the way the data are represented internally.

Most of the systems start with the definition of the geometric components, and then with the definition of the topological connection between them. Geometric components comprise points, lines, and planes by means of geometric equations while topology regards a point as a vertex that bounds a line to define an edge and it links together the components from which the geometric information has been stored. Topological information can be easily entered as a graph because it is dimensionless, while geometric information must be derived from one or more two-dimensional drawings. The geometric and topological data computed for each object must be unique in the system.

The simplest and most common way to define a new shape is by applying linear transformations to an existing shape. This procedure is called "instances." It may involve a translation and a rotation operation, or a dilation and a shear operation. This linear transformation affects only the geometry of the object but not its topology. Furthermore, it is common to supply the designer with frequently used shapes that might be defined by specifying a set of parameters from which the computer can represent the shape. These standard forms may be kept in a catalogue of parts for further uses. Finally, the language also allows shape operators to combine primitive shapes into more complex ones.

Statements in the definition language will be evaluated upon entry into an expanded internal form that contains more explicit shape information needed for the application programs. Shape representation consists of a complex spatial point set occupied by an object. Surface representation, which is different from shape representation, relies, however, on only a subset of the full point set for its characterization. The most general form of representing solid models is a set of continuous points in a three-dimensional space. In Cartesian space, the solids are represented by a three-dimensional array of adjacent cells. The cell size is the maximum resolution of the representation. A more concise approach is to represent only the points on the bounding planar surface that divides internal and external points. This approach is usually called the polyhedra representation.

The basic problem of solid modeling is describing a three-dimensional, continuous object in a one-dimensional, finite, discrete computer. The available geometric representation schemes[5–10,14] can be classified into six types. They are (1) primitive instancing, (2) spatial enumeration, (3) cell decomposition, (4) constructive solid geometry (CSG), (5) sweep representations, and (6) boundary representations.

Primitive Instancing. The primitive instancing representation scheme[8] contains a collection of families of solids, e.g., holes, cubes, etc. In each family the members are only distinguished from one another by a few parameters. The family of solids is called generic primitive. A particular object is represented by a tuple of specified parameters of a generic primitive, which is called the instance of the generic primitive. The advantage of this approach is conceptually straightforward and structurally concise. However, the representation domain is limited and representation of a complex object may be difficult because no appropriate combining operators exist to represent the object.

Spatial Enumeration. The spatial enumeration representation scheme[8] consists of an ordered set of three-tuples, which is called spatial array. Each three-tuples is the coordinates of a cube (voxel) which is occupied by the represented object. One of the advantages of this scheme is that it is easy to detect the overlapping of two objects. But the disadvantages of this scheme are (1) difficulties in obtaining information about faces, edges, and vertexes, and (2) difficulties in compromising the appropriate resolution with the needs of large amounts of memory space.

Cell Decomposition. The cell decomposition representation scheme[8] represents a rectilinear polyhedron by first decomposing it into tetrahedra which must either be disjoint or meet precisely at a common face, edge, or vertex, and then storing the cells and the decomposition relation. This scheme has limited application to objects with curved surfaces.

Constructive Solid Geometry (CSG). The CSG representation scheme represents solids through Boolean constructions or combination of solid components via the regularized set operations, such as union, intersection, or difference. The solid is represented as an ordered binary tree with the Boolean operations as its nonterminal nodes and primitive solid as its terminal nodes. The most commonly used primitive components are cubes, cylinders, cones, and spheres. Figure 4 shows a constructive solid geometry (CSG) representation. This method is conceptually straightforward. If appropriate operators are included to support more complex objects, the

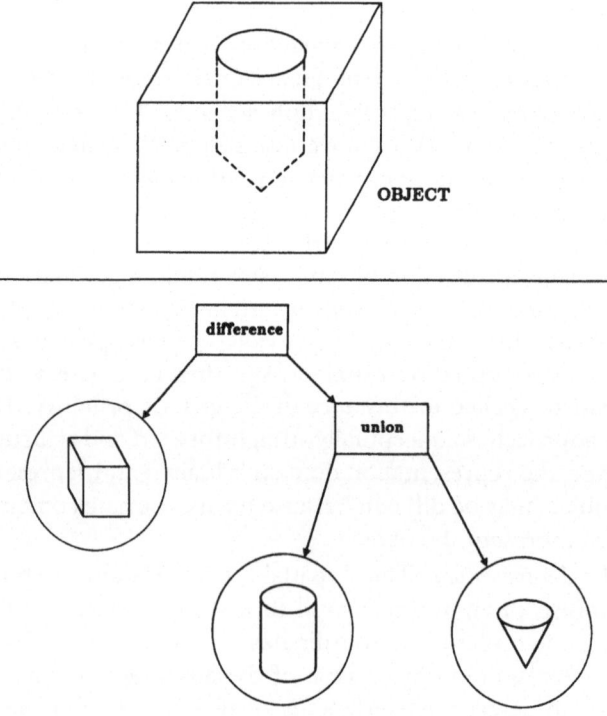

FIGURE 4. Constructive solid geometry (CSG) representation.

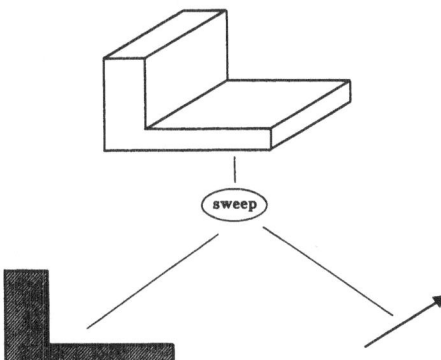

FIGURE 5. Sweep representation.

approach will become a powerful representation scheme in geometrical modeling, particularly in the robotics field. However, objects with sculptured surfaces and other complex contours may require extensive use of large amounts of computer time and user interfaces to construct them. This leads to other approaches to model objects with complex shapes.

Wesley[11] applied this approach to construct a geometrical knowledge database to meet the needs in the robotics field. The database is originally designed to represent the assembly environment for AUTOPASS,[12] a high-level robot programming language. The model has a graph structure where each vertex represents a volumetric entry (e.g., a part, a subpart, an assembly, etc.) and the edges represent the geometry or attachment relation between the nodes, including part-of, attachment, constraint, and assembly-component relations.

Sweep Representations. The sweep representation scheme is based on the concept that translating/rotating a plane along/above some axes in E^3 space will form a 3D solid. In this method, the representation of a 3D solid is reduced to the representation of a 2D one. Figure 5 shows that a solid can be represented as the Cartesian product of an *area set* and a *trajectory set*. Other operations in the sweep representation include *gluing*, where two previously created solids are joined at a common surface to produce a new unified model (Figure 6 provides an example); *tweaking*, which makes local changes to an overall shape; *Euler operations*, which are the dual operation of tweaking and change connectivity while the shape remains fixed[7] (see Fig. 6). The sweep representation domain is usually limited to those objects which have translational or rotational symmetry.

Boundary Representation. The boundary representation scheme[7,8] is widely used in computer graphics. A solid model is represented as a union of faces, where each face is represented in terms of its spatial surface

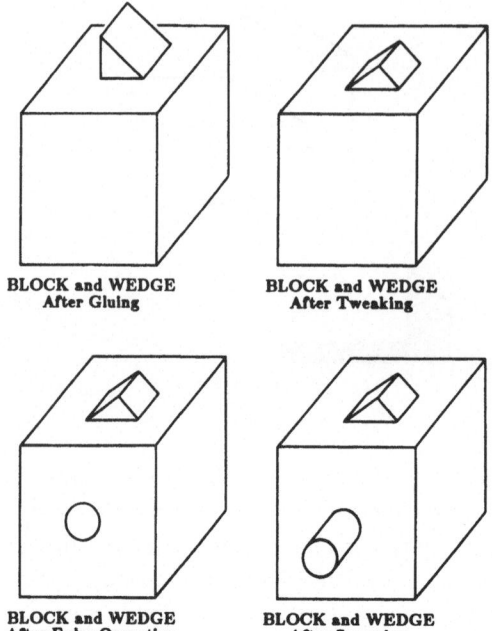

BLOCK and WEDGE
After Gluing

BLOCK and WEDGE
After Tweaking

BLOCK and WEDGE
After Euler Operation

BLOCK and WEDGE
After Sweeping

FIGURE 6.   Gluing, tweaking, and Euler operations.

boundary, and additional data define the surface in which the face lies. That is, each solid is represented by storing the geometrical location of its vertices (geometric information) and the constructional relations between any or all of these pairs: edges and vertices; edges and faces; faces and vertices (topological information). The greatest advantage of this technique is that it allows a great variety of appropriate operators. This is a very important feature for users having existing databases. This schema has application to mass properties, two-dimensional finite element meshes, and to numerical control tape generation. Figure 7 shows a schema for representing a solid in terms of its boundary.

This approach is widely used in CAD/CAM. However, the representation is tedious and difficult to construct manually because faces must satisfy elaborate geometric and combinational conditions. It is convenient to supply information about faces, edges, and vertices of a solid. But since its representation schema is so different from the user's view, a special input language is usually required to aid users in constructing and manipulating the geometric model.

In summary, the user interacts with a CAD system in constructing a geometric model of a product, analyzing its structure, performing kinematic studies, and producing engineering drawings.

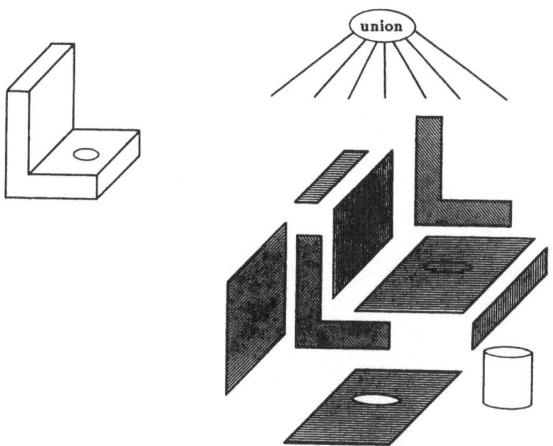

FIGURE 7. Boundary representation.

4. Computer-Aided Manufacturing (CAM)

After the products have been designed, computers are used to compute an optimized plan for manufacturing the product and to select the appropriate tools and fixtures necessary to adapt the manufacturing facility to the production of the specific products. Sparse work has been done in this area because of the complexity of the whole manufacturing process. Instead we shall discuss the use of computers for controlling and monitoring manufacturing operations and processes and as support systems for other manufacturing operations.[1–3,13]

Applications of computers in manufacturing operations and processes are very extensive and they can be separated into two categories:

1. On-line applications: computers are used to control and monitor manufacturing systems and processes in real time. In this process, machine tools, industrial robots, and production control are managed efficiently by small computer systems.
2. Off-line applications: computers are used as support systems in production planning, computer-assisted numerical control part programming, inventory and production control, cost analysis, company resources management, financial planning, and development of work standards.

In the following sections we will discuss some of these processes in manufacturing operations.

4.1. Process Monitoring and Control

This monitoring and control process involves direct application of computers in observing and collecting data from the manufacturing operations to determine the performance of the operations. Then appropriate decisions are made to control the devices to produce quality end products. In this context, computers used for these processes must possess three fundamental capabilities:

1. fast central processing unit (CPU) and input/output capabilities for performing any real-time computations and transactions;
2. large core memory for programs and large disk capacity for data storage; and
3. decision-making capability.

Besides monitoring the operating conditions of various manufacturing devices, computers also collect information for operation management, production maintenance, and manufacturing records. Monitoring for operation management involves recording current production levels and then comparing them with the last production rate to predict future production. Monitoring for production maintenance involves monitoring and recording the equipment utilization and efficiency in a production period which is concerned with machine breakdown, electrical power consumption, and the average of production time lost for each production period. As a result of this record keeping, the computer can alert the maintenance personnel to service, repair, or replace the equipment at appropriate times without disrupting the production schedule. Monitoring for manufacturing records is mainly done to keep track of products produced for government regulations or product warranty considerations. A common example would be when a certain engineering design change becomes necessary owing to safety reasons; then it may be necessary to know the serial numbers of the products that require recall.

4.2. Direct Digital Control

Traditionally, manufacturing processes are monitored and controlled by analog computers. Owing to the advancement of digital computers, time-shared computers are currently used to regulate the production operation on a sampled-data basis. Direct digital control (DDC) involves directly linking a time-shared computer to the process to continuously sample the process variables and using these variables to generate appropriate control signals for controlling the individual feedback control loop of the process. Figure 8 shows some basic components of a direct digital

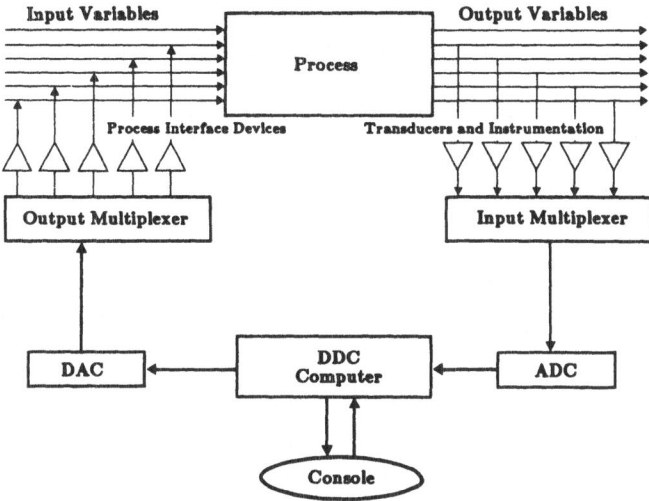

FIGURE 8. Direct digital control.

control system. Using multiplexers, the DDC computer can monitor and sample several output variables and generate appropriate control signals for performing feedback control of each process variable. Some advantages of DDC over analog computer control are (1) it is easier to program a more complex control strategy with DDC computer than with an analog computer because of its flexibility to be reprogrammed; (2) based on the time-shared concept, the cost of a digital control system is less than analog computer control for a process with several process variables and multiple feedback control loops; (3) the components used in a DDC system are generally more reliable than those used in an analog control system.

Although the hardware cost for DDC is decreasing, the increasing cost for computer programming is the main disadvantage of direct digital control.

4.3. Supervisory Computer Control

Supervisory computer control utilizes a digital computer to determine the appropriate reference output for each feedback control loop in the manufacturing process in order to optimize the performance objective function of the overall operation. This performance objective usually pertains to each process and might be maximum production rate and/or minimum cost per product. A supervisory computer has frequently been associated with process industries such as chemicals, petroleum, steel part making, and so on. Different control strategies of supervisory computer control in the manufacturing process are identified and briefly explained below.

Regulatory Control. In regulatory control, individual control loop set points and corrective actions for disturbances of the process are calculated by a supervisory computer in order to maintain the desired output. This control is analogous to feedback control except that the feedback control applies to individual control loops in the process while the regulatory control regulates the entire process. The performance is measured in terms of product quality and the supervisory computer will perform appropriate actions to maintain the product quality at that particular level.

Feedforward Control. The feedforward control compensates for any disturbances by computing the corrective actions in advance so that the desired output will not be affected. Figure 9 shows how disturbances can be measured by feedforward elements; then the necessary corrective actions are computed to anticipate the effect of the disturbances in the process. The control scheme requires an accurate mathematical model of the process.

Direct Numerical Control. In this control, the computer program generates a desired sequence of steps for machine tools to follow. Then the computer utilizes the feedback control concept to make certain that each step of the operation sequence is completed before proceeding to the next step.

Optimal Control. The computer is used to compute an optimal control law or function which optimizes an objective function of the process such that the process will operate in the optimum operating conditions. Typical performance objectives used are cost minimization, profit and production rate maximization, and quality optimization.

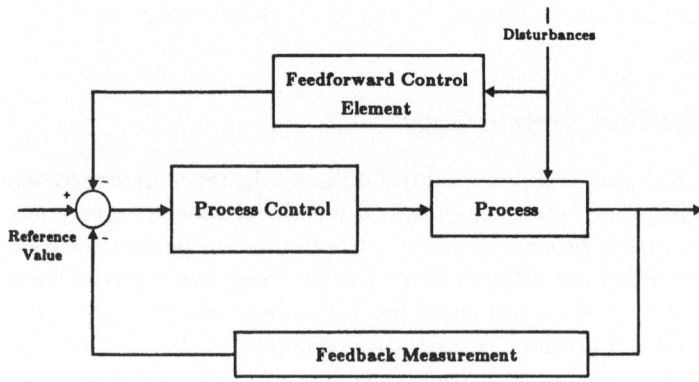

FIGURE 9. Feedforward control.

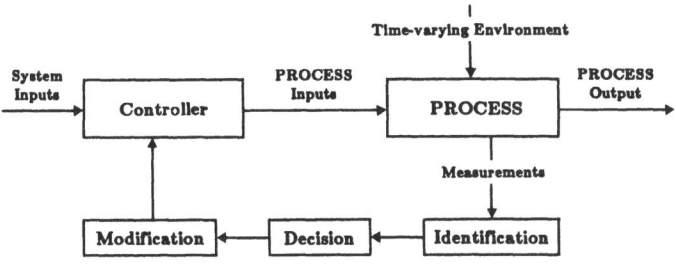

FIGURE 10. An adaptive control.

Adaptive Control. The computer monitors the performance of the process and redetermines the optimal operating conditions continuously. Adaptive control is different from a feedback control or an optimal control. It has the capability to deal with the time-varying environment. Large environment changes will not significantly affect the final result, because it evaluates the performance and makes necessary changes in its control efforts to improve and optimize the desired output. Figure 10 shows the control block diagram of an adaptive control for a process.

4.4. CAM for Manufacturing Support

Besides directly controlling and supervising manufacturing processes in real time, computers are also used extensively as support systems for reducing the production cost. The computers are utilized to provide data and information of the effective management of production activities. Major activities in this category are described in the following sections.

4.4.1. Computer-Aided Process Planning (CAPP). The computer is involved in this planning process to generate a near optimal operation sequence and describe resources and machines required to produce a new product. The computer generates these sequences of processing steps by coding parts based on its characteristics such as geometry and materials, etc. A general operation of an idealized automated process planning system is illustrated in Figure 11. In the first step, the user enters the part code number into the computer, and the CAPP program then searches the part family matrix file. If there is a match, both the standard machine routing and the standard operation sequence files exist, then the process plan formatter prepares the desired document. If the computer cannot find an exact match in the matrix file, it will continue searching the machine routing file and the operation sequence file for a similar code number. If the search fails, the user has to edit the new process plan for this new part; this operation sequence becomes a standard process for future parts of the same classification. Some other programs such as standard cost, setup

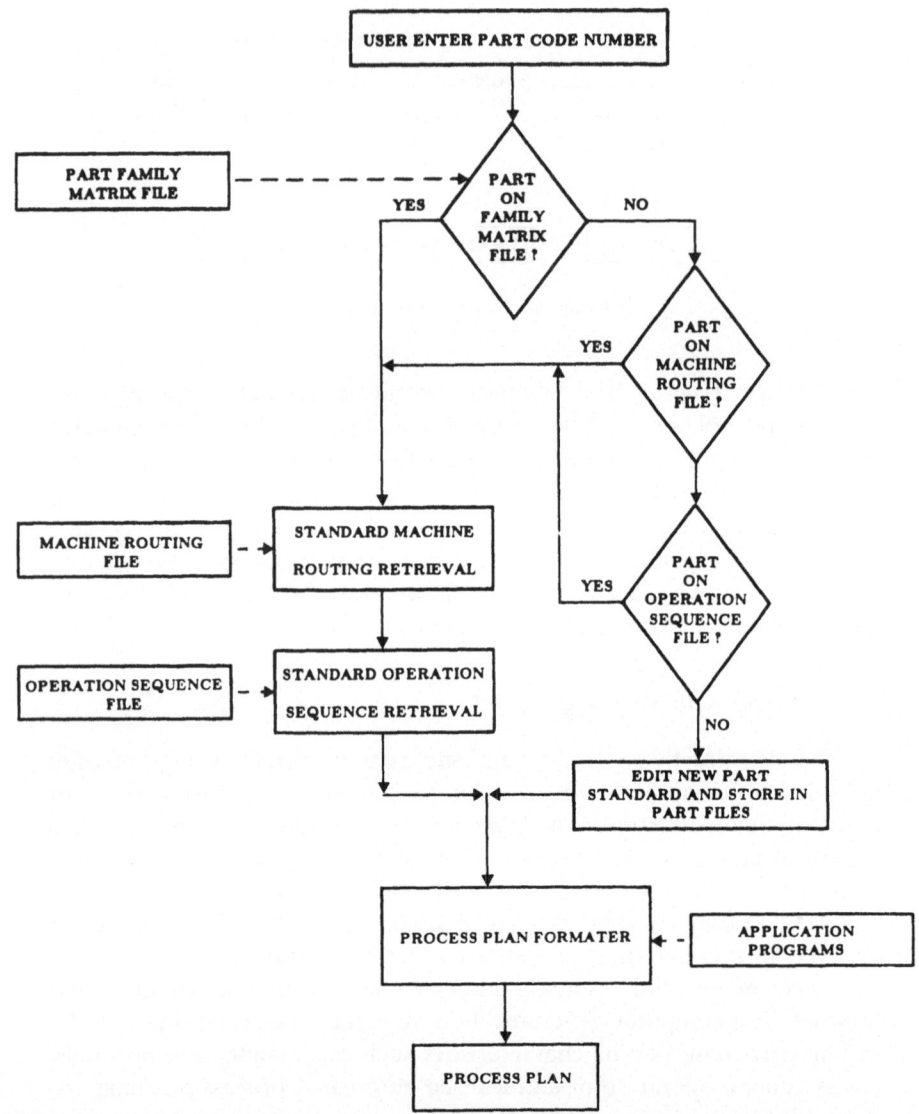

FIGURE 11. An automated process planning system.

times, and work standard could be used to make the process plan formatter. Some benefits derived from CAPP are process rationalization and
standardization, increased productivity of process planners, reduced turnaround time, and improved legibility on route sheets.

    4.4.2. Cost Calculation. Computers are also used in the cost analysis to estimate the production cost of a new product. This process is based on the actual cost analysis of previously produced similar products. In order to obtain more accurate cost analysis, time and cost of production are continously monitored and analyzed to estimate the production cost of the new product.

4.4.3. Computerized Machinability Data System. In order to increase the production rate and decrease the production cost, the best operating conditions for machine tools have to be determined, such as optimal operating speed and correct feeding schedule for the machines. All of these complex and time-consuming calculations are performed by the computer. Two computerized machinability data systems have been used quite successfully for the above purpose.

Database system. The database system maintains information which contains large quantities of data from laboratory experiments and shop experience. The information can be utilized for determining the proper cutting speed and feed rate of the machines.

Mathematical model systems. The mathematical model systems attempt to predict the optimum operating conditions based on the objectives of minimizing production cost or maximizing the production rate.

4.4.4. Computer-Assisted Numerical Control Programming. This is an efficient method that is used to produce complex part geometry for numerical control machines. The computer generates coded punched tape which is used for NC machines. The computer performs arithmetic calculations to define part geometry. For example, on tool path construction process, the computer subroutine will solve mathematics required to generate the part surface, considering the cutter offset factor. (The actual tool path is different from the part defined in the design because the tool path is the path taken by the center of the cutter.)

Several NC programming languages have been developed for producing the tapes for NC machines and the most widely used are as follows:

APT *(Automatically Programmed Tool)* is a contouring language developed by MIT in 1959. The modern version of APT can also handle both positioning and continuous-path programming for up to 5 axes.

AUTOSPOT *(Automatic System for Positioning Tools)* was developed by IBM in 1962 for point-to-point programming. The current version of AUTOSPOT can also be used for contouring.

SPLIT *(Sunstrand Processing Language Internally Translated)* was developed by Sunstrand for its machine tools. It can handle up to 5 axes of positioning and contouring.

COMPACT II was developed by Manufacturing Data System Inc. The language is quite similar to SPLIT.

ADAPT *(Adaptation for APT)* was developed by IBM under Air Force contract to be used in smaller computers. Although the language is based on APT, it is not as powerful as APT.

4.4.5. Production/Inventory Control. This process consists of maintaining production/inventory records for automatic reordering of stock items when inventory has been depleted to the reorder point. Computer systems with database management capability can check constantly the status of these inventories, generate order scheduling reports, maintain tool records, produce labor and machine utilization reports, and perform material requirements planning.

4.5. Robotics and CAM

Generally speaking, there are three types of manufacturing industry. The first type is continuous process industry, such as those in petrochemical and steel industry. This type of industry is highly automated. The second type is discrete-part manufacturing, such as those in automotive and appliance industry. This type of industry employs what is known as assembly line automation, which requires heavy use of fixture machines, NC machines for producing high-volume discrete-part product. The third type is called batch assembly or programmable automation, which handles the low-volume batch production of discrete parts. It has been estimated that a third of the manufacturing in the United States is done in batch assembly. Products are made in small batches on general-purpose machine tools with labor-intensive techniques. In most instances there has been little attempt to employ automation. Industrial robots find their best utilization in batch manufacturing and assembly tasks through "simple reprogramming" for various tasks. However, lately in a computer-aided manufacturing environment, industrial robots find their applications to various manufacturing tasks, such as loading and unloading NC machines, spot/arc welding, die-casting, forging, spray painting, deburring, and palletizing, etc.

An industrial robot is a general-purpose manipulator which consists of several rigid bodies, called links, connected in series by revolute or prismatic joints. One end of the chain is attached to a supporting base, while the other end is free and attached with a tool to manipulate objects or perform assembly tasks. The motion of the joints results in relative motion of the links. Mechanically, a robot is composed of an arm (or main frame) and a wrist subassembly plus a tool. It is designed to reach a workpiece located within its work volume. The work volume is a sphere of influence of a robot whose arm can deliver the wrist subassembly unit to any point within the sphere. The arm subassembly typically consists of three degrees of freedom of movement. The combination of the movements will place or position the wrist unit at the workpiece. The wrist subassembly unit usually consists of three rotary motions. The combination of these motions will orient the tool according to the configuration of the object to ease pickup. Hence for a six-joint robot, the arm subassembly is the positioning mechanism, while the wrist subassembly is the orientation mechanism.

The major obstacle in using manipulators as general-purpose assembly machines is the programming barrier. In the past, robots have been successfully used in the areas of arc welding, spray painting, etc. Those tasks that require no interaction between the robot and the environment can be easily programmed by guiding (or teaching by leading). However, with the advent of the idea of using robots to perform assembly jobs, more high-level programming techniques must be developed. This is because robot assembly relies heavily on sensor-based motions and guiding does not handle them well. Other limitations—such as lack of control structures, difficulty of modifying once programmed, and the fact that not all assembly tasks can be taught—further hinder using guiding in programmable assembly tasks.

Recently, a considerable amount of research has been focused on using textual robot programming languages[12,15-29] as an alternative. This effort is warranted because the manipulator is usually controlled by a computer, and for now the most effective way for humans to communicate with the computer is through textual programming language. Furthermore, using programs to describe assembly tasks, the robot can perform different jobs by simply executing the appropriate program. Thus, this increases the flexibility and versatility of the robot.

However, robot programming is substantially different from traditional programming. We can identify the following special features of robot programming:

- the objects we manipulate are three-dimensional and have physical properties;
- the robot operates in a spatial complex environment;
- the objects we describe are imprecise;
- sensory information has to be monitored and manipulated.

All these demand a different approach towards robot programming. Current approaches to designing robot programming languages can be classified into two major categories, namely, robot-oriented/robot-level programming and object-oriented/task-level programming.

In robot-oriented programming, the assembly task is explicitly described as a sequence of robot motions in terms of robot position and velocity. The robot is guided by the program throughout the entire task. This requires the user to know the basic operations of a manipulator in order to write an efficient program. On the other hand, object-oriented programming describes the assembly task as a sequence of goals or subgoals for the robots to achieve. No explicit robot motions are specified. The advantage of using object-oriented language is that it simplifies the programming task. The natural way to describe an assembly task is not in terms of robot motions, but in terms of the objects being manipulated. An object-oriented language allows the programmer to describe the task in a high-level language (task specification) and a task planner will then consult a database (world modeling) and transform the task specifications into robot-level programs (robot program synthesis) to accomplish the task.[19]

There are several high-level languages available for the control of manipulators. They include WAVE,[24] VAL,[27] AUTOPASS,[12] AML,[26] AL,[15,21,22] and others. Table 1 shows some comparisons among various existing robot programming languages. More evaluation and comparison of other robot programming languages can be found in Refs. 17,20.

5. Integration of CAD and CAM

The computerization of product design and manufacturing has resulted in productivity gains in their respective areas. Linking these product design functions, the various manufacturing functions, and the information processing functions, such as inventory control and cost analysis together to perform in optimized operating conditions presents a new challenge to manufacturing industry. The success of integration relies heavily on two aspects of software engineering: integrated database system and high-level languages for automation.[30-32] The integration of CAD and CAM into computer-integrated manufacturing (CIM) promises further gains in productivity by sharing information from a common centralized database system to optimize each step of manufacturing operation and streamline the entire manufacturing process.

Two fundamental aspects of manufacturing help to explain why integrated database system and high-level languages for automation are the important ingredients for linking these various disciplines in manufacturing. First, the entire manufacturing process, from product design through production to service support, is a monolithic, indivisible function. Second, the common denominator for all manufacturing operations is the processing of data. The creation, storage, analysis, transmission, and modification of data are key to all manufacturing operations. The integrated database system and high-level languages for automation directly address these two fundamental issues in manufacturing. The integrated database system links and integrates all manufacturing operations and processes

TABLE 1
Comparison of Various Robot Programming Languages

Language system	AL.	AUTOPASS	PAX/FSYS	PUMA/VAL	SIGMA/SIGLA	SRI AUTOMATION	REX	FREDDY	WAVE
Institution	Stanford A. I. Lab	IBM	Draper Lab.	Unimation	Olivetti, Italy	Stanford Research Institute	Jet Propulsion Lab.	University of Edinburgh	Stanford A. I. Lab.
Explicit or implicit	Mix	Implicit	Explicit	Mix	Explicit	Explicit	Explicit	Mix	Explicit
Compiler or Interpreter	Compiler	Compiler	Interpreter	Interpreter	Interpreter	Interpreter	Interpreter	Interpreter	Interpreter
Multiple arm control	Yes	No	No	No	Yes	Yes	No	No	Yes
Use of sensors	TV, force	Touch, Force, Acoustic Ranging	Force	None yet	3-DOF wrist force sensor	TVs, force	2 TVs, laser ranger, proximity sensor	2 TVs, force, torque	TV
Distributed system	No	No	No	No	No	Yes	No	No	No

through a common data link which reduces duplicate engineering drawings and improves product quality because all operations refer to the same data information. High-level languages for automation present a better solution for communication with the various users in different departments and as a means for communicating and integrating the various control programming languages used for machine tools and industrial robots. Because of the scope of this chapter, various aspects of integrated database system and languages for automation will be discussed in other chapters of this book.

6. Conclusion

Various processes and functions of CAD and CAM have been described. The productivity gains in these areas are due to the computerization of product design and manufacturing processes. In an idealized CAD/CAM system, the user interacts with the computer via a graphics terminal, designing and manufacturing a part from start to finish with information for each activity stored in the computer database. In a CAD system, the user constructs a geometric model, analyzes the structure, performs kinematic studies on the product, and produces engineering drawings in an iterative process. With the CAM portion of the system, the user creates NC instructions for machine tools, produces process plans for fabricating the complete assembly, programs robots to handle tools and workpieces, and coordinates plant operations with a factory management system. Other important characteristics of CAM are the computer monitoring and controlling of processes, direct digital control, and supervisory computer control of processes. These computer monitoring and controlling processes are currently well developed and implemented in industry. Industrial robots are finding their way into more advanced assembly tasks as more advanced high-level programming languages are developed to release the user from the burden of detailed programming of the robots.

Efforts to integrate these two disciplines into computer-integrated manufacturing are gaining momentum. This level of sophistication would lead ultimately to the automated factory. The success of the integration depends on the further developments of database system and high-level languages.

Acknowledgment

This work was supported in part by the AFOSR under grant no. F49620-82-C-0089.

References

1. M. P. GROVER, *Automation, Production Systems and Computer Aided Manufacturing*, Prentice Hall, Englewood Cliffs, New Jersey, 1980.
2. Y. KOREN, *Computer Control of Manufacturing Systems*, McGraw-Hill, New York, 1983.
3. R. S. PRESSMAN, *Numerical Control and Computer Aided Manufacturing*, Wiley, New York, 1977.
4. J. D. FOLEY, *Fundamentals of Interactive Computer Graphics*, Addison-Wesley, Reading, Massachusetts, 1982.
5. W. MYERS, An industrial perspective on solid modeling, *IEEE Comput. Graphics Appl.* **2**(2), 86–97 (1982).
6. A. BAER, C. EASTMAN, and M. HENRION, Geometric modeling: A survey, *Comput. Aided Design* **11**(5), 253–272, (1979).
7. C. EASTMAN and K. WEILER, Geometric modeling using the Euler operators, Proceedings of First Annual Conference on Computer Graphics in CAD/CAM System, April, 1979.
8. A. A. G. REQUICHA, Representations for rigid solids: Theory, methods, and systems, *Comp. Surv.*, **12**(4), 437–464, (1980).
9. A. A. G. REQUICHA and H. B. VOELCKER, Solid modeling: A historical summary and contemporary assessment, *IEEE Comput. Graphics Appl.* **2**(2), 9–24 (1982).
10. J. K. KROUSE, Ed., Solid models for computer graphics, *Mach. Des.*, **54**(12), (1982).
11. M. WESLEY, Construction and use of geometric models, in *Computer Aided Design Modelling, System Engineering, CAD-Systems*, J. Encarnacas, Ed., Springer-Verlag, New York, 1980, Chap. 2, pp. 79–136.
12. L. LIEBERMAN and M. WESLEY, AUTOPASS: An automatic programming system for computer controlled mechanical assembly, *IBM J. Res. Dev.*, **21**(4), 321–333, (1977).
13. *Mech. Eng.*, CAD/CAM: A special issue. November (1981).
14. *IEEE Comput. Graphics Appl.*, October and November (1983) issues.
15. T. O. BINFORD, The AL language for intelligent robots, in Proceedings of the IRIA Seminar on Languages and Methods of Programming Industrial robots, Rocquencourt, France, June, 1979, pp. 73–87.
16. D. D. GROSSMAN, and R. H. TAYLOR, Interactive generation of object models with a manipulator, *IEEE Trans. Syst. Man and Cybern.* **SMC-8**(9), 667–679 (1978).
17. W. A. GRUVER *et al.*, Industrial robot programming languages: A comparative evaluation, *IEEE Trans. Syst. Man and Cybern.* **SMC-14**(4), 321–333 (1984).
18. C. S. G. LEE, K. S. FU, and R. C. GONZALEZ, *Tutorial in Robotics*, IEEE Computer Press, New York, November, 1983, 573 pages.
19. T. LOZANO-PEREZ, Automatic planning of manipulator transfer movements, *IEEE Trans. Syst. Man and Cybern.* **SMC-11**(10), 681–698 (1981).
20. T. LOZANO-PEREZ, Robot programming, *Proc. IEEE* **71**(7), 821–841 (1983).
21. R. FINKEL *et al.* An overview of AL, a programming language for automation, Proceedings of the Fourth International Joint Conference on Artificial Intelligence, 1975, pp. 758–765.
22. M. S. MUJTABA, R. GOLDMAN, and T. BINFORD, The AL robot programming language, *Comput. Eng.* **2**, 77–86 (1982).
23. R. P. PAUL, *Robot Manipulators: Mathematics, Programming, and Control*, MIT Press, Cambridge, Massachusetts, 1981.
24. R. P. PAUL, WAVE: A model-based language for manipulator control, Technical Paper MR76-615, Society of Manufacturing Engineers, Dearborn, Michigan, 1976.
25. R. J. POPPLESTONE, A. P. AMBLER, and I. BELLOS, RAPT, A language for describing assemblies, *Artif. Intell.* **14**(1), 79–107 (1980).
26. R. H. TAYLOR, P. D. SUMMERS, and J. M. MEYER, AML: A manufacturing language, *Intl. J. Robotics Res.* **1**(3), 19–41 (1983).
27. B. SHIMANO, VAL: A versatile robot programming and control system," Proceedings of COMPSAC 79, Third International Computer Software Applications Conference, Chicago, Illinois, 1979.

29. K. TAKASE and R. P. PAUL, A structured approach to robot programming and teaching, *IEEE Trans. Syst. Man Cybern.*, **SMC-11**(4), 274–289, (1981).
30. Y. KALAY, A relational database for non-manipulative representation of solid objects, *Comput. Aided Design* **15**(5), 271–276, (1983).
31. G. M. E. LAFUE, Integrating language and database for CAD applications, *Comput. Aided Design* **11**(3), 127–130, (1979).
32. W. E. FISCHER, PHIDAS—A database management system for CAD/CAM application software, *Comput. Aided Design* **11**(3), 146–150, (1979).

20

PERCEPTUAL ROBOTICS: TOWARD A LANGUAGE FOR THE INTEGRATION OF SENSATION AND PERCEPTION IN A DEXTROUS ROBOT HAND

THEA IBERALL AND DAMIAN LYONS

1. Introduction

Controllers of industrial robot arms basically use positional information, generally programmed in a language which has no concept of an object other than as occupying a fixed position. They require a well-structured work environment before they can operate effectively. Three features characterize these robots: few degrees of freedom (DOF), little use of dynamic sensory information, and limited control architectures. Sensors to measure other physical features, and world-modeling robot systems, which present the user with a looser tie between object and position, are under active development.[1,5,17,23] However, we maintain that lack of sufficient sensory information is not the only problem, there is also the interfacing of sensory information to motor behavior.

Mammals, with their many DOFs, have evolved sophisticated sensors in the periphery and a highly integrated processing apparatus, the central nervous system (CNS). Movement and sensation are integrated in the action/perception cycle.[2,21] Actions are guided by expectations, and the organism perceives the consequences of these actions.

A *perceptual robot* not only has multiple sensory windows on its envi-

THEA IBERALL and DAMIAN LYONS • Laboratory for Perceptual Robotics, Department of Computer Science, University of Massachusetts, Amherst, Massachusetts 01003

ronment, but it also integrates sensory information into goal-oriented motor behavior. Sensors measure environmental features and provide sensations; these are integrated into perception, which could be described as the sensory state of the robot within a task context. Sensations may have different interpretations for different tasks, and to some tasks they may be irrelevant. Thus perception is integrated with the current task. We argue that this provides an easier and more robust way to program robots.

At the Laboratory for Perceptual Robotics (LPR), we are currently exploring one particular problem domain: that of the real-time control of a perceptual robot limb equipped with a *dextrous hand*. The control framework we are using is based on schema theory,[2] which defines schemas as units of motor control, acting within the action/perception cycle. A network of processors is being developed to realize a system of interfacing schemas to control the perceptual robot limb.

2. A Perceptual Programming Paradigm

We are developing a distributed programming environment[18] in which to implement our perceptual paradigm. A schema monitors feedback from the system it controls in order to tune its activities. Cooperating and competing schemas each monitor an aspect of an activity, and the overall behavior of the system is the combined behavior of all of its component schemas acting within a coordinated control program.[3] The set of all schemas defined in a particular robot system is similar to long-term memory in a human, or a program library in a computer. When schemas are instantiated, they are provided with an instantiation number, specific parameters, and a context; schemas do not execute until they are instantiated. An important property of schemas is that instantiations can be grouped in tightly coupled networks called schema assemblages; such an assemblage can in turn be treated as a single schema instantiation.

Direct sensory data are hidden from the system by schema instantiations which accumulate the data continuously and present it in a task-oriented form. For example, an instantiation of the Obstacle schema will provide information to the obstacle avoidance task, perhaps by culling its information from one sensor or from many, from other instantiations, or possibly by computing its information from internal representations. This information is updated constantly to represent the current world state. This eliminates "frame-oriented" sensory primitives from the robot language, e.g., primitives of the form TAKE PICTURE, etc. Instead the emphasis is placed on the task structure; how the robot manipulates objects, rather than how it recognizes objects.

Different tasks may call for different views of the same real-world entity; for example, several different versions of the Limb schema may be concurrently used for different tasks. Since they all refer to the same (real-world) object, an efficient implementation is performed by using the schema assemblage; it is, in effect, the total internal model of that object. However, unlike a more standard approach to modeling, multiple windows on the object (each a particular task-oriented view) are present, and the model can be easily piecewise extended. This representation clearly facilitates a hierarchical object representation, but does not force one. To illustrate the flavor of schema programming, we provide two simple examples.

Simple Obstacle Avoidance Example. We would maintain that the most basic property of any real-world object is its ability to act as an obstacle. Therefore, for any schema assemblage representing an object, there will be at least one instantiation of the Obstacle schema. Other schemas may identify other properties of the object (e.g., its use as a tool).

1. Let there be one instantiation of a Limb schema for each limb in the manipulator.
2. For each real-world object there is one instantiation of the Obstacle schema.
3. Each Obstacle schema instantiation outputs the position of the centroid of the object, the radius of a sphere tightly enclosing the object, and the radius of a "care" sphere around the object.
4. Each Limb schema will compare its position with the "care" sphere around nearby objects, and if its position comes within the sphere, it will cause the limb to have a velocity component away from the object, proportional to its distance along the radius of the "care" sphere.

The instantiation mechanism allows the above algorithm to be expandable in the number of obstacles and manipulator limbs. Also, the task can run independent of other tasks, allowing, say, a series of moves to be programmed without worrying about obstacle avoidance.

Task Triggering Example. An assembly subtask may be started when the principal component has been identified, or a corrective action taken if a nonstandard event should happen (e.g., a component is defective). Each task has a schema whose duty it is to recognize the main object components of the task and instantiate task-appropriate schemas into the assemblages for these objects. Once this has been done, the task can be triggered using the output of these schema instantiations. Let us consider triggering a mug grasping task[3]:

1. A GraspWatch schema is instantiated. It looks at Obstacle schema instantiations for hints that the obstacle may be a mug. If such clues are

present, a Mug schema is instantiated. The GraspMug schema is also triggered with a pointer to the instantiated Mug schema.

2. A Mug schema instantiation attempts to fit this object as a mug.
The relative success of this operation is indicated by the activation level of
the instantiation. This activation level is essentially the credibility of the
hypothesis that this object is a mug. Instantiations with activation levels
below a certain threshold just die away. When they die, so do any tasks
attached to them. If the activation level is super-threshold, then the
GraspMug schema attached also has this activation level.

3. The Advantages of Dextrous Perceptual Robot Hands

The industrial robot's end effector is constructed more like a hand
holding a tool than a hand itself, being designed for some special task capability. Specialized end effectors reduce the control problem because they
have exactly the DOFs needed for the task, and they minimize the sensory
information needed. Simple two-fingered grippers are easy to control,
although they are not very good at gripping awkwardly shaped objects.
They also may not easily be able to perform the variety of actions required
in an involved task, and are difficult to reprogram for a new task.

We are attempting to unify concepts in the robotics literature, in
order to make it reflect behavioral studies on human grasping and reaching. In this regard, we define the Grasp schema to reflect the steps used
by humans in grasping. Specifically, these steps are: the grasp preparation
(the preshape), the gripping of objects (the grip), and the manipulation.
This unifies what has been called static grasping (the grip) with dynamic
grasping (the manipulation).[6] A dextrous perceptual robot hand, based on
a model which resembles the human hand, would have the ability to shape
into the necessary tool for a given task, due to tactile and visual perception,
and could be used for direct manipulation of objects.

A dextrous perceptual robot hand has many advantages over a standard robot gripper. It can provide a range of stable grasps, since it can be
dynamically reshaped. With more DOFs in the fingers, it can also manipulate the object once it has been grasped. Finally, since it can be dynamically reshaped, it would be a better exploration tool than the limited industrial end effector.

4. The Human Hand as a Model

The human hand provides a rich model for a dextrous perceptual
robot hand. As a dextrous manipulator, it represents a general-purpose

tool that can be shaped into "the necessary tool for the task," allowing it to take on a variety of shapes. This, in effect, limits the number of available DOFs to match those needed for the task. Among its various movements are prehensile movements, which involve all the functions put into play when an object is grasped: intent, permanent sensory control, and a mechanism of grip.[26]

The involvement of the action/perception cycle is seen in the way humans use their hands in prehensile movements. As a human reaches for an object, the hand preshapes, using proprioceptive feedback (internal stimuli), into a shape suitable for the interaction (a shape initially perceived through the visual system). The action eventually produces stimuli against the skin as the hand touches the object; these stimuli modify further action as the hand encloses around the object. Anticipation of that contact allows smooth transition from proprioceptive feedback for shape movements to cutaneous feedback for contour following. Not only is a new special-purpose tool created in hand preshaping (the kinematic issue), but the tool is created in such a way as to optimize the use of the moments of force (the dynamics issue).

Cutaneous sensory information is provided by mechanoreceptors to the CNS about contact, pressure, and vibrations against the skin.[25] Spatial perceptions, such as two-point discrimination, depend on the resolution and location of those sensors. Temporal tactile perceptions involve movement of the hand across an object (active touch) or else movement of an object against the skin. Stereognosis, the perception of three-dimensional structure (object size, shape, and orientation), develops as a function of knowing which movements are being performed.[14] The sense of kinesthesia (position and movement) is a complex perception stemming from not only somatosensory stimuli (from cutaneous and proprioceptive sensors in the body) but from chemical receptors in the retina as well.[16,19] Nontactile perceptions of objects begin as sensations in the retina, which lead to perceptions of environmental features, object features,[10,25] and also the awareness of movement, or dynamic vision.[7]

The human hand has evolved an independent index finger, highly dependent ring and little fingers, and a saddle joint at the base of the thumb; this latter gives humans the ability to bring the thumb into opposition with the other fingers. The hand's basic architecture provides "functions for free" at the lowest level of a control hierarchy: the fingers, when flexed, also adduct (move inward); when extended, they also abduct (move outward). While there are over 25 DOFs, they interact in kinetic chains.[26]

In gripping an object, it is important to maximize contact (within a task context, of course), in order to maximize stability. The architecture of the hand forces a cylindrical object (such as a hammer) to be gripped obliquely across the palm, maximizing the skin surface contacting it. As the fingers and thumb flex in opposition, the fingerpads are brought toward the thumb pad, thus again maximizing contact.[26,15]

Preshaping can be done under somatosensory guidance and/or visual guidance. In experiments where subjects could not see their hands while reaching for various objects, Jeannerod showed that preshaping can be done under somatosensory guidance alone.[11,12] The shape that the hand took on was the same shape with or without visual guidance, whereas Jeannerod[13] studied a patient who received no sensory feedback from her hand, and found that she could not preshape her hand unless it was within sight.

We believe that tasks can be defined in functional, or virtual, terms. The preshaping of the hand has been shown to take on a virtual characteristic when attempting to grasp objects of different sizes.[3] In a task such as reaching to grasp a mug by its handle, we found that by varying the size of the mug handle, the shape which the hand takes on varies as well. As seen in Figure 1, if the mug has a small handle, the hand approaches the mug with only the index finger extended. If the handle is a bit larger, two fingers are extended; if it is a beer mug, three or four fingers are extended. The suggested mechanism behind this is a virtual finger representation, as a hierarchical substructure in hand control. This task would be represented in the higher levels of the CNS as a three-fingered task, since three basic forces must be brought to bear in this task: bringing a virtual finger (VF1) to a place on top of the handle, inserting a virtual finger (VF2) into the handle, and (perhaps) supporting the handle from below (VF3). The number of real fingers mapped into a VF does not affect the task definition, and, in reverse, a task such as reaching for a mug with a handle does not need multiple representations at higher levels of the CNS. A task can be defined in terms of the number of VFs needed to conceptualize the bringing to bear of the necessary forces, while the actual playing out of the movements during the task is done at a lower level across the real fingers and hand.

Hollerbach's[9] analysis of the task of handwriting is similar, defining functional degrees of freedom within the task. Preshaping movements, as well as gripping and manipulating movements, can be analyzed by the functional requirements of the task, as they are the movements which restrict the available DOFs to those needed in the final task.

Napier[20] has defined a taxonomy of hand shapes in terms of a power grip and a precision grip: power grips are prehensile activities involving the whole hand, whereas precision grips involve fractionated finger movements which use the tips of the fingers exerting little force.

Characteristics of the precision grip include: isotonic, dynamic, fingers flexed, wrist slightly extended, and thumb in opposition with the finger pads. The radial fingers (thumb, index, and middle), forming a

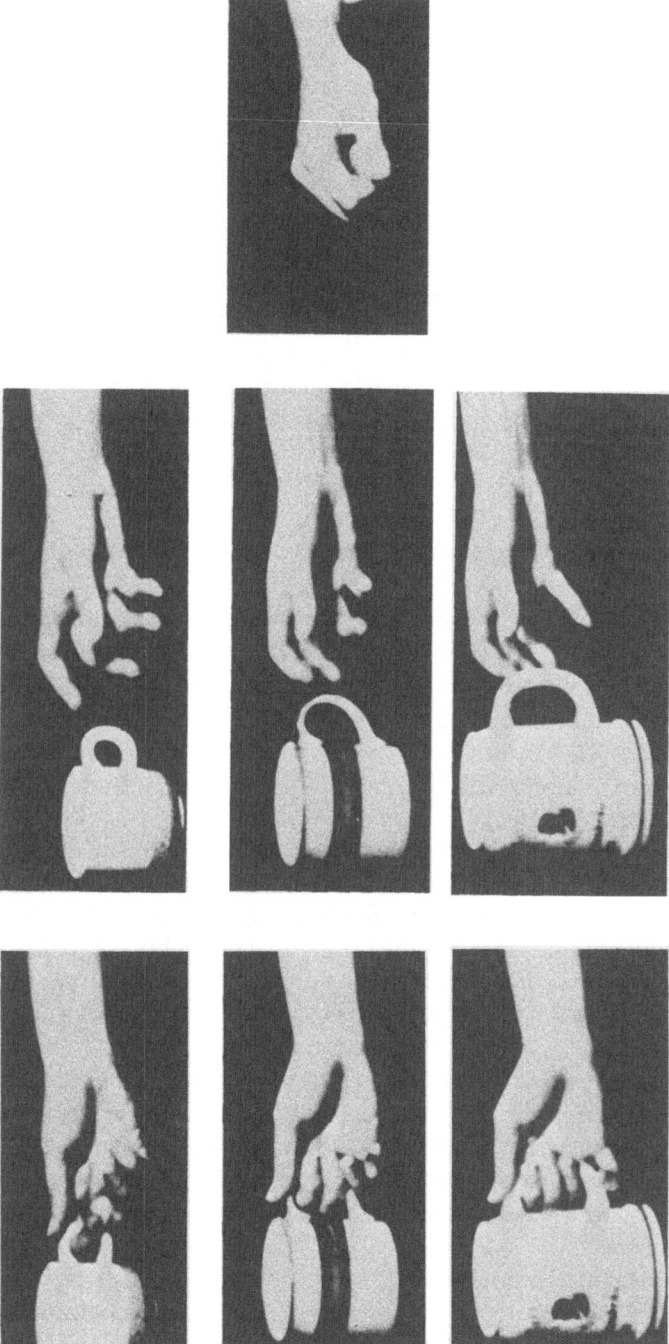

FIGURE 1. Virtual fingers. Different mappings for virtual fingers for different size objects.

dynamic tripod,[26] can be less flexed than the ulnar fingers (ring and little), as for example in holding a hammer precision style. Characteristics of the power grip include: isometric, static, wrist slightly extended, wrist deviation toward ulnar side, and thumb alignment longitudinally with forearm.[20] An abducted thumb provides some precision to it; an adducted thumb provides a stronger stabilizing influence. The intrinsic muscles (those within the hand) rotate the fingers to accommodate the shape of the object, optimizing contact between the finger pads and the object, and allowing the extrinsic flexors (those inserted into the forearm) to do the major part of the gripping. Even the little finger, although having the least strength in flexion, has a special role in the power grip, due to its peripheral position, its mobility, and capacity to press an object against the palm. In its unique position, the middle finger works well in both the power and precision grip: it can align with the ulnar digits for a power grip, or with the radial digits for a precision grip.[26]

5. Laboratory for Perceptual Robotics

The dextrous hand which the LPR is currently using in the design of a perceptual robot is the Salisbury hand.[24] The hand is an articulated three-fingered hand with nine DOFs, and it is currently being added to a four-DOF Cartesian robot arm. It is being provided with sensory devices such as CID cameras, force sensors, and tactile sensors. Current work is going on in developing sensors, and in integrating static and dynamic sensory information captured from the sensors.

Active research within the LPR has produced a variety of tactile sensors.[5,23] In the Overton tactile array sensor, conductive polymers measure resistance as proportional to the force applied to the surface. The Begej optical tactile sensor measures the intensity of the light reflected at a given location as proportional to force being applied there. These are similar to mechanoreceptors in the human skin, which sense the intensity of pressure applied.

In terms of vision, work is going on to incorporate image-processing programs developed at the University of Massachusetts for feature extraction, optic flow analysis, stereopsis, and object recognition.[4,8] Static vision processing, in terms of image segmentation, provides positional information, while movement of an image can be done using dynamic vision concepts.[4,7]

A schema-based controller for the integration of motor commands with sensory and perceptual processing is being built which involves a multiprocessor, hierarchical architecture. This will be further described in Section 8.

To develop models for grasping movements, we are conducting human skills studies to analyze how humans reach for and grasp objects. From these studies we are developing motor programs involving the activation of schemas.

To test various ideas in visual and tactile sensation and perception, we have constructed a graphic simulation of a generalized hand: specific models, such as a human hand or a Salisbury hand are inserted as desired.[18] We are also developing a programming environment based on our schema concept. Using these in conjunction, we can compare data from our schema-based model to data gotten experimentally from human subjects. In this way, we can both develop our model and fine tune it. We can also transfer our control concepts between robot and human models easily.

6. The Grasp Schema

A Grasp schema involves a preshape configuration, a grip configuration, and a set of associated DOFs (typically much less than the sum of all the DOFs in the hand). A grasp is chosen on the basis of three main environmental features: the function which the object is to be used for, the intrinsic object features (shape, weight, size), and the environment surrounding the object. Once chosen, the preshape is used to form the hand for anticipation of gripping the object. Once gripped, the object can be moved according to any of the functional DOFs associated with the grasp. These associated DOFs can be used for task-oriented object manipulations (e.g., rotating a screwdriver), or obstacle avoidance (i.e., the gripped object can, to some extent, be moved out of the way of obstacles using just the hand).

In order to shape the hand for grasping a particular object, a Preshape schema needs sensory feedback provided by the visual system or the somatosensory system. A prehensile classification for the human hand is seen as follows:

	Precision			Power	
Hand shape	VF1	VF2	VF3	VF2	VF3
I. Basic power	1			2345	
II. Modified power	1			234	5
III. Modified precision/power	1	2		34	(5)
IV. Basic precision/power	1	2			(345)
V. Basic precision	1	2	3		(45)
VI. Fortified precision/power	1	23			(45)

The real fingers (1 is thumb, etc.) are mapped into virtual fingers (VFn), within six basic grasps, which are not divided so much by precision vs. power as by the amounts of precision and power needed. It divides the human hand up into its precision and power sides, which seems more reasonable from the architectural study described in Section 4. Parentheses around real fingers indicate that the VF is either tucked into the palm (e.g., holding a dart), involved in the task (e.g., holding scissors), or even balancing outwards (e.g., the pinky in the air). For the three-fingered Salisbury hand, the shapes overlap, degenerating into two shapes, basically shapes I and IV. Parameters, which the Preshape schema needs in order to decide which grasp to use, might include degree of precision, degree of power, object size (how large a virtual finger is needed), object type (the shape of the handle), object orientation, and functional DOFs. Many objects have formal handles (e.g., mugs, hammers, bolt, etc.), and others have an area that can act as a pseudohandle (e.g., cylinder, ball, etc.). Preshaping for the task is performed by two subschemas, the Precision preshape schema and the Power preshape schema, activated in parallel. The actual behavior depends on the strategies used by these two schemas, which must cooperate, yet also compete for hand resources (e.g., control of the thumb, possible control of the index, etc.).

The Grip schema becomes active when the hand begins closing in on the object. The first part of it involves the anticipation of contact with the object, and the second part involves contour following, all within the virtual fingers defined by the preshape. With similar types of parameters, it would also need to know the expected contact locations.

We postulate the intended object usage as the primary index for grasps. The stereotyped configuration produced by this index is then particularized using the intrinsic object properties. Finally the existence of confining obstacles around the object is used to modify the grasp. The regulation of forces of a grip is important, varying due to weight, fragility, surface characteristics, and utilization. Continued sensory information is needed for compliance and maintaining contact.

7. Towards a Perceptual Robotics Language

We divide the robot architecture into a programming level and a real-time control level. This programming level will treat the grasping of an object, its manipulation, and releasing, as atomic operations. It sees therefore, not the total hand, but merely the DOFs associated with each grasp; the details of grasping and object manipulation are carried out at the real-time control level. The real-time control will move the gripped object to avoid any obstacles where necessary, using just the DOFs of the grasp.

We have looked at aspects of the action/perception cycle in human grasping movements, and can now begin to list the requirements of an automation language for a perceptual robot. First, it should describe movements in functional humanlike terms, instead of in typical input/output primitives seen in most programming languages. It should avoid the use of frame-oriented sensory primitives, and it should allow use of the hand in a task-oriented way and not force the programmer to worry about the specific DOFs of each joint. For example,

GRASP object (degree of precision, degree of power,
   object size, object type)

```
                 TWIST CW,
                 TWIST CCW,
                 PUSH AWAY,
IN_ORDER_TO      PULL TOWARD,   [ STRONGLY ]
                 PLACE,
                 SWING AROUND,
                 SWING DOWN
```

A possible task would be to use a screwdriver, TWISTing it against a screw. A large screwdriver would be grasped using shape I, with the wrist twisting during the task; a medium screwdriver would use shape IV. If the screwdriver was small, it would be grasped using shape V, where the twisting could in fact be done by finger rotation. Thus, by supplying parameters such as object size, object type, and the degrees of precision and power needed along with the functional DOFs, the Grasp schema will determine which hand shape is needed and which real DOFs to use.

8. The LPR Hand Control Network

Control of the Salisbury hand, equipped with various sensors, by a single processor is virtually impossible because of the real-time demands. Therefore, we have developed a logical network topology to control the hand. The network is logically divided into three levels.

Top Level: Hand Processor. The input to this level consists of high-level manipulation commands (our hand control language). At this level the basic grip appropriate to object and intended function are chosen. This produces an appropriate preshape configuration for the hand. Processed visual information is available for the selection of the grip, and for the changes necessary on the preshaped configuration due to the necessity of inserting the hand into confined locations, or the avoidance of local obstacles when the hand is in transit. More global obstacle avoidance must, of course, be done at a higher, plannerlike, level.

Middle Level: Finger Processors. The input to this level comes from the hand processor and consists of finger configuration commands. The finger processors are then responsible for the production of finger trajectories, and for the coordination of fingers amongst themselves. Sensory input to the finger level will come from tactile sensors, prototypes of which are being tested in our lab. Sensory input will also arrive from joint position and strain sensors. In addition, the finger level will do some of the kinematic transformation necessary to go from Cartesian space through finger-joint space to motor space.

Bottom Level: Motor Control Processors. For each motor in the hand there is one motor control processor. These processors will be grouped in fours, each under the direct supervision of a finger processor. These groups of fours will cooperate to perform the dynamic and kinematic transforms necessary to servo the motors of the Salisbury hand. Initially the LPR group intends to perform simple position control servoing; however, we shall then proceed to extend this to do more general stiffness control.

We are currently investigating possible hardware configurations for the hand control network. Although the initial configuration may statically distribute the functions at each level, the long-term plan is to dynamically schedule these functions. The intention is to base the network on the DEC T-11 processor. All programming will be done in FORTH, a particularly easy language to use for development.[22]

9. Acknowledgments

We would like to thank our co-workers at the Laboratory for Perceptual Robotics, including Steve Begej, Randy Ellis, Judy Franklin, Gerry Pocock, and T. V. Subramaniyan. We would like to especially thank Dr. Michael Arbib, as director of the lab, for his continued efforts, encouragement, and insights into this research.

This work was supported in part by grants nos. NS14971-65 from NIH and ECS-8108818 from NSF.

10. References

1. A. P. AMBLER, I. M. BELLOS, and R. J. POPPLESTONE, An interpreter for a language for describing assemblies, *Artif. Intell.* **14**, 79–107 (1980).
2. M. A. ARBIB, Perceptual structures and distributed motor control, in *Handbook of Physiology—The Nervous System, II. Motor Control,* (V. B. Books, ed.) American Physiological Society, Bethesda, Maryland, 1981, pp. 1449–1480.

3. M. A. ARBIB, T. IBERALL, and D. LYONS, Coordinated control programs for movements of the hand, in *Hand Function and the Neocortex*, (A.W. Goodwin and I. Darin-Smith, eds.), Springer Verlag, 1985, pp. 111–129. *COINS Technical Report 83-25*, Department of Computer and Information Sciences, University of Massachusetts, Amherst, Massachusetts, (1983).

4. M. A. ARBIB, K. J. OVERTON, and D. T. LAWTON, Perceptual systems for robots, *Interdiscip. Sci. Rev.* 9(1), 31–46 (1984).

5. S. BEGEJ, An optical tactile array sensor, *COINS Technical Report 84-26*, Department of Computer and Information Sciences, University of Massachusetts, Amherst, Massachusetts (1984). *Proceedings of the SPIE Conference on Intelligent Robots and Computer Vision*, Cambridge, Massachusetts, November, 1984.

6. R. S. FEARING, Exploration of the dextrous hand control problem, *General Electric Technical Report 82CRD337*, December, 1982.

7. J. J. GIBSON, *The Senses Considered as Perceptual Systems*, Houghton-Mifflin, Boston, 1966.

8. A. R. HANSON and E. M. RISEMAN, VISIONS: A computer system for interpreting scenes, in *Compter Vision Systems*, (A. R. Hanson and E. M. Riseman, eds.), Academic, New York, 1978, pp. 303–333.

9. J. M. HOLLERBACH, An oscillation theory of handwriting, *Biol. Cybern* **39**, 139–156 (1981).

10. D. H. HUBEL and T. N. WIESEL, Sequence regularity and geometry of orientation columns in the monkey striate cortex, *J. Comp. Neurol.* **158**, 267–294 (1974).

11. M. JEANNEROD, The timing of natural prehension movements, *J. Motor Behav.* 16(3), 235–254 (1984).

12. M. JEANNEROD and B. BIGUER, Visuomotor mechanisms in reaching within extrapersonal space, in *Advances in the Analysis of Visual Behavior*, (D.J. Ingle, M.A. Goodale, and R.J.W. Mansfield, eds.), MIT Press, Cambridge, Massachusetts, 1982, pp. 387–409.

13. M. JEANNEROD, F. MICHEL, and C. PRABLANC, The control of hand movements in a case of hemianaesthesia following a parietal lesion, *Brain*, **107**, 899–920 (1984).

14. E. R. KANDEL, Central representation of touch, in *Principles of Neural Science*, (E.R. Kandel and J.H. Schwartz, eds.), Elsevier/North-Holland, New York, 1981, pp. 184–198.

15. I. A. KAPANDJI, *The Physiology of the Joints*. Vol 1, *Upper Limb*, 5th Edition, Churchill Liningstone, Edinburgh, 1982.

16. D. LEE, Visuo-motor coordination in space-time, in *Tutorials in Motor Behavior*, (G. Stelmach and J. Requin, eds.), North-Holland, Amsterdam, 1980.

17. L. I. LIEBERMANN and M. A. WESLEY, AUTOPASS: An automatic programming system for computer controlled mechanical assembly, *IBM J. Res. Dev.* 21(4) (1977).

18. D. LYONS and M. A. ARBIB, A schema-based system for hand movement, *COINS Technical Report*, Department of Computer and Information Science, University of Massachusetts, Amherst, Massachusetts, (in preparation).

19. D. I. McCLOSKEY and S. C. GANDEVIA, Role of inputs from skin, joints and muscles and of corollary discharges, in human discriminatory tasks, in *Active Touch*, (G. Gordon, ed.), Pergamon, Oxford, 1978, pp. 177–187.

20. J. NAPIER, The prehensile movements of the human hand, *J. Bone and Joint Surgery*, **38B**(4), 902–913 (1956).

21. U. NEISSER, *Cognition and Reality: Principles and Implications of Cognitive Psychology*, Freeman, San Francisco, 1976.

22. T. A. NOYES, G. POCOCK, and J. FRANKLIN, Design of a FORTH-based robot control language, *Proceedings of 1983 Rochester FORTH Applications Conference*, June 1983, pp. 133–140.

23. K. OVERTON, The acquisition, processing, and use of tactile sensor data in robot control, Ph.D. dissertation, Department of Computer and Information Science, University of Massachusetts, Amherst, Massachusetts (1984).

24. J. K. SALISBURY, Kinematic and force analysis of articulated hands, Ph.D. dissertation, Department of Computer Science, Stanford University, Stanford, California 1982.

25. G. M. SHEPHERD, *Neurobiology*, Oxford, New York, 1983.

26. R. TUBIANA, Architecture and functions of the hand, in *The Hand*, (R. Tubiana, ed.), Vol. 1, W.B. Saunders, Philadelphia, 1981, pp. 19–93.

21

PRODUCTION SYSTEMS FOR MULTIROBOT CONTROL: A TUTORIAL

MARK C. MALETZ

1. Introduction

Current robot control systems and their programming languages are designed to handle completely specified tasks. They can be characterized by elaborate flow of control, primarily serial processing, and substantial syntactic constraint (e.g., AUTOPASS,[4] AL,[2] WAVE,[5] etc.). Such systems are poorly suited to the problem of providing robots with a local intelligence capability. As the complexity of the task that the system is expected to perform increases, there is a corresponding increase in the complexity of the robot programming and control system design. In direct contrast to current robot programming systems, which belong to the class of procedural systems, is the class of production languages. This chapter introduces production systems, which are highly parallel, do not require elaborate control constructs, and are particularly valuable for implementing heuristically specified tasks.

The application of the production system methodology to the area of multirobot control develops a formalism for pursuing aritifical intelligence research in the robotics area. Among the areas for possible research using the production system formalism are robot synchronization and collision avoidance, the introduction of expert advice, and adaptive modification of the underlying production using genetic algorithms.

Section 2 provides a general introduction to production systems. In Section 3, the application of production systems to robot control in multirobot environments is discussed. This discussion includes the description of a prototype multirobot environment. The application of production sys-

MACK C. MALETZ • Inference Corporation, 5300 West Century Boulevard, Los Angeles, California 90045

tems to robot control necessitates changes to the basic production system. These changes are also introduced in Section 3. Section 4 contains a detailed example of the way in which the production system-based multi-robot control system functions. Finally, Section 5 presents conclusions and areas for further research.

2. Introduction to Production Systems

Two classes of programming languages can be identified: procedural systems and production systems. Almost all traditional programming languages are procedural in nature. They are characterized by the importance of flow of control, which is often elaborate, and by an algorithmic specification of the tasks to be undertaken. Procedural systems are generally processed serially, although a degree of parallel processing is occasionally permitted (such as tasking). This parallelism, however, must be carefully specified and is subject to substantial syntactic constraint. Even where parallelism is permitted, flow of control concerns still dominate. In direct contrast is the class of production systems, which are highly parallel, lack elaborate control constructs, and are particularly valuable for implementing heuristically specified tasks.

The dependence of traditional programming languages on syntax requirements means that a change of even a single bit is likely to prevent the program from functioning correctly, perhaps even from functioning at all. This dependence on syntax makes automatic programming and the implementation of adaptive algorithms particularly difficult. The highly parallel nature of production systems makes their "instructions" less interactive than those of traditional programming languages. This results in the graceful degradation property which characterizes production systems and permits the alteration of individual "instructions" without significantly degrading system performance.

production system ::= {production _ set, message _ list$_m$}
production _ set ::= $\langle$production$\rangle^+$
$\langle$production$\rangle$ ::= $\langle$condition$\rangle^+$ /$\langle$action$\rangle$
$\langle$condition$\rangle \in \{0, 1, *\}^n \mid$ '$\{0, 1, *\}^n$
$\langle$action$\rangle \in \{0, 1\}^n$
message _ list$_m$ is a data structure that can contain a maximum of m
　　　messages, where each message is an n-bit string over $\{0, 1\}$.

Each production in the production set contains a condition part and an action part. The + superscript indicates that the superscripted element may appear one or more times. This means that productions are composed of one or more conditions comprising the condition part, followed by exactly one action in the action part. Conditions are n-bit strings over the $\{0, 1, *\}$ alphabet, where 0 and 1 are the ordinary binary digits, and * is a don't care character (i.e., it will match both a 0 and a 1). The actions are n-bit strings over the $\{0, 1\}$ alphabet. The relative simplicity of the message list belies a great versatility which will be developed.

Given the above specification, a production system functions as follows:

1. The system is first initialized by placing one or more messages on the message list (up to a maximum of m messages).

2. Each of the productions is checked against the message list to determine whether the messages on the message list satisfy that production's conditions.

A production's conditions are *satisfied* if and only if every condition matches a message on the message list, where matching is done on a bit-by-bit basis. The matching process is performed bit-wise for every condition in the condition part of the production, and therefore, the condition part is treated as a conjunction of conditions. Only the 0 and 1 digits in a condition must match the corresponding digit in a string on the message list since the don't care character (*) will match either a 0 or a 1 digit.

3. After every production has been checked against the message list, the message list is erased. Notice that the production checking procedure can be performed in parallel if the hardware permits. In fact, each production could potentially be processed by a dedicated microprocessor. Because of the message list erasure at the end of each time step, all messages have a single time step lifespan. To post messages for longer periods, repeated postings are required. A simple message renewal construct involves the use of a production whose condition and action parts are identically equal to the message to be renewed. This results in perpetual message renewal for the associated message once it appears on the message list, unless the production is not selected to post its message because more than m productions with distinct messages attempt to post their messages to the message list. A more general approach would be to use additional conditions in the condition part of the production to indicate when mes-

sage renewal should occur. Using this approach, when the message to be renewed appears on the message list it will match one of the conditions, and when the other conditions are matched signaling message renewal, the message is reposted.

4. A new message list is then constructed from the action parts of every production that was satisfied by the old message list. As long as no more than m productions with distinct messages are satisifed in step 3, every production can post its action part to the new message list.

> The action part of a production is *posted* to the message list by placing the n-length bit string that comprises the action part on the message list.

5. The processing procedure is iterated by returning to step 2.

As long as the number of distinct messages to be placed on the new message list in step 4 is not greater than m, all productions whose conditions are satisfied by the old message list may post their actions on the new message list. This results in a parallel system that is internally conflict free due to the fact that messages, from the message list viewpoint, are simply bit strings and cannot conflict with one another. Conflict can occur when the messages are interpreted if they generate contradictory interpretations (e.g., two messages might generate two robot control commands, one which instructs the robot to move to the right and the other which requires a left rotation). Interpretation conflict resolution is, however, exogenous to the production system. Another type of conflict is due to the fixed length message list. If more than m productions attempt to post distinct messages to the new message list, then m of the new messages must be selected. This problem can be resolved by removing the length constraint from the message list. This could be done by setting the message list length (m) sufficiently large, since the system itself imposes a maximum message list length equal to the number of distinct actions in the production set.

The production system processing can continue indefinitely. There are three possible terminal conditions:

1. During some iteration, it is possible that no productions will be satisfied by the current message list. This results in no messages being posted to the new message list. Once an empty message list appears, the system stops processing.

2. A message list cycle may appear. This represents a type of dynamic equilibrium in which the system cycles through a set of message lists. It corresponds to the control of repetitive tasks.

3. A special case of message list cycling occurs when the cycle length

is equal to one. In this case, the same message list appears during all future time steps.

An empty message list brings the system to a halt and should be avoided. Other control constructs for halting the system can be provided if necessary. A simple mechanism that will prevent an empty message list from occurring is the inclusion of a set of productions in the production set that covers the n-bit string space over $\{0,1\}$.

A *cover* of $\{0,1\}^n$, the n-bit string space over $\{0,1\}$, is a collection of strings in $\{0,1,*\}^n$ such that every string in $\{0,1\}^n$ is contained in a string in the cover. A string in $\{0,1\}^n$ is contained in a string in $\{0,1,*\}^n$ if and only if 0 and 1 digits in the $\{0,1,*\}^n$ string exactly match the corresponding digits in the $\{0,1\}^n$ string.

A simple example of this would be the use of a production whose condition consisted of the string ** . . .*, which would match any string on the message list. This universal condition is particularly useful in systems where the message list is a limited resource and not all productions whose conditions are satisfied can post their actions to the new message list. In such systems, the universal condition production could be assigned a low priority so that it would only post its action to the new message list if there were few other satisfied productions during the current time step. In a system where all satisfied productions are guaranteed a space on the new message list for their actions, the use of universal condition productions is discouraged since their actions would perpetually appear on the message list. In such systems, it is better to use a larger set of productions that cover the n-bit string space over $\{0,1\}$. The universal condition production is the smallest cover. The largest possible cover contains 2^n productions, where the condition part of each production is a distinct string in $\{0,1\}^n$.

A message list cycle can be beneficial in systems which are used to control cyclic or repetitive tasks. The steps of the task cycle would then correspond to the cycles of the message lists. The single time step cycle is a special, and generally degenerate, case of these cycles. Although cycles can be programmed using production systems, they are better handled by procedural programming systems because of the algorithmic nature of cycles and the advantages inherent in procedural systems for algorithmic programming.

The don't care (*) character has been permitted to appear in the condition part of productions, but not in the action part where it would function as a *pass-through* character. A pass-through on the action side of a production involves passing the corresponding bit of the message that matched the condition through to the action that is posted. If the corresponding character in the production's condition is a 0 or 1, then that digit is always passed-through. Usually, the condition character corresponding to a pass-through is the don't care character. In this case, the corresponding character in the message that matches the condition will be passed through. We will not use the pass-through character because of computational difficulties that result. The most prominent difficulty is the resolution of a potential conflict when more than one message matches the condition of a production containing pass-throughs in the action part. While pass-throughs will not be used, their functionality can be emulated by using more than one production and the don't care character.

Greater flexibility and programming power can be attained in production systems if communication with the outside world is permitted. This communication must be introduced without disturbing the parallelism or internal conflict-freeness of the production system. This is most easily accomplished using the message list by permitting messages to be posted from the outside world. The outside world can also be permitted to read the message list. In this way, production systems can be provided with a primitive I/O capability.

3. Robot Control Using Production Systems

The application of production systems to the problem of robot control in a multirobot environment will now be discussed. The robots in these multirobot control (MRC) systems will initially consist of a microprocessor-based robot arm, a television camera or other imaging device and an associated image processor, and a locomotion drive unit. Alternate robot configurations are possible with relatively minor alterations to the MRC system command structure. No modification of the production system processing is required when the robot configuration is altered. The robots are able to interact with their environment using the robot arm which can execute a predefined set of control subroutines, view portions of the environment, and move through the environment. Associated with each robot is a robot control processor which is responsible for local processing. Each individual robot processor communicates with its respective robot and with the system supervisor.

An example configuration is presented in Figure 1. The robots consist of PUMA robot arms mounted on a locomotion platform along with a television camera which can be rotated to the left or right but which is set at a fixed height. The robot processors are LSI 11/23 microcomputers. Each

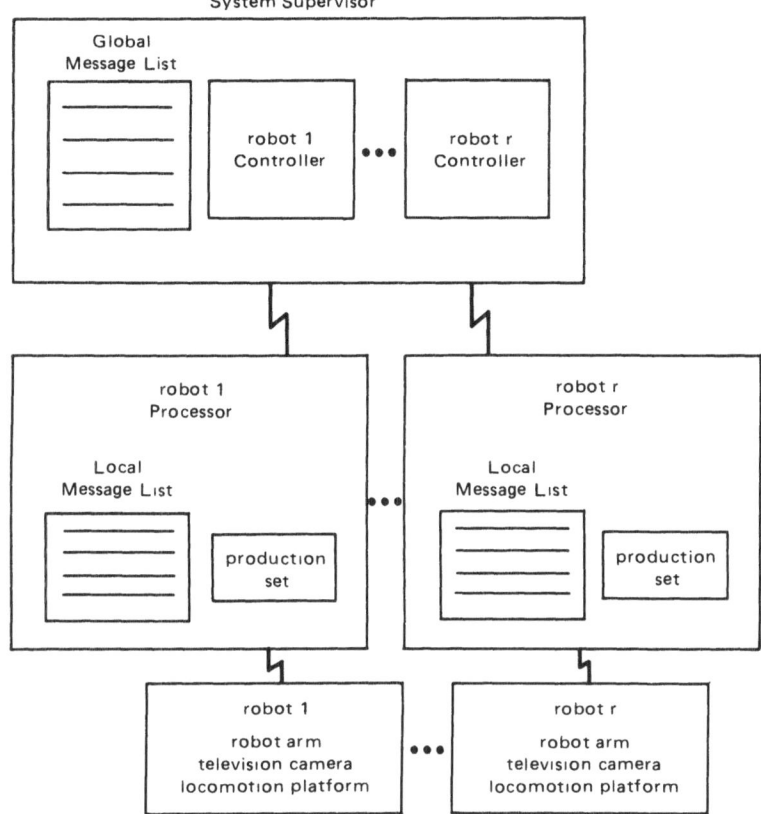

FIGURE 1. Example of the MRC system configuration.

LSI 11/23 is connected to the associated robot and to the supervisory computer a DEC VAX 11/780. Two channels from the VAX 11/780 to each robot processor are required. Therefore, the number of robots that can operate within the system is entirely determined by the number of available channels to the VAX 11/780.

Bidirectional communication between robot and robot processor permits robot control and television camera data transfer. The television camera has a local dedicated processor which converts images within the scope of its view into messages which can be placed on the message list. These messages are transferred along the communication channel to the robot processor. No other data originate at the robot end of the channel. Again, alternate robot configurations could be constructed which would require more extensive robot/robot processor data transfer. One example would be data generated by a wrist force sensor or other sensory devices. As with any alteration to the robot configuration, these changes would require MRC system modification. The robot processor sends robot control instructions to the robot. Initially, a limited set of robot control commands (RCC's) will be permitted. These initial commands are summarized in Table 1. Although relatively few commands are available, the set of Execute commands {En} permits the execution of any one of n robot control subroutines. These preprogrammed subroutines could be designed using task-specific robot control requirements. They therefore provide a large degree of flexibility in specifying the robot control capabilities of the system.

The set of RCCs has been divided into three general categories. The first five commands relate specifically to the vision vector. The next five commands are used to control the robot's locomotion, and the last two commands are used to manipulate the robot arm.

Implicit in Table 1 are several assumptions concerning the MRC system:

1. Each robot is equipped with a single television camera which corresponds to the vision vector, a robot arm which can grasp objects centered by the vision vector, and a primitive locomotion capacity which must be in the direction of the vision vector.

2. Robots move at a fixed speed in the direction of the vision vector. For this reason, the motion vector must always be coupled with the vision vector before movement (using the COUPLE command). The vision vector may be rotated during robot movements, but this will alter after the robot's course.

TABLE 1
Robot Control Commands

Command	Keyword	Description
{S}	SCAN	360° vision scan
{A}	ALIGN	Align vision vector with nearest object
{L}	LEFT	Vision vector left $\epsilon°$
{R}	RIGHT	Vision vector right $\epsilon°$
{T}	TURN-AROUND	180° turn-around
{M}	MOVE	Move forward
{H}	HALT	Stop moving
{C}	COUPLE	Couple motion vector with vision vector
{U}	UNCOUPLE	Uncouple motion and vision vectors
{G}	GRASP	Grasp object aligned in vision vector
{En}	EXECUTE	Execute robot arm subroutine n

3. The vision vector may only be uncoupled from the motion vector after the robot has been halted (using the UNCOUPLE command). This permits the robot to scan the environment by moving its camera while leaving its locomotion platform fixed. Because of this requirement, the SCAN RCC can only be performed after the robot halts.

4. The robot can grasp an object that has been aligned in the vision vector. A primitive subroutine execution facility {En} has been included in the RCC set that can be used to code robot arm subroutines, in addition to the other robot control subroutines that {En} may encode. Robot arm control EXECUTE RCCs may be performed either before or after grasping an object. They are presumably related to the task that the robot is to perform. The availability of multiple robot arm subroutines using more than one {En} RCC corresponds to multitask capabilities.

Two additional RCCs are provided for use by the system's "expert" users (the individuals who provide "expert" advice to the production system):

SHOW R Show the portion of the environment contained in robot R's vision vector on the television monitor.

IMAGE R Show all image related messages posted to the message list by robot R during the current time step. These messages are created by the image processor associated with the robot's television camera.

These commands will never be invoked by the production system. They are used exclusively by "expert" users to help them understand the internal processing of outside world image data by the production system.

3.1. Production System Modification

The class of production systems specified by {production_set, message_list$_m$} can be used to control a MRC system, but only if robots have complete autonomy or no autonomy. In the case of complete autonomy, each robot would have its own production system and would never need to communicate with other robots. In the no autonomy case, all processing and control would be performed by a single production system which would reside at the supervisor level. Clearly, both of these options are inefficient. Because the robots operate in the same environment, some implicit communication and cooperation among the robots is desirable. However, because the robot tasks may be vastly divergent, centralized control is inappropriate. This indicates the need for a modification of the production system definition.

To isolate robot control, each robot should be equipped with its own production set. This will permit differentiation of robots according to task and behavior specific requirements through the distinct production sets. To permit implicit communication and cooperation, a single message list will serve as the common data structure. In terms of the example system configuration, this means that each local robot processor (LSI 11/23) will have a production set which is processed and updated separately from all other robot's production sets. This will permit each robot's production set to evolve separately when adaptation is introduced. The single message list for the MRC system will reside on the system supervisor (VAX 11/780), and be accessible to all robot processors. These modifications to the production system definition are sufficient to permit control of a MRC system. The new production system is specified by the following $r + 1$-tuple:

$$
\text{production system} ::= \{\text{production \_ set}_1, \\
\text{production \_ set}_2, \ldots, \\
\text{production \_ set}_r, \\
\text{message \_ list}_m\}
$$

Where there are r robots in the multirobot environment.

Figure 2 illustrates the example MRC system, including the locations of production sets and the message list.

3.2. Message Differentiation Through Tagging

In the multirobot environment, each robot is characterized by its individual production set. These production sets are processed using the common message list during each time step. The shared access to the message list necessitates segmentation of the message list. The message list segments correspond to the various interpretations of messages on the message list, and are summarized as follows.

1. Messages can be generated externally to provide the MRC system with "outside world" information. In the example system, these messages can only be generated by the television camera image processor. The messages are therefore specific to the robot whose television camera generated them. For this reason, other robots need not use these messages in production processing. Future MRC systems may include an external source of outside world data. This external source could then make its data available to all robots. An example of this would be a motion detection device which could identify all robots that were in motion.

2. Internal processing messages can be generated by activation of a production belonging to a particular robot for processing associated

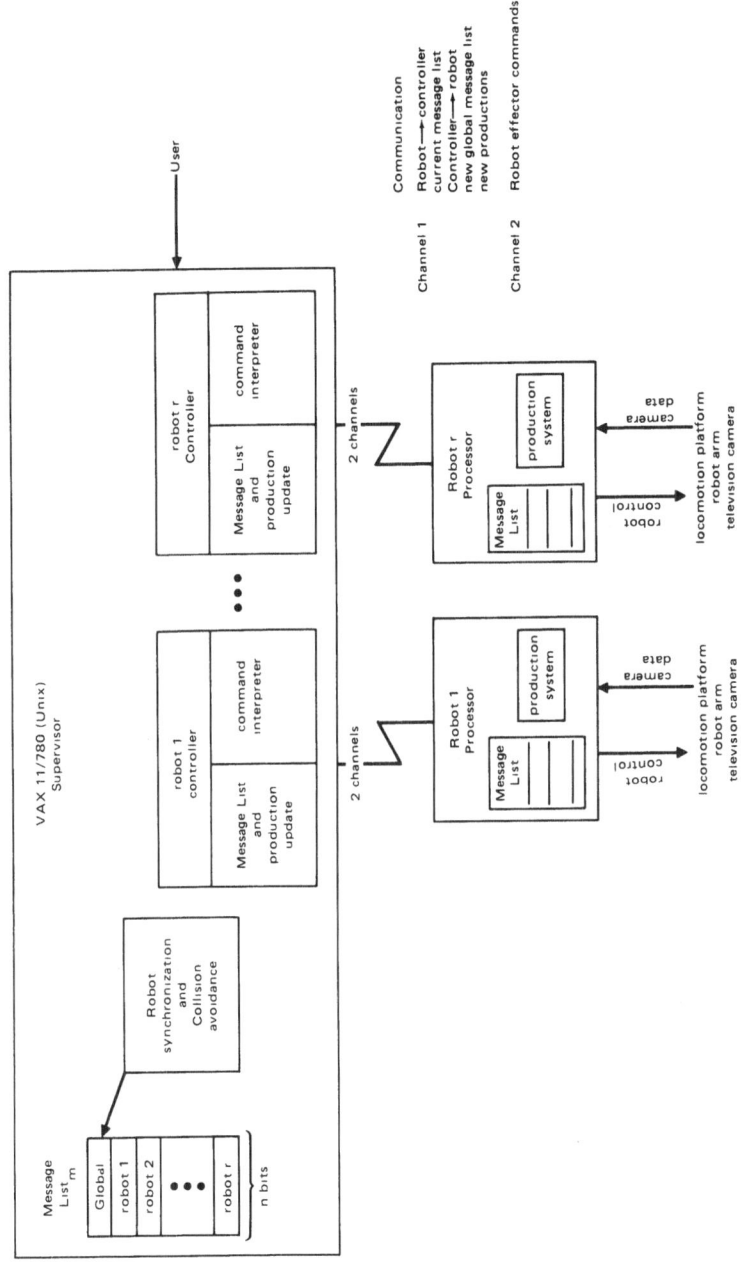

FIGURE 2. Example of the MRC system with production sets and message list.

uniquely with that robot. These are the messages that permit each robot's production processing to occur, and need not be interpreted by other robots.

3. Control messages can be generated internally by production activation for a particular robot. These messages can be used both for production system processing and for robot control. The robot control function is implemented by translating the message into a robot control command, which is then passed to the robot for interpretation. These control messages are also robot specific, and need not be interpreted by other robots.

4. Coordination messages are accessed by all robot production sets and are used to permit robot coordination and joint processing capabilities.

Figure 3 illustrates the segmentation of the message list into these four segments, and describes the degree of shared access to each segment

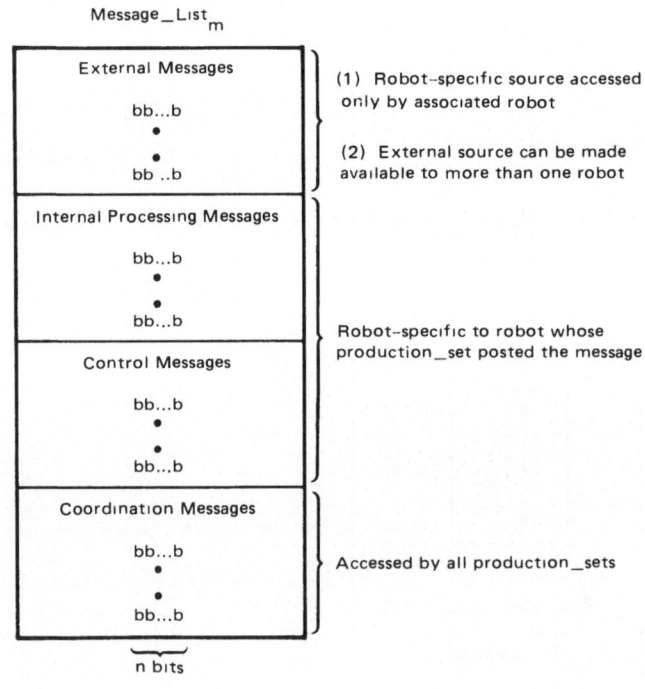

Figure 3. Four segment message list segmentation.

among robot processors. This message list segmentation will be further refined for MRC system usages. The first category of messages provides the production system with a primitive input capability. These are the messages that are generated by the image processor associated with each robot's television camera. Encoded in these message is information concerning the image currently contained in the vision cone. The third message category provides the production system with a primitive output capability, in the form of robot control commands which are executed by the robot. These commands permit the robot to interact with its environment. The final message category represents the potential for I/O among robots in the form of messages which are accessed by all robots.

Segmentation of the message list into the four categories of messages is accomplished through the use of message tags.

> A *tag* can either be a prefix or a suffix consisting of a fixed number of bits. These bits can be used to identify message list segments.

In this case, prefix tags will be used. The number of bits dedicated to the tag depends on the number of distinct message list segments desired. By subdividing the prefix bits into fixed length substrings, it is possible to provide subsegments for the message list. In this way, the message list may be segmented into principal components, and then each of these components may be further subdivided as necessary. For the MRC system, primary message list segmentation will be used to separate the global portion of the message list and the robot-specific segments. The global segment of the message list contains messages from category four. These messages are required for robot coordination, and must be accessible to all robot processors. Each robot-specific segment of the message list contains external messages, internal processing messages, and control messages for the associated robot. If there are r robots, then the message list must have $r + 1$ segments, which requires INT_CEILING $[\log_2(r + 1)]$ bits dedicated to the tag prefix (where INT_CEILING $[x]$ is the integer ceiling function, the largest integer greater than or equal to x).

No further subdivision is required within the global portion of the message list. The robot-specific segments, however, must be further divided into three subsegments according to message category. This further segmentation requires implicit segmentation of the tags.

> Each tag used by the production system will be divided into two components: a *primary tag* and a *secondary tag*. The *primary tag* is used to segment the message list into its primary segments,

global and robot specific. This requires $r + 1$ distinct tags, and hence INT_CEILING [$\log_2(r + 1)$] bits will be dedicated to the primary tag. The primary tag will always be represented by the initial bits in a message. A *secondary tag* is used to establish message list subsegments, within segments defined by the primary tag. The number of bits required for the secondary tag is computed in the same way as the number of bits required for the primary tag. The secondary tag bits immediately follow the bits dedicated to the primary tag. While only primary and secondary tags will be used here, higher-order tags can also be defined.

The first subsegment of the robot-specific message list segments identifies external messages. These messages correspond to images in the vision cone and can therefore be interpreted as more than just a sequence of bits for production system control. The second subsegment contains internal processing messages. The third subsegment contains control messages which can be converted into robot control commands and transferred to the robot. These control messages are also used by the production system for internal processing along with internal processing messages. By distinguishing between internal processing messages and control messages, a substantial amount of command conversion overhead is eliminated, since the command conversion routine need only process messages tagged as control messages.

The following tagging convention will be adopted:

Suppose that there are r robots in the MRC system, and that the robots have been indexed 1 to r. Then $i = $ INT_CEILING [$\log_2 (r + 1)$] bits are required by the primary tag, and two bits are required by the secondary tag.

1. The global segment of the message list will always be assigned the tag: $00...0$ ($i + 2$ bits).
2. Each robot-specific segment of the message list will be assigned an i-bit primary tag whose decimal value is equal to the index number of the associated robot.
3. The two secondary tag bits will immediately follow the primary tag bits and will be encoded as follows:

00	External messages
01	Internal processing messages
10	Control messages
11	Not currently used

The 11 secondary tag could be used for global messages posted by a particular robot. This could, for example, be used if image data collected by a particular robot were required by other robot processors. Figure 4 displays the complete message list segmentation used by the MRC system. This tagging scheme segments the message list in a manner that requires little system overhead. Distinct message list segments can be identified by using a bit mask to select the messages of interest. As an example, a bit mask of 00. . .011 (i bits) will select all messages specific to robot number 3, and a mask of 00. . .01110 ($i + 2$ bits) will select all control messages for robot number 3.

Tag manipulation using bit masks and assembly language is fast and efficient. Tags represent a low overhead mechanism for partitioning the MRC system's common data structure into both shared and dedicated components. In future applications, tags can be used to perform even

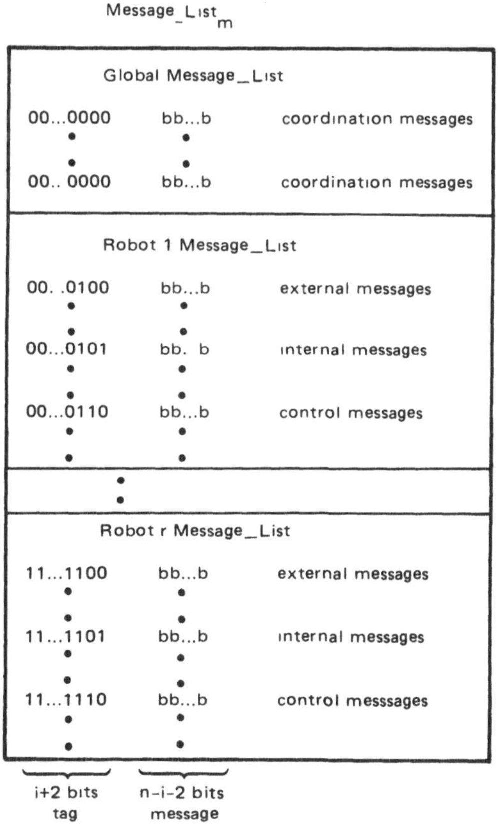

Message_List$_m$

Global Message_List		
00...0000	bb...b	coordination messages
•	•	
•	•	
00.. 0000	bb...b	coordination messages
Robot 1 Message_List		
00. .0100	bb...b	external messages
•	•	
•	•	
00...0101	bb. b	internal messages
•	•	
•	•	
00...0110	bb...b	control messages
•	•	
•	•	
•	•	
Robot r Message_List		
11...1100	bb...b	external messages
•	•	
•	•	
11...1101	bb...b	internal messages
•	•	
•	•	
11...1110	bb...b	control messsages
•	•	
•	•	

$\underbrace{\phantom{i+2 \text{ bits}}}_{\substack{i+2 \text{ bits} \\ \text{tag}}}$ $\underbrace{\phantom{n-i-2 \text{ bits}}}_{\substack{n-i-2 \text{ bits} \\ \text{message}}}$

FIGURE 4. Complete message list segmentation.

more sophisticated functions. One example involves message addressing using tags. In such a system, robot j might post a message on the message list, and address that message to robot k through the use of a tag. This tag would have two address components, one for the sender, and the other for the addressee. Although the message list is a relatively simple data structure, information coding using tags can permit sophisticated behavior, and the emulation of complex data structures.

4. MRC System Example

The example system will be configured with a single robot, consisting of a robot arm, a television camera, and a locomotion platform. The robot will reside in a contrived environment in which there are humans, ducks, rats, and a single elephant (Clyde, of course). The robot's desired behavior is to avoid the elephant, chase the rats, and ignore the ducks and humans.

The simplicity of the environment and task results in a relatively simple production system. The external messages generated by the television camera's image processor need only distinguish the four types of occupants of the contrived environment. This can be accomplished using two features: size and number of legs. A third feature which will prove to be useful is the range of the object, since the robot will react differently to the elephant depending on whether it is close or far away. The three features can be coded as follows:

size	0 = large	1 = small
# legs	0 = 2 legs	1 = 4 legs
range	0 = close	1 = far

Using these features, the external messages corresponding to the four residents of the robot's environment can be constructed. Each of these messages requires five bits: the two-bit 00 prefix identifies the message as an external message, and the last three bits correspond to the size, number of legs, and range features, respectively. The area inhabitants are identified as follows:

Humans	0000_
Ducks	0010_
Rats	0011_
Elephant	0001_

Note that in each case the range feature is left undetermined, since each of the area inhabitants can be either close or far away.

To obtain the desired behavior from the robot, a subset of the Robot Control Commands is sufficient. These RCCs are coded as control messages using the control message 10 prefix as follows:

10000	Halt
10001	Rotate vision vector $\epsilon°$
10010	Rotate vision vector 180°
10011	Move fast
10100	Align vision vector with nearest object

In this example, five RCCs were sufficient, and since the external messages required three bits for feature mapping, the RCCs could be coded using the message length determined by the external messages. In fact, one could have been coded up to eight RCCs using the five-bit message format. If more than eight RCCs were required, then it would have been necessary to expand the message length, and to pad the external messages with dummy features.

Because of the task's simplicity, a rough approximation of the desired behavior can be obtained using only seven productions:

00*0*/10001	Ignore humans and ducks by rotating the vision vector
00011/10000	Halt when the elephant is sighted at a distance
10000/10001	Rotate vision vector away from elephant after halting
00010/10010	Rotate vision vector 180° when elephant is close
10010/10011	Move fast to flee from close elephant
0011*/10100	Align vision vector on rats
10100/10011	Move fast to chase rats

The above production system fragment could be expanded to provide the exact behavior desired, but it is intended only to serve as an example and to help develop intuition about production systems. A particularly significant observation concerns the third and fifth productions. In both of these productions, a control message serves as an internal processing message. In the second production, the 10000 control message is posted to the message list and is interpreted as an RCC to halt the robot. In the third production, the 10000 control message serves as a condition to the production. The 180° rotation RCC (10010) found in the fourth and fifth productions functions in a similar manner.

5. Conclusions

A formalism for applying production systems to the area of multirobot control has been developed. The example system demonstrated the use of such a system in a single-robot environment. More sophisticated multirobot environments can be developed, in which the robots perform robot-specific tasks and also communicate with each other through the global message list.

In the future, the formalism developed here will be applied to several problems in the area of robot control:

1. Robot Synchronization. The fact that all robots access the global message list can be used by the system supervisor to handle such robot synchronization problems as collision avoidance. One simple approach to this problem would be to divide the environment into quadrants, and use the global message list to Lock and Unlock these quadrants so that only one robot would be permitted to be in motion in any quadrant.

2. Production Differentiation. Assuming that the production system does not completely correspond with the desired behavior (owing either to faulty coding, inaccurately specified behavior, or changes in the task itself), it is helpful to identify productions that contribute positively and negatively to the robot's completion of the specified task. This will involve assigning a strength parameter to each production, and modifying the strength based on the production's contribution to robot behavior.

3. Adaptive Modification of Production System. Once the productions have been assigned strength parameters, it is possible to use genetic algorithms to modify the production system so that the robot's behavior will converge to the desired task.

4. Introduction of Expert Advice. An expert can be asked to evaluate the production system's performance. This can be used to modify strength parameter values. It can also be used in conjunction with genetic algorithms to modify the task assigned to the robot.

These areas of future research will involve the development of simulations of the MRC systems, and then prototype systems using actual robots with simulated input data.

References

1. S. BONNER and K. G. SHIN, A comparative study of robot languages, *Computer* 15(12), 82–96 (1982).
2. R. FINKEL, AL, A programming system for automation, Stanford Artificial Intelligence Laboratory Memo AIM-243, Stanford University, Palo Alto, California, November, 1974.

3. J. H. HOLLAND, *Adaptation in Natural and Artificial Systems,* the University of Michigan Press, Michigan, 1975.
4. L. I. LIEBERMAN, and M. A. WESLEY, AUTOPASS, An automatic programming system for computer controlled mechanical assembly, Computer Sciences Department RC 5925 (25653), T. J. Watson Research Center, Yorktown Heights, New York, March, 1976.
5. R. P. C. PAUL, Wave: A model-based language for manipulator control, *Ind. Robot* **4,** 10–17 (1977).
6. K. TAKASE, R. P. PAUL, and E. J. BERG, A structured approach to robot programming and teaching, *IEEE Trans. Syst., Man, Cybern.* **11**(4), 274–289 (1982).
7. P. H. WINSTON, *Artificial Intelligence,* Addison-Wesley, Reading, Massachusetts, 1977.

22

A PROGRAMMING LANGUAGE FOR SOPHISTICATED ROBOT SENSING AND SENSOR-BASED CONTROL

MING-YANG CHERN

1. Introduction

The versatility of a robotic system depends heavily on its sensing capability.[1-4] Without appropriate sensing provisions, the robot is not able to react to unpredictable events in its working environment. It can only succeed for fully prearranged tasks, and thus the tasks it can perform will be quite limited. To get beyond this restriction, powerful sensing capabilities are indispensable. With the advance of sensing technology, more and better sensors will be developed while the cost is reduced. Robotic systems equipped with more sensors are expected for the near future. The use of sensors also relaxes the absolute accuracy requirement of the robot mechanical constructs.[5] And this makes the robots even more cost-effective and attractive for use.

The enhancement of robot sensing capability is not just a matter of equipping the robot with more sensors. The robot programming language has a profound effect on the whole issue.[6,7] It affects the human effort required in programming the given robot as well as the extent to which the system capability can be exerted. Quite a few robot programming languages have already been developed.[8,9] Among them, some are implemented on systems without sensors. In many robot systems, mostly with a small number of sensors, there is no general, systematic way of accessing

MING-YANG CHERN • Department of Electrical Engineering and Computer Science, Northwestern University, Evanston, Illinois 60201

and handling sensory information in the programming language provided. The PAL language[10,11] can specify robot end-effector positions, but only simple force, torque, and touch compliances can be referenced. VAL[12] allows the signal detection and setting through the use of IFSIG, SIGNAL, and REACT instructions; yet this provision is for single-bit channels only. The design of AUTOPASS aimed at a very high-level language, and as a result, the handling of sensor data is transparent to the applications programmer.[13] AML was lately developed at IBM.[14,15] It includes aggregate data objects and operations, and allows aggregate parameters for subroutine calls. It is a well-structured, semantically powerful language for robot programming. However, only a few sensor-related commands (system subroutines) are provided.

In the language RAIL,[16] there is a flexible way of defining and accessing input or output lines, either as single- or multiple-bit numbers. Geschke presented a programming system which can easily specify servo processes for sensor-based robot manipulators.[17] Through a simple declaration, the values of built-in (software) sensors may be converted from those of external sensors. Recently progress has been made[18] which combines the advantages of the above two designs, yet is still mainly for unisensors.

For the more complicated case of array sensors, few systems or languages have been developed so far. AML/V[19] is an extension of AML. It offers some basic subroutines for image acquisition, region analysis, and image arithmetic. In another system which integrates the PUMA robot with the MIC vision, again, it calls subroutines for image processing.[20] MCL offers some convenience for setting up two-dimensional models for part inspection.[21] Yet the programming language for vision and image processing in the general aspects is not available. Currently, most systems with vision or other array sensing offer some provisions in software to facilitate their specific applications. The high-level programming language for general array sensing and processing is not well developed yet. Nor is an easy way for low-level data access (to individual data in the sensor array).

In the recent robot programming language design, more attention has been paid to the aspects such as process synchronization, exception handling, real-time execution, and concurrent process control.[22,23] So are the tools for testing and debugging. Yet there are still several goals the current robot languages can hardly achieve.[24] These include the ease of expansion, robot independence (portability), completeness without much complexity, and multirobot concurrent operations.

In this chapter, some design principles toward a robot programming language for sophisticated sensing and sensor-based control are presented. The inclusion of sensor variables and control variables in the language

The cooperative use of these two variables also allows an easy access to the data of array sensors. Some arithmetic and logical combinations of sensor variables are permitted in this design. This facilitates more abstraction of sensory information and simplifies the specifications of sensory conditions. In addition, one way of specifying the simultaneous monitoring of multiple sensory conditions is suggested. Although this chapter emphasizes the language design, the execution efficiency is not neglected. An efficient scheme for monitoring sensory conditions in real time is also referred.

2. Sensory Programming Problems

In an advanced robot system, there may be various kinds of sensors used. Examples of these sensors include force-torque sensors, proximity sensors, touch sensors, slip sensors, tactile sensors, and vision imagers.[1-4] The hardware constructs of these different sensors may be quite varied. And the information delivered by these sensors depends on the sensor type and the physical arrangement. To the programmers' concern, it is important that sensor data can be accessed in a simple and consistent way from within the robot programming language. It is also desirable that these data be represented in a convenient format, i.e., with proper data type and using the unit most convenient to human programmers. The portability of robot control programs has been a major issue in recent years.[6,7] The difficulty in portability is largely caused by the difference in the system I/O configurations. We hope that the sensors referenced in the program can be easily mapped onto the system configuration. This has a definitely positive effect on the portability of the robot control program. The programmers need not look into the system's I/O driver routines to understand how the sensor data are obtained nor modify the I/O drivers to accommodate the new system or sensor I/O configuration. With all the above features, the readability of the robot control program would be much enhanced. In other words, this means that the language will be easy to learn and easy to use.

For the simplicity of robot programming, we desire a high-level language. Nevertheless, we still need to have access to all the low-level facilities. Some sensors require a certain control from the system to set their parameters such as the threshold or the position. The sensory programming may also require timing information for process synchronization or the measurement of speed. These require some ingenuity in the language design. The incorporation of sensory access and programming capability must not too much increase the complexity of the programming language. That is, we desire the programming language to be complete yet still simple.

To further simplify the program and reduce the programming effort, we will need more abstraction of sensory information. It may be necessary to define high-level software sensors and to have some simple language rules for specifying the monitoring of various sensory conditions. The continuous monitoring of sensory conditions during robot operations may cause a certain system load. Since the execution of a robot control program must satisfy the real-time operation requirement, an efficient scheme for monitoring sensory conditions is desired as well.

3. Defining Sensor Variables for Various Sensors

There have been several designs using program variables to reference sensor values directly.[16,17] The reason for using variables to represent the sensor inputs is quite straightforward. For example, instead of the sequence

$$\text{Fvalue = Sensor\_input (FORCE)}$$
$$\text{if (Fvalue > limit) goto Alert;}$$

we may like to simply write

$$\text{if (FORCE > limit) goto Alert;}$$

However, extra information must be provided to the system such that it knows where to access the sensor data and how the data should be treated.

In spite of the difference among various sensors, they look similar at the I/O interface level. At the input ports of the system, all sensor data are digital and of either one bit or several bits. The analog signals from sensors will be converted through A/D converters. The major differences between different sensors at these input ports are the number of bits and semantic interpretation of the bits comprising the sensor data. On the other hand, in order to be user-friendly, the sensor data in the program should be expressed in terms of the unit most convenient to human programmers. And this is also related to the data type of the variables assigned to store those sensor values. If the internal representation of the sensor data is not an integer or not with exactly the same unit as used at the sensor input port, some translation processes will be required. In the extreme case, a table look-up may be needed.

One approach to deliver the above sensor setup information to the system is to have a special system service routine for defining sensors. We may call a system function

DEFSENSOR (Name, Index, Port, Start_bit, Nbit, Type, Offset, Scale)

where Index contains the information whether there are several sensors under the same name. If Index = 0, there is only one sensor. If there are several sensors, Index will be larger than 0 and this number will be used as the index under the array name to represent the specific sensor.

The problem with the above approach is that it is not user-friendly enough. It takes a certain effort for programmers to memorize the meaning of the argument at each position. A novice in this programming language, without looking into the system manual, will have difficulty in guessing the meaning of the arguments in this sensor setup function call. For sensor data which do not need any transformation, we still have to fill out the Offset and Scale terms. For a variable other than an array, we still have to set Index = 0. The program expression based on this approach is not concise to the best extent. In some cases, there is only one bit of sensor input. Yet we use two arguments to deliver this information. On the other hand, a sensor may be input through more than one port. We can hardly inform the system using the above function call without modifications.

A better approach to set up the variables for sensors is to have these variables specified in a fashion consistent with the usual programming language rules. And this is the approach we adopt in the language design. Here we call the variables representing sensor data the *sensor variables*. A sensor variable is a read-only variable whose value is determined by a sensor rather than through internal assignment. To declare the sensor variables, a sensor-declaration section will be added to the program. In this special section, a variable is declared as a sensor variable while the sensor setup information is delivered. Similar to the language appearing in Ref. 17, we may include the sensor data transformation as a mathematical expression in one declaration statement. The data type of the sensor variables must be declared, before this section, just like the usual program variables. They may be declared as integer, real, Boolean, array, etc. The programmer may view sensors simply as program variables which happen to receive their values from an external source rather than through internal assignment. Some examples are shown in the following:

integer shaft, TOUCH[2], position;
real FORCE;

    sensor shaft := #21[7–3],
      TOUCH[1] := #12[7],
      TOUCH[2] := #12[6],
      position := #21[2–0]–#22,
      FORCE := A * float(#15[6–0]) + B;

where *A* and *B* are constants or previously defined variables. The meaning of the above declaration is easily understandable. The sensor variable "shaft" takes value from port #21, the seventh to the third bit; while "position" obtains its value by concatenating the three least significant bits of port #21 and the whole eight bits of port #22. Note that in defining FORCE, we can specify in the same declaration a transformation as well as the sensor input connection.

  To describe the declaration of sensor variables more exactly, we define the declaration in the following general form:

sensor ⟨sensor_variable⟩ := ⟨sensor_expression⟩

where ⟨sensor_variable⟩ is a variable (or an array element) with its data
   type already defined.
   ⟨sensor_expression⟩ ::= ⟨port_select⟩ | ⟨op⟩ ⟨sensor_expression⟩
          | ⟨sensor_expression⟩ ⟨op⟩ ⟨expression⟩
          | ⟨expression⟩ ⟨op⟩ ⟨sensor_expression⟩
   ⟨port_select⟩ ::= #⟨port_number⟩
        | #⟨port_number⟩ [⟨bit_number⟩]
        | #⟨port_number⟩ [⟨high_bit⟩ − ⟨low_bit⟩]
        | ⟨port_select⟩ − ⟨port_select⟩

  Most of the above terms are self-explanatory. The ⟨op⟩ is the arithmetic, relational, or logical operator allowed in the programming language. And ⟨expression⟩ is the program expression consisting of the usual program variables, constants, and operators. For the convenience of programming, we may further extend the definition of ⟨sensor_expression⟩ by including previously defined sensor variables as operands. For example, we may define

sensor ANY_TOUCH := TOUCH[1] *or* TOUCH[2] *or* TOUCH[3],
   Breakable := BEND_FORCE > limit,
   TOTAL_FORCE := FORCE[1] + FORCE[2];

With the high-level "software" sensor variables, the conciseness of the program can be further enhanced and the processing of sensory information can be more abstract. The advantage of using these high-level variables in specifying the sensory conditions for exception handling is obvious and will be shown in a latter section.

4. Control Variables

The operation of some sensors requires a certain control. For example, a sensor which gives a binary value may need the setting of its threshold. A complicated sensor such as the TV camera may need controls regarding its position or orientation. To achieve the sensor control, a robot programming system must have some ways of sending the control parameters to the proper output ports. Similar to the case of sensor variables, we may program the control output in a simple way. As a dual of the previous case, we call the variables which activate the output of control parameters to the designated I/O ports as the *control variables*. The data type of the control variables must be declared before the control-declaration section. Examples of defining control variables are shown below:

> real threshold;
> interger position_x, position_y;
> *control* position_x := #21[7–4],
> position_y := #21[3–0],
> (threshold − 2.8) * 3.7 := #24[6–0];

On the left-hand side of the declaration statement, only one variable is allowed. And this variable is the control variable we want to specify. This design allows a certain scaling and offsetting for the control variables such as "threshold." The control data will be translated into integers before being sent to the output port.

To be more specific in describing the control declaration, we define the declaration format in terms of the following:

control ⟨control_expression⟩ := ⟨port_select⟩

where ⟨control_expression⟩ ::= ⟨variable⟩ | ⟨op⟩ ⟨control_expression⟩
| ⟨control_expression⟩ ⟨op⟩ ⟨constant⟩
| ⟨constant⟩ ⟨op⟩ ⟨control_expression⟩

and the ⟨port_select⟩ is as defined in the previous section.

The control variables must be writable. In an appropriate design, the interface port designated for a control variable should have both input and output capability. Thus the current setting in the output register may be read back. This allows the setting of position_x not to affect the position_y in the previous example. The system reads the port #21, changes only the first four bits, and then outputs the new setting to the same port. In case the interface hardware does not support the output read back, then the software system must keep record of the current output values of control variables. In some instances, the sensors or devices offer specific status information. The status, if input through the same port, may be read from the control variable itself. We may also define a sensor variable to access this information.

Unlike sensor variables, the control variables cannot be easily abstracted without losing this generality. However, we may extend the definition of control variables to the aggregates of previously defined control variables. For example, we may define the control variable imager_position based on the other two control variables.

$$control \text{ imager\_position} := (\text{position\_x, position\_y});$$

This offers the programmer a chance of writing a more concise program. More examples will be shown in the next section.

The usage of control variables is not necessarily limited to the sensor control. It may be generalized to control the movement of the individual robot joints. Yet this extension is only good for the low-level motion programming. The major advantage of this programming concept may still be on the control of sensors and auxiliary devices. The use of control variables also offers a concise way of programming the timing control in the robotic system. The timer can be treated as an output ports for its initial time setting and mode programming. For example, we may program the timing as follows:

$$Control \quad \text{Timer} := \#50,$$
$$\text{Tmode} := \#51;$$
$$Sensor \quad \text{time} := \#50;$$

.

Timer = 100;
Tmode = code_of_countdown;
DO action1 UNTIL time <= 0;

.

Another way to obtain the timing control is to define a sensory condition:

$$Sensor \quad \text{TIMEUP} := \text{time} <= 0;$$

and declare TIMEUP as an exception in the program. When the timer is counted down to zero, an interrupt will be triggered. The timing control is thus achieved.

5. Access of Array Sensors

There are various array sensors used in robotic systems. The vision imagers and the tactile array sensors are examples. The data of the array sensors can easily be accessed through an appropriate use of the sensor variables and control variables mentioned in the last two sections. Taking the example from a tactile array sensor,[25] we can acquire the sensor data at any point in the array as simple as the following:

control select_x := #15[4–0],
       select_y := #16[4-0],
       select := (select_x, select_y);
sensor tacti := #17;

.

.

select = (i, j);
tactile[i][j] = tacti;

.

where the lower five bits of ports #15 and #16 are connected to the address lines of the tactile sensor, while the port #17 is used to input the addressed sensor data.

Another example of array sensor data acquisition shown here is for the vision input through a 320 × 240 image digitizer (frame grabber) commercially available. The digitizer itself has four full frames of memory to store the image. This memory can be accessed, via the system bus, just like the access of an I/O port. The initiation of grabbing an image and the addressing of the image memory are achieved through other I/O ports. The data access of this imager is shown below. In this way, the imaging and data acquisition processes are easy to learn and program.

Sensor Acqflag := #61[6],   /* acquisition flag */
       Busyflag := #61[7];   /* busy flag */
control Image := #60[5-0],   /* image data register */
       Imode := #61[5-0],   /* operating mode */
       Origin := #63[1-0]–#62,   /* image origin line address */
       Vsync := #63[5-4],   /* video synchronous source */

```
Image_Hpos := #65[0]–#64,    /* image access addresses */
Image_Vpos := #67[1-0]–#66;
```

```
/* Initialization */
Imode = 0;
Origin = 0;
Vsync = 0;
Image_Hpos = 0;
Image_Vpos = 0;
```

```
/* digitize an image frame into the memory */
Imode = 4;                 /* activate a snap shot */
Imode = 0;                 /* once activated, cancel the request */
WAIT UNTIL (Busyflag = = 0);
```

```
/* data acquisition from the computer system */
/* the image data will be transfered to an array A[320][240] */
Imode = 1;                 /* Imode = post_increment_read */
for (i = 0 to 319)
    for (j = 0 to 239)
        A[i][j] = Image;
```

6. Sensory Conditions and Exception Handling

An important use of sensory data in robot programming is to initiate or terminate motions. A specific value or combination of sensor data, which is used as the condition for an action or related decisions, is called a sensory condition. In a robot system, the sensor inputs may change from time to time owing to the changing environment or the motion of the robot itself. Some sensory conditions may need to be monitored continuously. This is rather similar to some system status or error checking in a computer system. The checking must be performed continuously, and on detection of the special status or errors, an interrupt is usually evoked. In some programming languages, these interrupt-causing status or errors are called exceptions. Processing of the interrupt caused by exceptions is called exception handling. Several programming languages, such as Ada,[26] have provided users this function. In robot programming, we may regard the continuously monitored sensory conditions as exceptions and the action taken to process the interrupt, evoked by the "matched" sensory conditions, as exception handling.

In VAL,[12] a few program instructions are provided to facilitate sensor-based detection and exception handling. The IFSIG instruction checks whether the indicated external input signals exactly match the states specified. For example, given "IFSIG 2, –3,, THEN 15″, the system will check if the input signal line #2 is high and line #3 is low. If this is true, the program will branch to the step labeled "15"; otherwise the next program step will be executed. Other instructions include REACT, REACTI, and IGNORE. The REACTI initiates continuous monitoring of external signal at the specified input channel and has the format: REACTI ⟨channel⟩, [⟨program⟩] [ALWAYS]. The ⟨program⟩ is optional. On detection of a matched channel status, the current program step will be immediately aborted and the system jump to either the subroutine ⟨program⟩ or the next program step depending on whether the ⟨program⟩ is specified. If the optional argument ALWAYS is specified, the given channel will be monitored for the rest of the program execution until a corresponding IGNORE is encountered. The REACT instruction has the same format and performs almost the same function as REACTI except that the interrupt takes effect only after the completion of each program step. The IGNORE ⟨channel⟩ [ALWAYS] instruction is used to terminate either REACTI or REACT. If ALWAYS is specified, the "reaction" is permanently disabled; otherwise it is temporarily disabled until the completion of the next motion instruction. With these provisions, the continuous monitoring of sensor data, in a program step or over a segment of program, can be easily specified. However, the above design is only good for single-bit sensory data. For multiple-bit sensor data and complicated sensory conditions, there is a need to extend the language.

In our design, we propose to replace the single-bit channel by one or more than one sensor variables. With the use of UNTIL ⟨sensory_condition⟩, the continuous monitoring of the given condition during a single program step has been implied. Thus we may use the REACTI as if it is with ALWAYS in the VAL language. Here the format of the REACTI is redefined as REACTI ⟨Boolean_sensor_variable⟩, [⟨Boolean_sensor_variable⟩, ... ,] ⟨label⟩, [⟨program⟩]. We allow several Boolean sensor variables to be listed in one instruction. When all of them are true, the sensory condition is matched and an interrupt follows. Once a REACTI is issued in the program, the monitoring of the specified sensory conditions is initiated and will be continued until a corresponding IGNORE is encountered later. The positive integer ⟨label⟩ is added here for the convenience of reference in IGNORE. The IGNORE can either reference the sensor variables directly or only the ⟨label⟩ number to terminate the corresponding sensory monitoring.

The following example program illustrates how the continuous monitoring of sensory conditions and the exception handling can be specified in our language. This program directs a robot manipulator to move to an object, grasp it, and then move to another location.

— Assume that PowerFail, PROX, TOUCH, PRESS, and SLIP
— have been declared as sensor variables earlier.
—

sensor PROXIM := PROX[1] < 2.5 *or* PROX[2] < 2.5,
       TOUCHED := TOUCH[1] *or* TOUCH[2],
       TIP_HIT := TOUCH[3] *or* TOUCH[4],
       GRASPED := PRESS[1] >= 3 *and* PRESS[2] >= 3,
       SLIPPED := SLIP[1] *or* SLIP[2],
       DROPPED := *not* (TOUCH[1] *and* TOUCH[2]);

— The executable program starts here.
— The sensor variable TIMEUP and control variable Timer
— are as defined in the previous section
—

REACTI   PowerFail, 1, BackUp;

Timer = 100;
REACTI   TIMEUP, 2, Alert;
MOVE TO (x1, y1, z1) UNTIL PROXIM;
—

— The robot manipulator is commanded to move to a certain position
— or stop when proximate to some object.
— If this movement cannot finish within the prescheduled
— time interval, the exception TIMEUP will trigger an interrupt.
—

— After the above motion, we do not need to check the timing.
— The exception TIMEUP is turned off.
IGNORE 2;
—

— We open the end-effector and move down for 3 inches.
— Meanwhile, we add one more sensory condition to monitor.
REACTI   TIP_HIT, 3, Retreat;
OPEN;
MOVE DOWN (3);
—

— After moving down, we no longer need to check TIP_HIT.
IGNORE   3;
CLOSE UNTIL GRASPED;
—

— After grasped, we should make sure the object is not lost
— while moving it to another position.
—

— We check two possible conditions:
— (1) The object is NOT GRASPED tightly enough and SLIPPED,
— (2) The object is DROPPED.
—

REACTI —GRASPED, SLIPPED, 4, Repick;
REACTI   DROPPED, 5, Repick;
MOVE UP (3);
MOVE TO (x2, y2, z2)

— The exception handling subroutines BackUp, Alert, Retreat,
— and Repick may be attached anywhere following the main program.
BackUp:
{   BATTERY_ON;
      SIGNAL;
      return; }

In the above design, we do not adopt the approach of setting the scope of monitored sensory conditions by placing the exception handling routine at the end of each active range. This allows the full flexibility in specifying the scope of each monitored sensory condition and sharing the service routines.

7. Real-Time Monitoring of Sensory Conditions

The values of sensor variables may vary while the robot is moving or the environment is changing. Thus, it is important that the sensor data be collected and the involved sensory conditions be evaluated with adequately high frequency such that no significant error will occur. Taking an example, if a robot manipulator is moving at a speed of 0.5 m/sec in executing the command MOVE UP UNTIL TOUCHED, the "TOUCHED" condition must be detected and put into effect within 5 msec in order to keep the error smaller than 2.5 mm. Yet in a sophisticated operation, the sensory conditions may be quite complicated. The evaluation of them may probably involve floating-point or even exponential computations, and hence can be very time-consuming. As shown in the previous example program, several sensory conditions may need to be monitored at the same time. Without a special provision, the sensory condition monitoring may largely increase the load on the robot control computer and deteriorate the performance of the whole system. The situation is especially severe for a robotic system with multiple manipulators and numerous sensors performing concurrent operations.

The use of interrupts may be advisable for some simple I/O designs. For a sophisticated sensor-based robotic system, however, many sensory data may not be just Boolean values. A series of arithmetic calculations, comparisons, and logical operations are often required to determine if a sensory condition is true or not. Hence, the direct use of interrupts is difficult in these cases. Another approach to cope with this situation is to interrupt the system at a fixed time basis so as to invoke a polling of sensor data and an evaluation of the currently concerned sensory conditions. However, the frequency of the interrupt must be well tuned. If the frequency is too low, some sensory conditions may not be processed in time at some critical moment. If the frequency is too high, the system speed for other tasks or computations will be slowed down too much. The fixed-time interrupt arrangement is not appropriate especially when the system computational load is heavy and certain complex sensory conditions are to be monitored.

To solve the above problem, an efficient scheme for monitoring sensory conditions using an auxiliary processor has been proposed.[27] Under this scheme, all the unisensors in the system are connected to the auxiliary (sensory) processor. While the host computer controls the robot manipulator, the auxiliary processor collects the sensor data and monitors sensory conditions. This processor receives from the host computer a checking list once at the compile-time and later the information indicating which sensory conditions are to be monitored during run-time. A key point in this approach is that we check only the threshold values of the raw sensor data which make some sensory conditions true. In this way, we may avoid the time-consuming floating-point computations probably required for sensor data scaling or translation in the continuous monitoring. The sensory processor triggers an interrupt to the host computer whenever a sensory condition is found true. On the other hand, the host computer can request sensor data and the matched (interrupt-causing) condition numbers from the sensory processor. This scheme allows the sensory processor to be as simple as a microprocessor without floating-point capability. Yet it is still efficient enough to monitor numerous sensory conditions in real time. As analyzed in Ref. 27, this scheme can handle a broad range of sensory conditions possibly programmed in a robot system. The details of this scheme are described in that paper.

8. Conclusions

This chapter highlights some important ideas toward designing a programming language for a robot system with sophisticated sensing. The use of sensor and control variables makes the easy access and control of both unisensors and array sensors. The mapping between the hardware I/O ports and these variables, as designated in the program declaration, will reduce the human programming effort and enhance the portability of the robot control program to a large extent. With the combined use of these variables and the exception-handling provision, the timing control and process synchronization can also be achieved. The control variables presented here may be generalized to program the low-level robot movement. Although the efficiency of sensory condition monitoring is not the major emphasis of this language design, an efficient system configuration and scheme have been developed to secure the monitoring of multiple sensory conditions in real time.[27]

References

1. A. C. Sanderson and G. Perry, Sensor-based robotic assembly systems: Research and applications in electronic manufacturing, *Proc. IEEE* **71**(7), 856–871 (1983).
2. L. X. Nguyen and S. N. Dwivedi, A critical study of sensory feedback in industrial robotics, Proceedings of the 1983 Robotic Intelligence and Productivity Conference, pp. 76–81.
3. G. G. Dodd and L. Rossol (Eds.), *Computer Vision and Sensor-Based Robots,* Plenum Press, New York, 1979.
4. A. K. Bejczy, Sensors, controls and man–machine interface for advanced teleoperation, *Science* **208**, 1327–1335 (1980).
5. R. P. Paul, The off-line programming of robots and the need for sensors, Proceedings of COMPSAC, 1983, pp. 498–499.
6. C. C. Wagner, A need analysis for intelligent robot languages, Proceedings of the 1983 Robotic Intelligence and Productivity Conference, pp. 45–50.
7. S. N. Dwivedi and K. N. Karna, High level programming languages for controlling robots, Proceedings of the 1983 Robotic Intelligence and Productivity Conference, pp. 58–65.
8. T. Lozano-Perez, Robot programming, *Proc. IEEE* **71**(7), 821–841 (1983).
9. W. A. Gruver, B. I. Soroka, J. J. Craig, and T. L. Turner, Evaluation of commercially available robot programming languages, Proceedings of the 13th International Symposium on Industrial Robots and Robots 7, 1983, pp. 12–58 to 12–68.
10. K. Takase, R. P. Paul, and E. J. Berg, A structured approach to robot programming and teaching, *IEEE Trans. Syst. Man Cybern.* **11**(4), 274–289 (1981).

11. R. L. Paul and S. Y. Nof, Human and robot task performance, in *Computer Vision and Sensor-Based Robots*, S. G. Dodd and L. Rossol (Eds.), Plenum Press, New York, 1979, pp. 23–50.

12. Unimation Inc., *User's Guide to VAL: A Robot Programming and Control System*, 1980.

13. L. I. Lieberman and M. A. Wesley, AUTOPASS: an automatic programming system for computer controlled mechanical assembly, *IBM J. Res. Devl.*, 21(4), 321–333 (1977).

14. R. H. Taylor, P. D. Summers, and J. M. Meyer, AML: A manufacturing language, *Int. J. Robotics Res.* 1(3), 19–41, (1982).

15. R. H. Taylor and D. D. Grossman, An integrated robot system architecture, *Proc. IEEE*, 71(7), 842–856 (1983).

16. J. W. Franklin and G. J. VanderBrug, Programming vision and robotics systems with RAIL, Proceedings of SME Robots VI, 1982, pp. 392–406.

17. C. C. Geschke, A system for programming and controlling sensor-based robot manipulators, *IEEE Trans. PAMI* 5(1), 1–7 (1983).

18. M. Y. Chern, M. L. Chern, and T. G. Moher, A language extension for sensor-based robotic systems, Proceedings of the IEEE Workshop on Languages for Automation, 1983, pp. 11–16.

19. M. A. Lavin and L. I. Lieberman, AML/V: An industrial machine vision programming system, *Int. J. Robotics Res.* 1(3), 42–56, (1982).

20. H. Lechtman, T. Sutton, R. N. Nagel, N. Webber, and L. E. Plomann, Connecting the PUMA robot with MIC vision system and other sensors, Proceedings of SME Robots VI, 1982, pp. 447–466.

21. B. O. Wood and M. A. Fugelso, MCL, The manufacturing control language, Proceedings of the 13th International Symposium on Industrial Robots and Robots 7, 1983, pp. 12–84 to 12–96.

22. B. E. Shimano, C. C. Geschke, and C. H. Spalding III, VAL-II: A new robot control system for automatic manufacturing, Proceedings of the IEEE 1st International Conference on Robotics, 1984, pp. 278–291.

23. W. T. Park, The SRI robot programming system (RPS), Proceedings of the 13th International Symposium on Industrial Robots and Robots 7, 1983, pp. 12–21 to 12–41.

24. B. I. Soroka, What can't robot languages do?, Proceedings of the 13th International Symposium on Industrial Robots and Robots 7, 1983, pp. 12-1 to 12-8.

25. W. D. Hillis, A high resolution image touch sensor, *Int. J. Robotics Res.* 1(2), 33–44 (1982).

26. G. Goos and J. Hartmanis, Eds., *The Programming Language Ada Reference Manual*, Springer-Verlag, New York, 1983.

27. M. Y. Chern, An efficient scheme for monitoring sensory conditions in robot systems, Proceedings of the First International Conference on Robotics, 1984, pp. 298–304.

23

COMPUTER-AIDED DESIGN OF DIGITAL SYSTEMS: LANGUAGE, DATA STRUCTURES, AND SIMULATION

CHRISTOS A. PAPACHRISTOU AND PEI-CHING HWANG

1. Introduction

Hardware description languages, HDLs, are widely used for simulation and synthesis of digital systems. They have also been used as the input for generating microcode, as the input language for system partitioning, design verification, performance evaluation, test generation, and automatic generation of logic design equations and diagrams. In general, HDLs are useful to express the *structural* and *behavioral* information of digital systems serving as a medium of communication between hardware and software engineers.

Presently, design techniques for complex digital systems are hierarchical, i.e. top-down from the initial specification to the final circuit implementation. Although there are many layers of abstraction in this design hierarchy, it is convenient to distinguish four main levels, namely, the system, register-transfer, logic, and physical design levels. HDLs have been employed at all main levels of digital design, and some successful cases are briefly surveyed below.

One major attempt at the system level description is the PMS (processor–memory–switch) notation introduced by Bell and Newell.[1] The

CHRISTOS A. PAPACHRISTOU • Computer Engineering and Science Department, Center for Automation and Intelligent Systems Research, Case Western Reserve University, Cleveland, Ohio 44106   PEI-CHING HWANG • Phoenix Data Systems, Inc., 80 Wolf Road, Albany, New York 12205

PMS uses a number of primitives and allows them to form structures representing complex digital systems. Although this notation is powerful in its descriptive capability, it is not formal and hence it is unsuitable for CAD purposes.

Most HDLs are written at the register-transfer level. Here, the description is given by means of sequences of information transfers among the system storage elements, mainly registers. This approach was introduced by Bartee et al.[2] implementing an algorithmic control graph of a system as sequences of microoperation transfers. Schorr improved this approach by providing a formal language for register-transfers.[3] Further improvements were obtained by Chu's CDL language,[4] Duley and Dietmeyer's DDL,[5] and Hill and Peterson's AHPL.[6] Alternatively, an instruction-set interpretation algorithm may also be expressed in a language of register-transfer microoperations. This approach was taken by Barbacci for the ISPS language[7] and, similarly, by Rose et al. for the ISP'.[8] An interesting variation is used by Zimmermann[9] in the Mimola system.

There are many languages at the logic design level for gate modeling and simulation. Good example languages among numerous such systems are the following: FST,[10] APDL,[11] CASD,[12] LALSD,[13] LOGAL,[14] and more recently FTL[15] and PHPL.[16]

The main problems at the physical design level are the automatic partitioning, placement, and routing. Although these problems were first defined in the PC board framework, they now find direct application to the VLSI chip environment. Essentially, a description language at this level can express logic design equations in a form that can be efficiently compiled into layout diagram.[17] Most notable examples of such layout languages are the Mead and Conway CIF,[18] LAVA,[19] LGEN,[20] CHISEL,[21] and SLIM.[22]

Depending on the design objectives, there are several other levels that may be interleaved into the four main levels of the preceding hierarchy. Quite important among them is the microprogram design level which is related to the register-transfer level, in some sense. The use of description languages to generate microcode for microprogrammed architectures is one of recent research developments. Compilers for such languages have been constructed by Sheraga and Geiser,[23] and by Ma and Lewis.[24] However, more work in this area is needed.

The main purposes of this paper is to discuss a high-level description language with the following important objectives:

1. to define the structural and behavioral informaton of a digital system;
2. to generate efficient microcode for the system;
3. to be used as a functional simulation language.

The language is easy to use and can facilitate the development of design tools. It can provide the description of the system instruction-set at the register-transfer level. This language is the essential ingredient of a design automation system, MDSS, developed at the University of Cincinnati.[25]

Other important objectives of this work are to develop techniques and tools required for efficient microcode generation of horizontally microprogrammed processors and, further, the simulation and evaluation of the processor system performance. For the simulation of the system, the stimuli will be the assembled program stored in the system memory together with the interactive commands issued by the user to control the simulation process and display the results.[26]

The overall problem associated with this design automation system may be divided into three major steps, directly corresponding to the above three objectives. First, to devise a method for expressing and collecting the structural and behavioral information of a target machine architecture. The hardware description language that fulfills this task will be presented in Section 2. Second, to establish a software translation process that will convert input descriptions based on the language constructs into target microcode. Finally, one must introduce a methodology for extracting the information useful for target machine simulation. The microcode generation and system simulation techniques will be discussed in Sections 3 and 4, respectively.

2. The Hardware Description Language MDSL

The hardware language MDSL (microcode development and simulation language) is suitable for the descripton of microprogrammed microprocessors at the register-transfer level. It contains the facilities to describe the structure and behavior information, simultaneously, for the complete specification of a microprogrammed digital system.[27]

The structure information is defined in the *structure section*, which is divided into a number of subsections. An important subsection type is the *element*, distinguished into the storage, the link, the input/output, and the functional element types. The link element definitions contain appropriate information regarding their control point locations. Similarly, the functional element definitions contain the information concerning the register-transfers and their control requirements. There are three additional subsection types to be defined in the structure section: the sequencing scheme the special function routine, and the control format.

The behavioral information is defined in the *behavior section*, which contains the instruction-set definition of the register-transfer microoperations.

The control requirements are defined explicitly in the functional element definitions, and implicitly in the link element and instruction-set microoperations definition. The control word format declared in the format subsection can be used to map the control requirements of the instruction-set microoperations into microcode. The concurrency of the microoperations can be checked to determine the control word combinations.

In the following, we will discuss the structural and behavioral description methodology in detail and give some examples to demonstrate their usage.

2.1. Structure Description Section

The structural description contains the following subsections:

1. Declaration of storage elements for data storage and retrieval, including the declaration of input/output terminals of a processor;
2. Declaration of functional elements, such as ALU, multiplexers, selectors, etc.;
3. Declaration of linking elements between the storage elements, i.e., the data flow and data paths;
4. Declaration of the control part of a predefined sequencing scheme;
5. Definition of the control word format for the control memory;
6. Definition of special gate circuit functions or element functions without explicit control conditions.

The storage elements include specification of registers, subregisters, memory, and bus. Specification of the input and output terminals is also provided. The length of each element also needs to be defined. An example follows:

$$\begin{aligned}
&\text{Register: } A\langle 7{:}0\rangle, \text{ FLAG}\langle 3{:}0\rangle, \text{ WZ}\langle 15{:}0\rangle; \\
&\text{subreg: } W\langle 7{:}0\rangle = WZ\langle 15{:}8\rangle, \\
&\qquad\qquad Z\langle 7{:}0\rangle = WZ\langle 7{:}0\rangle, \\
&\qquad\qquad CA = \text{FLAG}\langle 3\rangle, \\
&\qquad\qquad ZO = \text{FLAG}\langle 2\rangle; \\
&\text{memory: } M[1023{:}0]\langle 7{:}0\rangle, CM[525{:}0]\langle 40{:}0\rangle \\
&\text{bus: } \quad\; DBUS\langle 7{:}0\rangle, ABUS\langle 15{:}0\rangle; \\
&\text{input: } \quad INT; \\
&\text{output: } INTA;
\end{aligned}$$

In the example above, each identifier with no bit length definition requires a single bit.

The storage elements declaration defines all the elements used in the subsections to follow as the operand of register-transfer operations. All elements declared in the storage subsection are global variables that need to be allocated machine storage in order to simulate the described system. In parsing the transfer operations of the subsequent subsections, the compiler will check this section to see if the storage elements have been defined or not. If not, the compiler will indicate a definition error.

The functional element subsection is to define the data transformation elements between the storage elements. It has to specify the source and destination storage elements, the transform operations, and their control conditions. The definition of timing requirements is optional. If timing is not defined, then single phase timing conditions are assumed. An example follows:

element:
ALU [phase 1] : ⟨4 bit⟩
  b'1000': AO ← A+T,
  b'1001': AO ← A−T,
  b'1010': AO ← A+T+CA,
  b'1011': AO ← A−T−CA,
  b'1100': AO ← A & T,
  b'1101': AO ← A | T,
  b'1110': AO ← ≪A,
  b'1111': AO ← ≫A;
MUX [phase 2] : ⟨3 bit⟩
  b'001': DBUS ← B,
  b'010': DBUS ← C,
  b'011': DBUS ← H,
  b'100': DBUS ← L,
  b'101': DBUS ← SPH,
  b'110': DBUS ← SPL;

In the example above, ALU and MUX are functional elements, having 4 and 3 control lines, respectively, as shown in the description and in Figure 1. The statement b'1000': AO ← A + T, means that under the control condition binary 1000 for ALU, the data transformation and register transfer will take place among the declared registers. The ALU is connected to phase 1 clock whereas the MUX is connected to phase 2. The allowed operators for data transformations will be discussed later (see Table 1).

In case of having more than one functional element between the storage elements, then one can use two different ways for their definition. The first way is to insert a pseudostorage element between the functional ele-

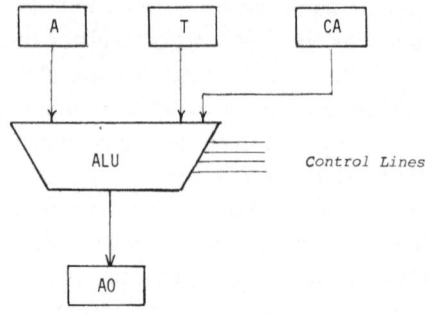

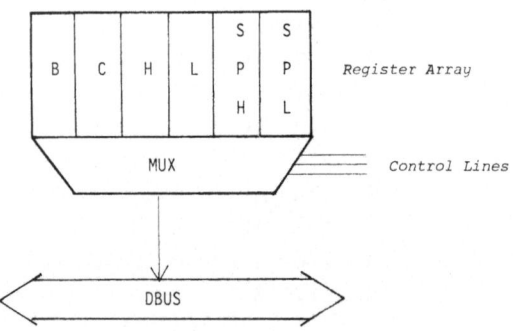

FIGURE 1. Functional element declaration example.

ments. The second is to combine the two functional elements together and describe the combination as one element (see Figure 2). The combination approach is preferred to the insertion approach since it can reduce the total number of defined microoperations.

The linking element declaration is used to defined the data flow or data paths of the system, for example,

link:
  LINK1 [phase 2] : A ← DBUS,
  LINK2: MDR ← DBUS,
  LINK3: DBUS ← MDR;

The control conditions in the link declaration are implicit. When the data path is open, it is '0' condition; it is '1' for closed path situation. The link-

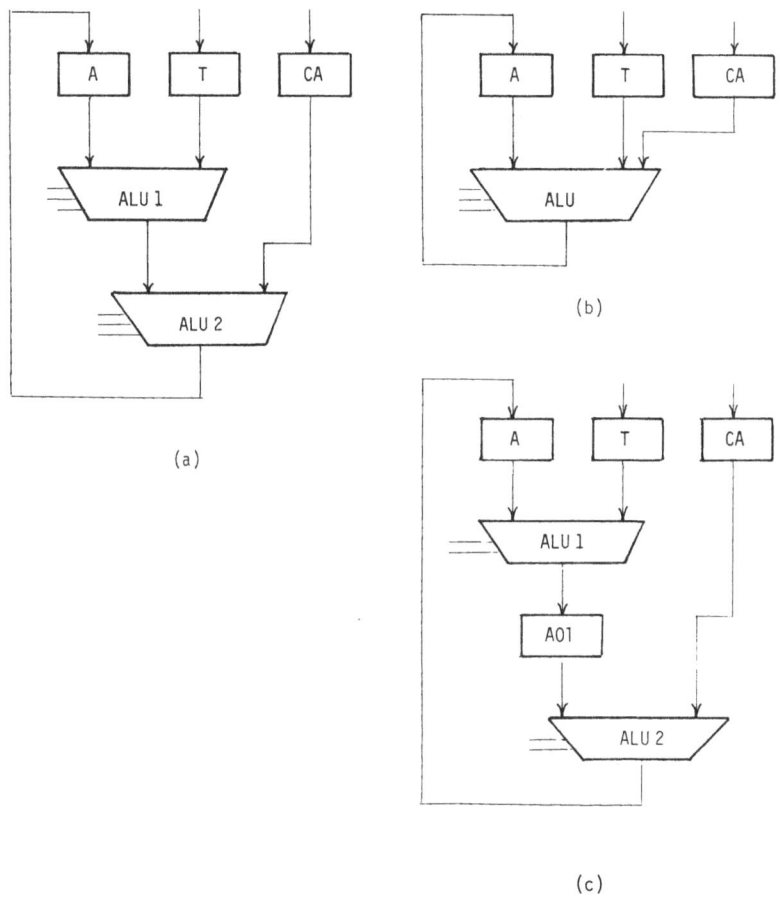

FIGURE 2. Two ways of defining more than one functional element between storage elements: (a) the original circuit; (b) the combinational approach; (c) the insertion approach.

ing element can also be connected to the control clock, Figure 3, as LINK1 is in closed path situation only at phase 2. The bidrectional link LINK2-LINK3 has no timing definition, as it is assumed to be at phase 1.

In register-transfer languages, there is a need to provide some sequencing information, such as branch test conditions, branch addresses, or next addresses. In this language, the user only needs to provide information regarding the branch test conditions. The microcode generation system will generate the other sequencing information (next-address) automatically based on the predefined sequencing controller architecture, to be discussed shortly. The microcode generation system will compute the next address information from the behavioral description section and

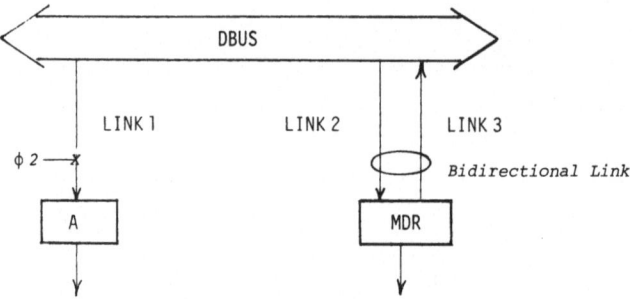

FIGURE 3. Link element example.

insert it into the control memory. The specification of the branch test is similar to the element definition. For example, if the status register contents are used as branch test conditions, then a four-bit, status register, branch test specification could be

> sequencing:
> BRTEST: ⟨8 bit⟩
> /* x' ' means hexidecimal number */
> x'01': if P = 0,
> x'02': if P = 1,
> x'04': if CA = 0,
> x'08': if CA = 1,
> x'10': if S = 0,
> x'20': if S = 1,
> x'40': if ZO = 0,
> x'80': if ZO = 1;

The status flags P, CA, S, and ZO should be declared in the storage subsection. Since there are eight conditions to be tested, the branch test field is chosen to be eight bits wide. Branching is effected when there is matching between the branch test condition in control memory with the corresponding status flag.

The predefined sequencing scheme is the so-called "variable format sequencing".[28] The reasons for choosing this scheme are its flexibility and its effectiveness for control memory compaction.

The control word format definition is used for mapping the control requirements into the control memory. It will be used to form the resultant microcode. The information regarding the element names and their corresponding bit locations in the control memory needs to be provided. The ADDRESS is the default name that is reserved for specifying the next sequencing address. The dual format controller using format-0 for microoperation control and format-1 for sequencing control is illustrated in the following example:

> format: CM⟨40:20⟩ : =
> format0:
> ⟨40⟩: LINK1,
> ⟨39⟩: LINK2,
> ⟨38:36⟩: DMUX2,
> ⟨35:33⟩: MUX2,
> ⟨32:29⟩: DMUX1,
> ⟨28:25⟩: MUX1,
> ⟨24:21⟩: ALU,
> ⟨20⟩: 0;
> format1:
> ⟨40:33⟩: BRTEST,
> ⟨32:21⟩: ADDRESS,
> ⟨20⟩:1;

In the above example, format0 defines the memory mapping for the element controls whereas format1 refers to the sequencing.

There are some gate connection circuits betwen the storage and/or functional elements. These complex circuits may not be controlled by the control memory. The language provides the facilities to define these networks. For example, the gate circuits between the status register and the ALU are used for determining and gating the status register contents after the ALU operations have been performed (see Fig. 4). The circuit can be defined using the special function technique in the following way (the operators are defined in Table 1):

```
ALUTOF( )
 P ← AO[7] ^ AO[6] ^ AO[5] ^ AO[4] ^
    AO[3] ^ AO[2] ^ AO[1] ^ AO[0],
       /* ^ is EXOR operator */
 S ← AO[7],
 CA ← AO[8],
 ZO ← ! AO⟨7⟩ & ! AO⟨6⟩,
 ZO ← ! AO⟨5⟩ & !ZO,
 ZO ← ! AO⟨4⟩ & !ZO,
 ZO ← ! AO⟨3⟩ & !ZO,
 ZO ← ! AO⟨2⟩ & !ZO,
 ZO ← ! AO⟨1⟩ & !ZO,
 ZO ← ! AO⟨0⟩ & !ZO;
   /* the nand operator can be derived from
      De Morgans Law */
```

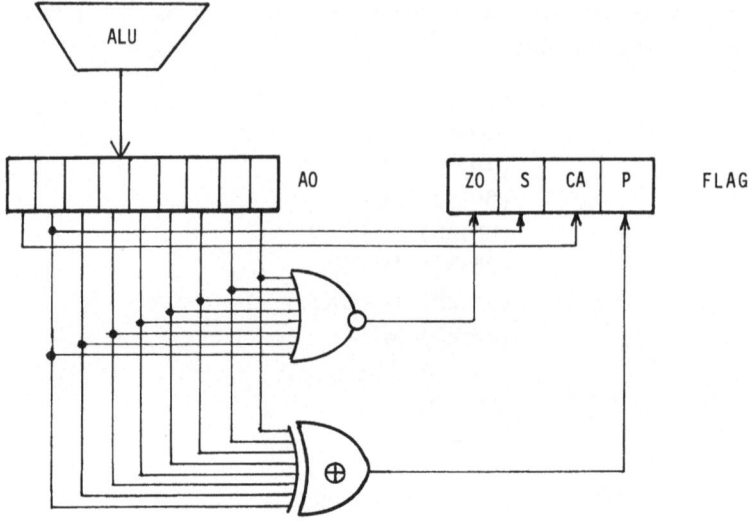

FIGURE 4. Gate circuit connection example.

The ZO status bit is derived in multiple statements using the "not" and the "and" operator to achieve the "nor" operation.

If the special function is activated by the control memory, the function name can be used to substitute the transfer operation in the element definition. Of course, it can be used to define the machine behavior whether it is controlled by the control memory or not. No matter how it is controlled, the simulator will still need to know the actual data transformation defined in the function.

The functional element DECODE can be defined using a special function. It is possible to specify the label constant which is defined as address in the function definition part, as illustrated below:

```
DECODE( )
  if IR = 0 than → NOP fi, /* → means goto */
  if IR = 1 THEN → MOVAB fi,
  . . . . .
  if IR = 255 then → IN fi,
  → error;
```

2.2. Behavioral Description Section

The behavioral description is used to define the machine instruction set operations at the register-transfer level. The special functions in the previous section can also be used to define such operations.

The language also allows the macro definition to specify frequently used register-transfer block, for example the memory location computation block, in order to facilitate the behavioral description. The macro definition uses the following format:

macro macro-name:
microoperations;

The microoperations within the macro can be simple transfers, compound transfers, multidestination transfer operations, calling another macro that is previously defined, or calling functions. To call the macro definition we have

call macro-name;

The macro definitions must be collected at the beginning of the instruction set description section.

The language operators are listed in Table 1. These operators have the same meaning as in the element and function definition of the structural description section.

TABLE 1
List of Language Operators

Operator category	Operators	Meaning
Assignment	$\leftarrow$	Data transfer
Relational	$>$	Greater than
	$<$	Less than
	$=$	Equal to
	$>=$	Greater equal
	$<=$	Less equal
	$!=$	Not equal
Arithematic	$+$	Add
	$-$	Subtract
	$++$	Increment
	$--$	Decrement
Logical	$\&$	And
	$\vert$	Or
	$!$	Not
	$\char"005E$	Exor
	$\ll$	Shift left
	$\gg$	Shift to right
Conditional	If then	
	If then else	
Branch	$\rightarrow$	Go to

The compound transfer statement, not allowed in the element definition, is allowed here to define the behavior. The compound statement

$$MAR \leftarrow ABUS \leftarrow PC$$

is translated into the following statements:

$$ABUS \leftarrow PC,$$
$$MAR \leftarrow ABUS;$$

If there is more than one destination in the transfer, i.e., a multidestination transfer, the "&&" operator can be used to define it. The multidestination transfer statement

$$MAR \ \&\& \ PC \leftarrow ABUS$$

is translated as follows:

$$MAR \leftarrow ABUS,$$
$$PC \leftarrow ABUS;$$

The comma (,) is used to separate the microoperation statements; the semicolon (;) separates sequences of microoperations comprising microoperation blocks.

In addition to the microoperation transfer statements, there are two kinds of sequencing statements. First, the "if-then-(else)-fi" statement for defining the conditional branch sequencing statements. Second, the "→ label" statement to define the unconditonal branch sequencing. The conditional branching statements can be nested to construct a more complicated sequencing structure.

Shown in Figure 5 is an example of some instruction description of a typical microprocessor. In Figure 5, labels are used as microoperation block names to be referred to by the sequencing controller as the corresponding block symbolic addresses or simply as the branch targets of the goto statements.

3. Microcode Generation Methodology

The MDSL (microcode development and simulation language) discussed in Section 2 is an efficient tool to develop microcode and simulate the hardware operations of microprogrammed processors. A software translator for MDSL has been developed for this purpose. An important

```
instructset:
macro NEXTLOC:
ABUS <- ++PC,
MAR && PC <- ABUS,
Z <- DBUS <- MDR <- M[MAR],
ABUS <- ++PC,
MAR && PC <- ABUS,
W <- DBUS <- MDR <- M[MAR],
ABUS <- WZ,
MAR <- ABUS;
macro NEXTMEM:
ABUS <- ++PC,
HL && PC <- ABUS,
MAR <- ABUS <- HL,
T <- DBUS <- MDR <- M[MAR];
FETCH: MAR <- ABUS <- PC,
       IR <- DBUS <- MDR <- M[MAR],
       PC <- ABUS <- ++PC,
       DECODE () ;
NOP:    -> FETCH;
ADDM: call NEXTMEM,
       A<- DBUS <- AO <- A+T,
       ALUTOF (),
       -> FETCH;
JC:   call NEXTLOC,
       if CA=1 then
       PC <- ABUS <- MAR,
       -> FETCH,
      else -> FETCH fi ;
MOVAB: A <- DBUS <- B,
       -> FETCH;
STA:   call NEXTLOC,
       M[MAR] <- MDR <- DBUS <- A,
       -> FETCH;
LDA:   call NEXTLOC,
       A <- DBUS <- MDR <- M[MAR],
       -> FETCH;
error: -> error;
/* if error happens stuck here */
```

FIGURE 5. A typical microprocessor behavioral definition.

aspect of this translator is that, in lieu of object code, it generates data structures by parsing MDSL input descriptions. In this section, we first discuss the data structures that are employed to hold the information generated by the MDSL compiler and then our approach to optimize the microcode generation.

In the process of generating efficient microcode, the element description can be translated into a microcode template table. Each template contains information concerning the source and destination storage devices, and the control requirements. This template table can be used to map the behavioral description statements into microcode, provided that the behavioral information can be translated into data structures resembling the template.

3.1. Parsing the Behavioral Information into Link List Structures

Before the translation of the behavioral description, the compound and multidestination transfer statements and the call macroname statements of the microoperation description need to be preprocessed to a simpler form.

The compound transfer statement will be translated into a block of simple transfer statements according to the rules given in the previous sections.

The calling of macro definitions needs special care because there are no corresponding templates provided in the template table for the call macroname statements. There are two applicable solutions. The first is a software, the other is a hardware solution. The software solution is to generate the statements that the macro represents each time the call macro statement is encountered. In this way, the generated microprogram will be longer, but no additional hardware is needed in the microcontroller. The hardware solution is to add the additional hardware stack in the microcontroller to save the return addresses. This will save some control memory space, but will make the control circuit somewhat more complicated. In order to facilitate the description and maintain the predefined controller scheme, the software solution is employed in the system design.

After these translations are made, the instruction set description can be transformed into a link list node structure of the form shown in Figure 6a to represent microoperations, or Fig. 6b to represent the sequencing statement.

For the microop node in Fig. 6a, the addr field is computed by the compiler and inserted into the list. The information in the destination (dest) and source fields is somewhat different. For register-transfer operations, the fields are used for destination and source(s) storage, respectively. The arithmetic and logic operators are allowed in the source field to represent the data transformation operations. For the special function call, dest holds the function name, and source is null (no information). The elem field contains information regarding the linking element in executing the node's microoperation. This information can be inserted by the translator by searching the template table for that particular microoperation. The status field signifies whether the node's microop is controlled by the control memory or not, since the special functions are not required to be controlled by the control memory. The link field is a pointer to the next node.

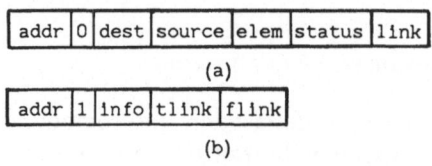

(a)

(b)

FIGURE 6. Link list to hold the microops and sequencing statement.

The sequencing controls are defined in the conditional and goto statements. For the "→ label" statement, the info and flink fields are assigned "→" whereas the tlink contains the address of "label".

For the conditional statement

$$\text{if id} = \text{cond then op1}$$
$$\text{else op2}$$
$$\text{fi}$$

the info field will contain the coding of "id = cond," the tlink a pointer to op1 node, and the flink a pointer to op2 node. If no "else" condition is specified in the statement, the flink will be null.

The second field of the node distinguishes the microoperation from the sequencing node. Thus, bit value 0 corresponds to format0 as defined previously, i.e., the mapping into the microop template. Bit value 1 corresponds to format1, i.e., the mapping into the sequencing template.

3.2. Compacting the Microoperations

It is possible that the sequential microoperations have disjoint data path device sets. These operations can be executed in their original order or in reverse order. And since they are hardware independent, they can be executed at the same time with equivalent results. The collection of parallel microoperations into microinstructions can reduce the length of the control memory and accelerate the execution.

The algorithm for the detection of parallel microoperations is based on the following principle.

For the microops MO, $(i = 1, 2, \ldots, n)$ in a straight line microprogram, MO_i and MO_j, with MO_i preceding MO_j, are parallel if

$$I_i \cap O_j = O_i \cap O_j = C_i \cap C_j = \phi$$

where I, O, C represent the source, destination, and link fields, respectively.

If MO_i and MO_j are parallel, then they are combined by taking the union of their respective fields, as follows:

$$UO_i = O_i \cup O_j$$
$$UI_i = I_i \cup I_j$$
$$UC_i = C_i \cup C_j$$

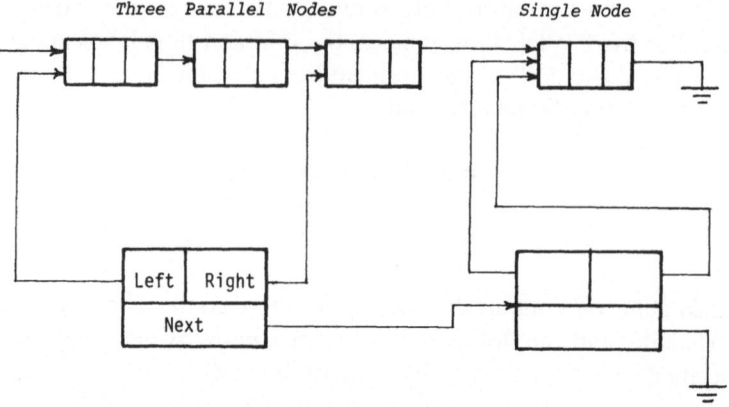

FIGURE 7. Parallelism list structure.

and then proceeding to the next MO to check further. The new MO_j will be added to the parallel set if

$$UI_i \cap O_j = UO_i \cap O_j = UC_i \cap C_j = \phi$$

The microoperations link list can be pointed to by a generalized link list representing the parallel microoperations block. This link list is called "parallelism list." Each node of the parallelism list has three field pointers to the leftmost, the rightmost microop node, and the next parallelism node, respectively. The structure shown in Figure 7 is an example to illustrate the use of parallelism list to point to three parallel nodes and the single node that cannot be executed in parallel with the other microoperations.

Using the structure of Fig. 7, constructed for every microoperation in the instruction set, the microcode generator will be able to produce the microcode by applying the template table to each node. For the parallel nodes, the resultant microcode will be composed of the union of each node's microcode.

4. System Simulation Methodology

The simulation of descriptive systems, as mentioned in the previous section, requires stimuli and interactive simulation commands. The simulation approach in this paper is executed at the register-transfer level. The stimuli are the machine codings assembled from a source program.

At the initilization stage, the declared storage elements are allocated with the host machine memory and are initialized to the values provided by an initialization data file.

There are two types of declared storage in the description. The register, subregister, and memory declarations are "static" since they hold the information as long as they are not changed. The memory allocated to bus and input/output declarations are "dynamic" because the information there will endure only temporarily. Every time the register (or subregister) information has been changed, the corresponding subregister (or register) needs to be adjusted accordingly.

Listed in Table 2 are the simulation commands that the simulation system contains. The "reset" command activates the initialization procedure and loads the initial conditions of the declared storage into the allocated memory. The "output to file" command will load the simulation results into the file specified by the user. The "output" command without file name will display the results at the terminal. The "execute–s" command executes the simulation step-by-step and displays the results after each step, while the "execute–f" command executes the simulation until the specified simulation time is up. The simulation ending time can be specified by the "simulate until" command. To start the actual simulation process, the "go" command is needed. For the step-by-step mode, each step of the simulation will end after one machine instruction is executed. To resume the next step, just type any character except "q." The "q" command is the exit command from the simulation process.

The simulation program uses the information associated with the declared storage subsection, the special function definition subsection, and the instruction-set definition section in the system description. After the compilation, the tables constructed for the storage declaration section, the list structures constructed for the special function, and the instruction-set sections will be used for the simulation purpose. Each node of the list structure will be seen as a subcommand for the simulation and will activate the corresponding data transfer or data transformation, and change the corresponding allocated memory location contents.

TABLE 2
Simulation Commands

Command	Usage
Reset	Initialize the initial conditions of processor
Output to file	Save the simulated results in file
Output	Display the simulated results
Execute −f	Execute the simulation process in whole
Execute −s	Execute the simulation process step-by-step
Simulate until	Specify the simulation ending time
Go	Starting the actual simulation process
q	Exit from the simulation process

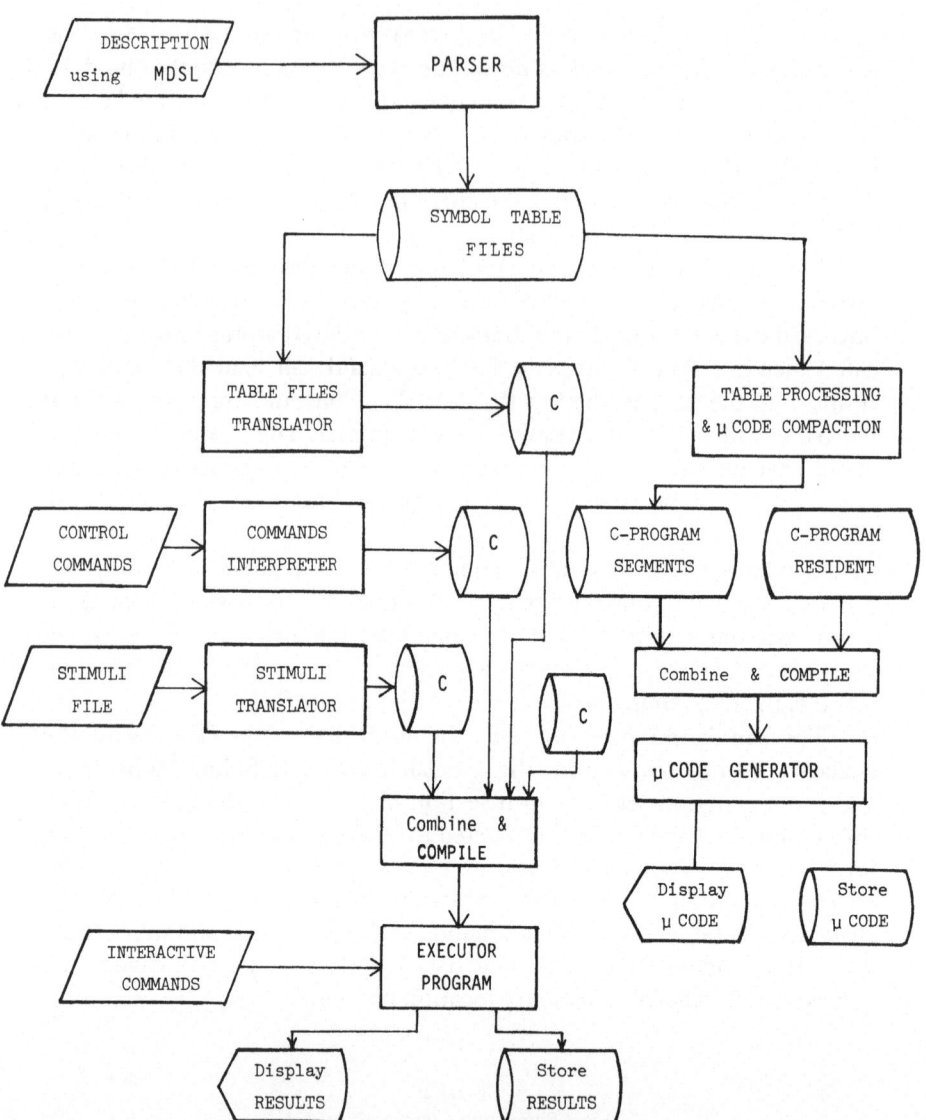

FIGURE 8. Design automation system diagram.

The compiler mode simulation is employed in designing the simulation. The simulation commands are translated into C language statements and subroutine functions. The declaration, the special function, and the behavioral sections of the MDSL description are also translated into equivalent initialization declarations, subroutines, and statements, respectively.

Shown in Figure 8 is the design automation system block diagram. The system description has to be translated into system tables and list structures using the compiler writing tools, i.e., the LEX and YACC utility programs, of the UNIX† operating system. Appropriate lexical and syntax rules and action rules provided to the utility programs will generate the needed tables and list structures. This information is saved on file for further usage in simulation and microcode design programs. More details on the simulation process including the syntax rules are in Ref. 25.

5. Conclusion

In this paper we have described a computer-aided design tool for microprogrammed digital systems such as LSI/VLSI and microprocessors. The full system provides the user with an efficient design automation system to operate in an interactive environment. The whole package includes (1) a hardware description language to specify structural and behavioral aspects of the system; (2) a processor simulator at the instruction-set level; (3) a design program that operates at the register-transfer level to perform microprogram controller design. All these are essential ingredients of a hierarchical design automation system. Lower-level designs should include standard gate-level simulators and other layout tools for VLSI designs.

The advantage of this system is that it uses higher-level description language constructs and it relieves the user from the tedious and error-prone tasks of lower-level design, including the design of the microprogram control scheme, with the aid of a host computer. The goal of this paper is to provide a top-down functional simulator and design automation environment for microprogrammed digital system design.

Acknowledgment

This work was partially supported by the Army Research Office under contract No. DAAG 29-82-K-0106.

References

1. C. G. BELL and A. NEWELL, The PMS and ISP descriptive system for computer structure, *Proc. AFIPS SJCC*, 1970.
2. C. T. BARTEE, L. I. LEBOW, and S. I. REED, *Theory and Design of Digital Machines*, McGraw Hill, New York, 1962.
3. H. SCHORR, Computer aided digital system design and analysis using a register transfer language, *IEEE Trans. Electron Comput.* **EC-13**(12) 730–737 (1964).

†UNIX is a trade mark of AT&T Bell.

4. Y. CHU, An ALGOL-like computer design language, *Commun. ACM* **8**(10) 607–615 (1965).
5. J. R. DULEY and D. L. DIETMEYER, A digital system design language, *IEEE Trans. Comput.* **C-17**(9), 850–861 (1968).
6. F. J. HILL and G. R. PETERSON, *Digital System: Hardware Organization and Design*, 2nd ed., Wiley, New York, 1978.
7. M. R. BARBACCI Instruction set processor specifications for simulation, evaluation, and synthesis, Proc. 16th Design Automation Conference, June, 1979.
8. C. W. ROSE, G. M. ORDY, and P. J. DRONGOWSKI, N.mPc: A study in university–industry technology transfer, *IEEE Design Test* Feburary 44–56 (1984).
9. G. ZIMMERMANN, MDS—The MIMOLA design method, *J. Digital Syst.* **4**(3), 221–239 (1980).
10. E. A. FRANKE, Automated Functional Design of Digital System, Ph.D. dissertation, Case Western Reserve University, November, 1967.
11. J. A. DARRINGER, The Description, Simulation, and Automatic Implementation of Digital Computer Processor, Ph.D. dissertation, Carnegie-Mellon University, May, 1969.
12. E. D. CROCKET *et al.*, Computer-aided system design, Proc. AFIPS FJCC, 1970.
13. M. B. BARAY and S. Y. H. SU, A digital system modeling and design language, Proc. 8th Design Automation Workshop, 1971.
14. J. LUND, LOGAL—Logic algorithmic language, Univac Technical Memo A00317, March 5, 1973.
15. S. HIRSCHHORN *et al.*, Functional simulation in FAN-SIM3—Algorithms, data structures, and results, Proc. 18th Design Automation Conference, June, 1981.
16. H. ANLAUFF, P. FUNK, and P. MENIN, PHPL—A language for logic design and simulation, Euromicro Symposium on Microprocessing and Microprogramming, 1977.
17. D. J. ULLMAN, *Computational Aspects of VLSI*, Computer Science Press, Rockville, Maryland, 1984.
18. C. MEAD and L. CONWAY, *Introduction to VLSI Systems*, Addison Wesley, Reading, Massachusetts, 1980.
19. R. MATTHEWS, J. NEWKIRK, and P. EICHENBERGER, A target language for silicon compiler, IEEE Compcon 1982, pp. 349–353.
20. S. C. JOHNSON, Code generation for silicon, Proc. 10th ACM Symposium on Principles of Programming Languages, 1983.
21. K. KARPLUS, CHISEL, An extension to the programming language C for VLSI layouts, Ph.D. dissertation, Standford University, 1982.
22. J. HENNESSY, SLIM: A simulation and implementation language for VLSI microcode, *Lambda*, Second Quarter, 20–28 (1981).
23. R. SHERAGA and J. L. GIESER, Automatic microcode generation for horizontally microprogrammed processor, Proceedings of the 14th Workshop on Microprogramming, October, 1981.
24. P. R. MA and T. G. LEWIS, On the design of a microcode compiler for a machine independent high-level language, *IEEE Trans. Software Eng.* **SE-7**(3), 261–274 (1981).
25. P. C. HWANG, MDSS: A design automation system including a hardware language, a microcode generator and a functional simulator, Ph.D. dissertation, University of Cincinnati, 1984.
26. C. A. PAPACHRISTOU and P. C. HWANG, A functional simulator for digital systems, Proceedings of the 16th Asilomar Conference on Circuits, Systems and Computers, November, 1982.
27. C. A. PAPACHRISTOU and P. C. HWANG A language for digital system specification and design, IEEE Workshop on Language for Automation, November, 1983, pp. 229–237.
28. M. ANDREWS, A firmware engineering development tool, Chapter 6 of *Principle of Firmware Engineering in Microprogram Control*, Computer Science Press, Rockville, Maryland 1980, pp. 210–274.

MANAGEMENT AND AUTOMATION

IMPLICATIONS OF COMPUTER-INTEGRATED MANUFACTURING FOR CORPORATE STRATEGY, CAPITAL BUDGETING, AND INFORMATION MANAGEMENT

MARTIN L. BARIFF AND JOEL D. GOLDHAR

1. Innovations in Manufacturing Strategies

A central theme of this chapter is that a combination of computer-integrated manufacturing (CIM) of computer-aided design (CAD), computer-aided manufacturing (CAM), and flexible automation has implications for capital budgeting, corporate strategy, and information management. The CIM factory has capabilities not found in traditional facilities:

- zero learning curve;
- zero setup time;
- zero inventory.

This leads to an operation with a potential economic order quantity (EOQ) of one-at-a-time, and one-of-a-kind, on-demand production. It is capable of low-volume production of a wide variety of designs on flexible manufacturing equipment at cost per unit equal to the cost of producing the same design in large quantities on special purpose facilities. Conversely,

MARTIN L. BARIFF and JOEL D. GOLDHAR • Stuart School of Business Administration, Illinois Institute of Technology, Chicago, Illinois 60616

the flexible, computer-controlled factory can produce a high variety of low-volume designs at a far lower cost than these same volume/variety mix could be manufactured on traditional manufacturing technology. Traditional learning curve benefits are derived from the production of repetitive, large batch size, single products using labor-intensive technology. In contrast, CAM, having little or no labor component, consistently applies the best manufacturing techniques through programmed machinery to produce output with a zero learning curve.[6,7] The initial experience with CAD/CAM methodology, however, will create a one-time learning curve effect. CIM also significantly reduces or eliminates job setup time and production schedule uncertainty which permits adoption of a just-in-time inventory policy. This leads to the idea of an opportunity cost to the firm for not investing in the new technology.

Two examples of new systems emphasize this point:

> Hughes Aircraft recently completed an FMS in El Segundo, California where nine machining centers plus a coordinate measuring machine, a tow-line conveyor system and supervisory computer control do the work of 25 stand-alone machining centers.[2] Compared to traditional technology, the FMS was built for 75% of the investment cost, operates at 13% of the labor cost, and will generate the required production at 10% of the machining-time cost. The system was started with five different part designs in the 4,000–7,000 pieces per year volume range with plans to add ten more designs in the near future.

> Vought Corporation in Dallas is currently installing a $10 million flexible machining center that was expected to begin work in July, 1984[3] The FMS will produce 600 different designs, one unit at a time in random sequence. The cell is expected to save $25 million in machining costs for these parts by performing in 70,000 hours work that would take 200,000 hours by conventional methods.

These examples and many other installations currently being reported in the manufacturing literature illustrate the essential difference between flexible and conventional systems. They exhibit "economy of scope" in place of traditional "economy of scale" that our conventional production systems thinking is based upon.[6] The combination of flexible tools, computer-aided manufacturing techniques for managing the information flows in the factory, and computer integration of both the knowledge work and the physical conversion of materials that we generally refer to as the "factory of the future" challenges all of our traditional thinking about manufacturing systems, marketing options, capital investment decisions, and corporate strategy. The new economics of production offers new possibilities for competitive strategies not possible in traditional factories.[7] It also requires new strategies and a new type of capital budgeting algorithm if a firm is to realize the full benefits of its investment. This requirement for a new strategy is reinforced by the current trends in the world marketplace.

1. Products are being designed for "manufacturability" as well as product function.
2. Consistent high quality is being recognized as productivity and a source of profit improvement.
3. Product life cycles are becoming shorter and new product designs more frequent.
4. More sophisticated customers are demanding reliability, high quality, and some degree of uniqueness in the products they purchase.
5. Many products are becoming more complex and technologically sophisticated with each succeeding generation—requiring more sophisticated and complex manufacturing techniques and systems.

Underlying these observable trends is a set of fundamental changes in the underlying science base of manufacturing, as well as the increasing technological complexity of many products—leading to a closer coupling between product design and process design than has been traditionally true in the "piece parts" manufacturing sector of our economy. Effective management of decision support aids and integrated information systems is critical to achieving improved product design-manufacturing process performance.

Manufacturing is now becoming a scientific activity. This trend starts with an increased level of understanding at the molecular level of the behavior of materials in the solid state under varying process conditions. If we ask why the chemical industry has been so productive and so willing to invest hundreds of millions of dollars in relatively untried new technology process plants over the years, we could argue that it is because within chemistry, the behavior of matter in the fluid state at the molecular level is understood well enough to design and optimize such a plant on paper, with a calculator, and in a computer. By the time major investments are made in facilities and process equipment, the level of uncertainty has been reduced considerably. We do not yet have the equivalent of a Reynold's number for mechanical factors. We rarely know whether a big flexible manufacturing system is going to work until we build it, debug it, and make it work over two, three, or four years. Slowly, however, that is changing. We are learning more about the fundamental behavior of materials in the solid state from research in materials science, and we certainly know more about materials science than we are using now in the design of manufacturing process technology. And this knowledge bank will continue to grow.

The sciences of metrology and measurement, development of sensors, improved control theory, computers, electronics, communication, information science, and artificial intelligence all form an increasing science base from which continuing new generations of technological capability and manufacturing technology are going to develop. They are not there yet. Things that we now consider the state of the art are very advanced and new for most factories, but they are in fact elemental compared to what we can expect to see in the early and middle 1990s. So, the good news is that the science base is being replenished and the technology of the far future is going to be very exciting. The bad news is that the technology of the near future that is available today calls for levels of expertise that very few of us have.

2. Design and Operating Characteristics of the Factory-of-the-Future

The applications of networked computers and software and the resulting economy of scope requires that we understand the factory with a very different philosophy.[14]

Consider the CIM factory as a combination of the following three ideas:

1. a continuous flow of product—like a chemical plant—but with economy of scope allowing the production of a variety of similar products in random order in addition to the economy of scale derived from overall volume of operations;
2. an integrated system with programmable machine tools, robots, and other process equipment as new "peripherals" to complement printers, plotters, terminals, and disk packs;
3. a response to the demand for greater variety, customized designs, rapid response, and "just-in-time" delivery.

We are gradually switching from the production of large volumes of standard products produced on specialized machinery to systems for the production of a wide variety of similar products in small batches (perhaps as small as one). These small batches will be produced on standard, but flexible machines that are reconfigured by their software to the required process for each different product design.

These combinations of computer systems and chemical plant flows with their attributes of scope, flexibility, close-coupling, predictability, and speed will allow U.S. industry to respond profitably to market pressures for increased variety and customization of products, close-coupling/minimum inventory linkages between suppliers, and the "demassification" of the marketplace as described by Alvin Tofler in *The Third Wave*.[12]

Following the analogy to a chemical plant and the logical consequences of economy of scope production and low-cost computing and memory, we can suggest a number of design characteristics that will differentiate the "factory of the future" from traditional manufacturing:

1. High fixed costs approaching 100%.
2. Relatively flat learning curve for a specific product configuration after the software is debugged. Emphasis shifts to an "experience curve" for a family of similar products over time in the entire process from design to production to distribution.
3. Short-run average costs approach long-run average costs.
4. High levels of investment in computer software, communications capability, and maintenance staff.
5. Extensive and expensive "preproduction" activities requiring a reallocation of management attention.

These characteristics lead to a factory that operates very differently from our traditional factories and that will require marketing and competitive strategies that are "counterintuitive" to our successes in the past. The operating capabilities of the factory of the future can be summed up as follows:

1. Economic order quantity approaches 1.
2. Variety has no cost penalty at the production stage.
3. Costs per unit are highly sensitive to total production volume because fixed costs approach 100%.
4. Joint cost economics—the value of the system is a function of the "bundle" of products it produces.
5. Rapid response to changes in product design, market demand, and production mix are possible.
6. Unmanned and continuous operation is standard.
7. "Close-coupled" and highly integrated production systems as well as "supplier-user" linkages resulting in minimal inventory levels and little slack for errors in timing.
8. Consistent high levels of quality and accuracy and repeatability introduce higher levels of certainty into the production planning and control activity allowing for higher levels of process optimization and product quality.

These operating characteristics lead to a factory capable of one-at-a-time, one-of-a-kind, irregular production systems capable of closely following changes in both demand and style. This is a very different factory and will lead to a new set of production decision algorithms and industrial/ manufacturing engineering concepts and techniques.

We can summarize the changes from tradition to computer integrated manufacturing as follows:

Traditional technology can be described by

- economy of scale;
- experience curve;
- task specialization;
- work as a social activity;
- separable variable costs;
- standardization;
- expensive flexibility and variety;

leading to factories that exhibit characteristics of

- centralization;
- large plants;
- balanced lines;
- smooth flows;
- standard product design;
- low rate of change and high stability;
- inventory used as a buffer;
- "focused factory" as an organizing concept;
- job enrichment;
- batch systems.

In constrast the CIM factory is described by

- economy of scope;
- truncated product life cycle;
- multimission facilities;
- unmanned systems;
- joint costs;
- variety;
- profitable flexibility and variety;

leading to factories that allow for

- decentralization;
- disaggregated capacity;
- flexibility;
- inexpensive surge and turnaround ability;

- many custom products;
- innovation and responsiveness;
- production tied to demand;
- functional range for repeated reorganization;
- responsibility tied to rewards;
- flow systems.

3. Information Management for Factory Automation

3.1. New Challenges for Information Managers

Improvements in manufacturing performance require better coordination among the product design, manufacturing engineering, production planning operations and control, and distribution functions. Such coordination may be achieved by shared data through both local and wide area networks.[14] The content of these shared data is scientific, operational, and financial. To complicate, further, this challenge, many applications and databases have been implemented previously with little consideration for future integration.

Factory systems integration, however, is essential for productivity improvements. Manufacturing operations' bottlenecks and breakdowns must be recognized, diagnosed, and resolved. The use of robots and numerical control tools often requires the participation of manufacturing engineering staff. Quality control exceptions may require involvement from the product design group. Production scheduling and capacity planning for a variety of variable batch size products, sharing some common tooling, also requires cross-functional deliberations. In many situations, communicating through networks rather than face-to-face discussion will occur. Thus, minimization of the time between planning/control analyses and implemented action is critical to maintain manufacturing resource productivity.

In discrete process manufacturing, information systems previously had been considered a support resource, i.e., useful but not critical. Now, low-cost production in a flexible manufacturing resources factory requires effective information systems integration to maintain or improve a firm's competitive position. Thus, the information system function becomes strategic.[9] Accordingly, appropriate attention must be directed towards developing a systems architecture (e.g., logical data models, loosely/tightly coupled applications, and communications networks) to better support the business strategy and factory automation requirements.

3.2. Product and Manufacturing Design

Within these preproduction activities, information systems supported CAD and computer-aided engineering (CAE), for product design and modification. Additionally, support for CAM configuration planning, robot action menu design, and numerically controlled tool programming is provided. These classes of semistructured applications represent decision/design support in contrast to the more structured activities on the shop floor.

CAD/CAE systems have been used traditionally for product design and analysis functions. Automated and integrated shop floor operations, however, offer an opportunity to test the impacts of product introductions and modifications on shop floor operations and performance in a computer-based simulation. Thus, CAD/CAE workstation clusters would be linked through a local area network (LAN) to a larger scale processor. Proforma resource requirements, planning schedules, output levels, and quality and productivity measures would be generated for evaluation by manufacturing management.[14]

Similarly, recommended initial and revised CAM configurations could be simulated. Tradeoffs between flexibility and rigidity in CAM from proposed product and process innovations could be assessed. Off-line testing of robot clusters can be achieved. CAD/CAE/CAM adopted plans can be implemented electronically through LANS or remotely through wide area networks (WAN). The use of a WAN may be chosen when the scarcity of technical staff (e.g., numerical control programmers) requires a distributed, rather than decentralized strategy.

These CAD/CAE/CAM decision/design support systems (DSS) require information systems staff support skills which differ from the traditional transaction system development. DSS requirements place greater emphasis on effective person–system interface design, flexible data structures, adequacy of modeling tools, and embedded intelligence to alert or guide users in their problem finding/solving/design creation activities.[10]

The high intensity of person–DSS interaction requires flexible and "user friendly" interface options for users.[11] For example, menus may be functional for novices, yet inhibitive for experienced users. Output display choices should include image format (table, graph, text, picture, or combination), color, contrast, screen scrolling speed, joystick zoom, and windowing choices. Additional ergonomic aspects of the workstation, user's chair, and environmental lighting and noise should be considered. These human factors may be chosen by the end user, perhaps in consultation with a DSS support analyst.

For some activities, e.g., CAD/CAE, DSS will be application specific software. For some aspects of CAM modeling a DSS generator may be chosen.[11] This generator is a general-purpose DSS package (e.g., ADR's EMPIRE or EXECUCOM's IFPS) by which a specific DSS (e.g., inventory planning and control, distribution planning) can be designed by an end user with possible help from a DSS analysit/builder. Some of the DSS, whether business, operations, or scientific oriented, become institutionalized in a shared model base. Thus, appropriate development and documentation procedures must be monitored by the information systems group.

Even if a DSS is not shared, often the data inputs and/or information outputs will be shared among users. Thus, appropriate data integrity and security practices for data bases located on LAN file servers and larger-scale mini/mainframes must be maintained. Since these classes of applications represent decision support with ad hoc queries, flexible query capabilities dominate the loss in storage and retrieval efficiencies. Since CAD/CAE and CAM staff will communicate with each other on common problems, a central or distributed data dictionary is required. This is a difficult development and maintenance task especially if numerous applications already exist. A stepwise approach to building the dictionary, however, may be chosen. As more CAD/CAE/CAM expertise becomes formalized, documented in DSS, and validated, data bases will contain rules in addition to traditional text and numerics.

The preproduction activities all require some computer-based modeling for structural analysis, product design, and manufacturing configuration/numerical control tool evaluation.[1] This model base should have abstracts of each model which discuss potential uses, model assumptions, and data requirements. Such standards also should pertain to user developed models placed in public software libraries. The use of some specialized models may require access to external data bases. Access instruction should be documented within the system or a DSS analyst should be available. For organizations which initially introduce CAD/CAE/CAM methodologies, a manufacturing "information center" may be developed during the transition to end user competency with these systems.

The flexibility and power of these DSS, however, at times may be counterproductive. A number of studies with both students and professionals have identified a set of information processing biases and decision heuristics.[8] If users are unaware of the benefits and risks related to these mental shortcuts and biases, the performance consequences from chosen actions may be less than anticipated. Thus, DSS designers should build alertors into systems which flag such situations to individuals who may accept or override the displayed advice. As the factory environment becomes more integrated the consequences from inappropriate decision assumptions and processes become more critical.

3.3. Manufacturing Planning, Operations, and Control

The recent use of computer-based support (e.g., MRP-II) mostly is related to more labor-intensive operations. Thus, much human judgment is used for machine/tool operations, resource deployment, planning and control, shop scheduling, and bill-of-materials maintenance. CIM, however, provides for many computer-driven (i.e., data-driven) procedures to replace human participation.[12] As previously noted, manufacturing is dominated by capital intensity and investment in computer-based management and execution. Significant, but fewer, human responsibilities exist for planning, control, and maintenance. Thus, the smooth and effective functioning of manufacturing activities is dependent upon the quality of integrated information systems support.

Production scheduling and capacity planning modeling software is now more complex. Machine selection, maintenance schedules, and capacity all are dependent upon the type of discrete batch to be produced. Thus, production mix planning must consider an expanded set of variables for flexible manufacturing systems.

Once a plan, now having a shorter time horizon, is adopted, robots, numerically controlled tools, and programmable controllers all must be readied expeditiously by local or central staff technicians. Both LANs and WANs facilitate rapid response to adopted plans or control report-triggered adjustments. CAM software may be able to cope with recognized bottlenecks or stoppages in production; otherwise human intervention is required. Although these operations' activities both draw upon and generate data for transactions data bases, the shop floor planning and control needs suggest a relational data model for these analyses. Thus, the production environment information systems applications and data bases will require both DSS planning/relational data model and transaction (operations)/structured data (e.g., tree or network) support. Both operational and financial data will be captured.

The manufacturing operations systems located on the shop floor require ruggedized protection to service the environment. Plastic badge, wand scanner, and smart cards (computerized work orders) may complement or replace keyboard input. Voices recognition requirements will be more difficult on the shop floor. Computer vision and tactile sensing is used by robots for production and quality control.

Manufacturing resources procurement will include network links to suppliers to place orders and monitor delivery status. For non-long-term contract vendors, those suppliers may electronically update product quality and pricing information in the buyer's information system. Further, CAD/CAE specifications may be transmitted electronically to suppliers. These decreases in communication time should provide more effective purchasing performance and inventory control. New, interorganizational information system access security considerations, however, must be addressed.

Many of the information technology components for factory automation exist today. Some have been implemented. Others will be introduced shortly. Development of other components continues. Productivity improvements, however, require the careful management of user expectations retraining of staff, and senior management's understanding of the longer-term benefits from CIM.

4. Productivity and Capital Budgeting

These changes must lead us to a new and more broadly based definition of productivity. We must move away from manufacturing's traditional focus on cost per unit and to a system that evaluates the factory in terms of its ability to contribute to the profitability and long-term competitiveness of the firm as a whole. The increasingly high level of integration that the computer makes possible requires organizational, financial, and strategic integration as well as close coupling technology and production steps.

The new criteria for factory "goodness" will have to include

1. minimum changeover costs and time;
2. maximum flexibility and quick turnaround capability;
3. minimum downtime for maintenance;
4. maximum product "family" range;
5. ability to adapt to variability in materials and process conditions;
6. ability to handle increasingly complex product designs and technology;
7. ability to integrate new process technology into the existing system at minimum cost.

These are the new variables by which we will evaluate the factory of the future. That is, we will look for a factory's ability to provide a competitive weapon for the market environment of the future in place of the narrow focus on cost per unit and schedules that has led to long runs of standardized products that no one seems to want to buy. Because our usual capital budgeting techniques are based upon traditional scale, economies and cost per unit productivity criteria will have to change in order to reflect the opportunity cost and strategic nature of the CIM factory. Bela Gold discussed these changes in detail in the *Harvard Business Review*.[5]

From one of the author's research on industrial decisions relating to CAM it appears that managers have not yet learned to analyze CAM's prospective contributions to their companies. Managers still

> Do not realize that CAM is a special kind of technology, whose adoption requires a broader level of analysis than that normally applied to the purchase of equipment and facilities.

> Rely on bottom-up processes for generating new equipment proposals.

> Depend on traditional capital budgeting techniques to evaluate such proposals.

The real promise of CAM technology lies not in its use as yet another, perhaps fancier than usual, machine tool located at a single point in an otherwise unchanged production process. CAM's promise lies, by contrast, in its ability to integrate adjacent operations with each other and with overall control systems. Because it offers a systemic—not a "point"—capability, neither its purchase nor performance should be evaluated in the traditional way. Yet, as a senior executive of an electrical machinery company observes, "I am sure that most companies that, like our company, have been making capital allocation decisions for many years already have established methods for developing and evaluating new proposals which have proved thoroughly satisfactory over the years."

Indeed, it is precisely such established methods that managers often apply to decisions about CAM. Their thinking influenced by the (here inappropriate) assumptions that underlie most companies' equipment proposals, these managers ignore the broad and distinctive capabilities of CAM systems.

These traditional capital budgeting systems assume that the investment being analyzed is a direct replacement for the old equipment, has a narrowly defined impact, and is a fixed technology that will not change. They assume that costs and revenues can be accurately estimated and that this evaluation can best be done by sepcialists in the narrowly defined application.

We know that these assumptions are not true for CAM or CIM. Yet our typical approach to the capital budgeting problem for a new machine is to ask what the production costs would have been if we had been using the machine last year and to evaluate the difference (less we hope) versus the investment cost and the time value of money. We neglect, however, to realize that last year's product mix was optimized for the volume/variety constraints of conventional production facilities. Thus, when we evaluate a system with high flexibility and do not consider the value of integration, it is not surprising to see low returns on investment.

These low ROI numbers are very misleading because they do not recognize the CIM is an ongoing process of progressive advances in technological capabilities. The benefits of CIM require a long-term point of view that assumes continuing improvements in both the technology and its utilization. The largest benefits of CIM are hard to define because they come from doing new things not possible with the equipment being replaced and from new strategies that emphasize the strengths of CIM derived from its economy of scope.

Gold summarizes the requirements for a new analytical framework as follows[4]:

1. penetrate more deeply into the bases for the estimates of inputs, outputs, and costs that are fed into the capital budgeting model;
2. include estimates of probable carryover effects on successively broader sectors of operations;
3. encompass estimated effects over an array of successively larger time periods.

Finally, we must change the objective function from "what is the ROI of this particular investment proposal?" to "what will be the ROI and the competitive position of my firm in 5 or 10 or 20 years if we do not make this investment?"

This will require a new set of competitive strategies based on the realities of changing markets and CIM's capabilities for low-cost variety production.

5. Changing Markets and Counterintuitive Strategies

The key trends in the marketplace can be summarized as follows:

- short life cycle products;
- high variety product lines;
- customized products;
- fragmented market segments;
- demand for quality and reliability;
- new and more complex technology products;

- new users and uses and wider variety of customers;
- reduced customer loyalty—more sophisticated customers;
- worldwide production, distribution, markets, competition, and innovation.

If you accept the twin premises that (a) the marketplace is changing as described above, and (b) the "factory-of-the-future" technology is real and available and can deliver high variety production at reasonable cost, then the challenge to U.S. industry becomes one of developing competitive strategies that play to the strengths of the factory of the future and minimize the strengths of competitors.

These strategies that must be considered in the capital investment decision will include deliberate efforts to proliferate the product design, deliberate efforts to truncate and shorten the product life style, deliberate use of distributed production locations closer to customers, and an emphasis on quality and reliability as a measure of value. All these things are counterintuitive to the way in which we have run our manufacturing systems and our manufacturing-intensive corporations in the past.

Many of our marketing strategies are based on a set of unspoken assumptions about what existing factories will do well. And we know that what they do well is long runs of standard products over a long period of time in which change is barred at the door. Deliver the product that you were told to deliver, on time, at acceptable quality, on spec, at the lowest possible cost. You do not want a change in the accounting system—you do not want anything that is radically different or the traditional factory no longer offers the best approach.

On the other hand, suppose you deliberately fragment the market into pieces so small that nobody can come in and say, "aha—if I focus on that market segment and I build a nice, well engineered factory, there is enough scale and stability in that market segment so that I can get down on a learning curve and compete as a low cost producer. I can capture just that little market away from a large firm, because they've got to be dealing in 7, 8, 10, or 25 market segments." On the other hand, if you fragment the market so that no segment is large enough to allow that kind of "cherry-picking," and choose to compete broadly in the market as a whole, then you will require advanced manufacturing capabilities. You can build a competitive edge based on your strengths in factory-of-the-future technology that minimizes the strengths of less agressive firms or off-shore competitors.

The old marketing theories about "positioning," "segmentation," "penetration pricing," and so on need to be reviewed and possibly reinvented or changed. Entirely new marketing tactics must be invented to take advantage of the capabilities of flexible and computer-integrated manufacturing.[7]

All of the above requires a fundamental change in attitude and philosophy regarding the role and capabilities of manufacturing at all levels and in all functions and by the top level general management of the firm. The very principles that served us well in the past and upon which the current generation of management built their track records of success is wrong for the future. But it will be very difficult for most experienced and successful managers to reverse their tracks. That is the challenge for our business education establishments, for corporate boards of directors, for bankers and investment analysts who need to recognize the new potential, and for each individual manager. And that is the weak spot in our march toward more competitive U.S. industry.

References

1. A. ARBEL and A. SEIDMAN, Performance evaluation of flexible manufacturing systems, *IEEE Trans. Syst. Man Cyber.* **14**(4) 606–617 (1984).
2. *Am. Mach.* May, 109–111 (1983).
3. *Am. Met. Mark./Metalwork. News* April 25 (1983).
4. B. GOLD, *Improving Managerial Evaluations of Computer Aided Manufacturing,* National Research Council, National Academy Press, Washington, D.C., 1981.
5. B. GOLD, CAM sets new rules for production, *Harvard Bus. Rev.* November-December, 88–94 (1982).
6. J. D. GOLDHAR and M. JELINEK, Plan for economies of scope, *Harvard Bus. Rev.* November-December, 141–148 (1983).
7. R. H. HAYES and S. C. WHEELWRIGHT, *Restoring Our Competitive Edge: Competing Through Manufacturing,* Wiley, New York, 1984.
8. D. KAHNEMAN, P. SLOVIC, and A. TVERSKY (Eds.), *Judgment Under Uncertainty: Heuristics and Biases,* Cambridge University Press, London, 1982.
9. F. W. MCFARLAND, J. L. MCKENNEY, and P. J. PYBURN, The information archipelago—Plotting a course, *Harvard Bus. Rev.* January-February, 145–156 (1983).
10. W. RAUCH-HINDIN, Expert system to plan PC board assembly in the factory, *Sys. Software* August, 77–80 (1984).
11. R. H. SPRAGUE, JR. and E. D. CARLSON, *Building Effective Decision Support Systems,* Prentice-Hall, Englewood Cliffs, New Jersey, 1982.
12. E. TEICHOLZ, Computer integrated manufacturing, *Datamation,* March 169–174 (1984).
13. A. Toffler, *The Third Wave,* William Morrow, New York, 1980.
14. F. W. WEINGARTEN, *Computerized Manufacturing Automation: Employment, Education, and the Workplace,* OTA-CIT-235, U.S. Congress, Office of Technology Assessment, April, 1984.

25

DECISION SUPPORT SYSTEMS AND MANAGEMENT

C. R. CARLSON

1. Introduction

The concept of decision support systems is not new. Managers have always used decision support systems of one type or another. To many managers it means asking a question, getting an answer, then using the newly found information to ask yet another question, get yet another answer, and so on until an issue or potential situation is clear enough in an executive's mind for an informed decision to be made. To these managers, the advent of computer-based decision support systems (DSS) has meant primarily that they can now get answers to questions faster than they could manually. In other cases, computer-based DSSs have provided managers with decision-making opportunities which, for all practical purposes, were not available to them before. Without the help of a computer, these decision makers frequently were forced to give up in the middle of the decision process once they realized that insufficient information and decision tools were available.

Advances in computer technology have made possible the development of decision support systems designed to be used directly by noncomputer-oriented executives, planners, and managers. Surveys [1-5] of decision support systems have identified numerous working systems which provide the decision maker with access through a user-friendly desktop terminal to a totally integrated data base, along with the interactive means of analysis necessary for decision-making support and problem solving. Some decision support systems help managers by expediting access to information that would otherwise be unavailable. Others contain explicit models that

C. R. CARLSON • Computer Science Department, Illinois Institute of Technology, Chicago, Illinois 60616

provide structure for particular decisions. Some DSSs are primarily tools for individuals working more or less in isolation, while others serve primarily to facilitate communications among people or organizational units whose work must be coordinated.

A commonly used definition of decision support systems describes them as interactive computer-based systems that help decision makers utilize data and models to solve unstructured problems. This describes a broad spectrum of computer-based systems. The following examples, taken from Ref. 6, provide us with further evidence of this diversity.

> A U.S. mining company was faced with a major capital budgeting decision. Should it develop a new copper mine in Africa? Assessing the potential payback of such a 20-year project required hundreds of assumptions about cost, price, and demand variables. To handle this task, the company developed a decision support system that could handle all the variables and the "what if" questions necessary to assess the project.

> With more than 100 operating units keeping separate financials over different periods of time, a large multinational electronics company found it almost impossible to compile current financial data that reflected overall corporate performance. The company solved the problem with a decision support system that allowed it to produce a consolidated financial statement within hours of closing the books on worldwide operations.

> A large money-center bank wanted to develop its portfolio management business without significantly increasing its staff. The bank also wanted to assist its portfolio managers in evaluating a wide range of investment strategies for its clients, a very time-consuming task using traditional methods. The bank addressed both problems by implementing a decision support system that allowed bank personnel to evaluate the status of portfolios, investigate alternative investment strategies, and improve client presentations.

2. Management View of DSS

DSSs are based on the belief that computer technology can help management run the corporate enterprise. Well-conceived strategic information made available for sophisticated and timely analysis can serve as the instrument of change that management needs to move their organization in the strategic direction they have chosen. Effective utilization of information can minimize the risk in strategic planning, guide investment to the optimal opportunity, and identify problem operations for remedial action.

Keen and Morton have identified several significant benefits that an effective DSS can provide [7]. They include

- increase in the number of alternatives examined;
- better understanding of the business;
- fast response on "what-if" problems;
- ability to carry out ad hoc analysis;
- improved communication between levels of management;
- better control of information;
- cost savings through more economical use of information systems;
- better information upon which to base decisions;
- more effective teamwork;
- time savings.

Not all executives share this optimistic view of decision support technology. Underlying their view is a negative and even mistrustful perception of information processing which has proliferated masses of data throughout the organization whose value senior executives cannot ascertain but whose cost of production has increased significantly. As late as 1982, *Sloan Management Review* reported: "The president of a medium-sized manufacturing company remarked, 'I receive about the same information today as was provided 30 years ago before our computers. Only now I spend millions to get it.'" In the same article,[8] the chairman of a $3 billion conglomerate is said to have commented repeatedly, "I get nothing from our computers."

In assessing the role of decision support systems in an organization several lessons from the past need to be considered.

(1) Decision support systems are quite different from Management Information Systems (MIS). MIS are typically menu-driven COBOL programs, the highly structured result of a long development process. The user is the passive recipient of predetermined, routinely reported information. On the other hand, DSSs are designed to support unstructured and seldom repetitive decision processes.

(2) A 1982 Booz-Allen & Hamilton study is reported[9] to have shown that about $30 billion of DP funds goes to support $450 billion of managerial/professional activities, while $50 billion of DP dollars goes for support of $150 billion of clerical activities. That is a 15:1 ratio for management vs. 3:1 for nommanagement. This says that most of DPs resources are directed at such clerical assistance as on-line data entry, word processing, and large-scale transaction processing, with a disproportionate

amount going to higher-level utilization and analysis of data. It comes as no surprise that some managers have a dim view of information processing technology since much of the information they receive is a by-product of data which was originally generated for individuals in lower-level positions in the organization by a computer-based system whose rationale was cost reduction of clerical level activities.

(3) The lack of communication between two very different cultures within an organization often thwarts the cost effective utilization of computer technology. Executives are interested in politics and/or results, not techniques. They are worried about their next promotion and how to reduce costs, increase revenues, improve customer service, and gain better information. Computer professionals, on the other hand, are craftsmen—in the finest eighteenth century use of the term—who understand and value high technology, tools and techniques. Unlike computer professionals, executives often do not see themselves as pioneers in search of the next area of computer automation.

(4) The rationale for decision support systems is that they improve the effectiveness of employees in making decisions. Thus, it is not surprising that decision support technology has gained its initial acceptance in industries and areas of organizations where timely decision oriented information is important and its impact on the bottom line is obvious. Further, this argues that executives "get what they ask for" from computer-based information processing systems.

(5) Information has become a corporate asset—as vital as men, money, machines, and material. The rapid pace of change in the dynamic business world is dispelling the vestiges of uncertainty about the place of decision oriented information in the corporate environment. Upheavals in the economy, foreign competition, shifting demand for goods and services, innovative technology, and new ways of doing business—some of them attributable to the expanding role of information systems themselves—are reshaping the traditional concepts of corporate thinking and planning. To meet these challenges, executives are increasingly convinced that the availability of a broad base of clearly defined, readily accessible, decision oriented information is a must. The view that information processing systems are simply tools to be used in "running" the business is shrinking. Management now needs tools to help "plan" the business.

3. General DSS Requirements

Surveys of DSSs show them to exhibit a wide diversity of functional requirements. The following general requirements illustrate this diversity.

3.1. How Executives Use Them

In the past, senior executives relied on functional staff for information on which to base decisions. However, today's improved technology, together with a heightened analytic orientation among top managers, has, in some companies, moved responsibility for using this database support into the executive suite. Rockart and Treacy[10] divide DSSs into two categories based on how they are used by executives: for access to the current status and projected trends of the business and for "personalized analysis" of data.

Status-access systems provide read-only output of the latest data on the status of key company variables. Executives using these systems essentially peruse data and have very little, if any, data manipulation ability. "In industries in which market conditions change rapidly, in which there are many factors to watch or in which hour-to-hour operational tracking is important, status access can be very useful. This approach also provides an easy, low-cost, low-risk means of helping an executive become comfortable with a computer terminal."

With personalized-analysis systems, executives use the computer not only for status access but also as an analytic tool. Some merely compute new ratios or extrapolate current trends into the future. Some graph trends of particular interest to gain an added visual perspective. Some work with elaborate simulation models to determine where capital investments will be most productive. All, however, enjoy a heightened ability to look at, change, extend, and manipulate data in personally meaningful ways.

3.2. Types of Decisions

R. N. Anthony[11] describes DSSs in terms of various types of decisions. Each type of decision corresponds to a class of decisions with which every enterprise must be concerned.

Strategic Planning: decisions related to setting policies, choosing objectives, and selecting resources.

Management Control: decisions related to assuring effectiveness in acquisition and use of resources.

Operational Control: decisions related to assuring effectiveness in performing operations.

Operational Performance: decisions that are made in performing the operations.

Each of these four decision types are found in manufacturing/office processes. With the current focus on computer support for the automation of these processes, system developers are presented with a strategic opportunity to develop integrated manufacturing/office systems that feature both operational and decision-making support. Decision support for each of these decision types needs to be built into, rather than retrofited, future applications of computer-based automation. A DSS architecture will be discussed in Section 4 which facilitates the integration of decision support with the operational component of such systems. This modular design approach accomodates the diversity of processing requirements necessitated by such applications.

3.3. Problem Domain

Most DSSs have been developed in the context of a specific problem area, providing accounting, simulation, or optimization procedures developed for well-structured problems. As a result, they offer the decision maker limited flexibility regarding the problems to which they can be applied. However, some systems have been designed to support multiple problem domains. This is possible if the domain-specific knowledge resides in a database instead of being coded within the decision procedures. Much of the research advancing this approach falls under the terms "automated reasoning systems" or "expert systems."

3.4. Decision Phases

A DSS should support all phases of the decision-making process. A popular model of decision making is given in the work of Simon.[12] He characterized three main phases in the process:

Intelligence: Searching the environment for conditions calling for decisions. Raw data are obtained, processed, and examined for clues that may identify problems.

Design: Inventing, developing, and analyzing possible courses of action. This involves processes to understand the problem, to generate solutions, and to test solutions for feasibility.

Choice: Selecting a particular course of action from those available. A choice is made and implemented.

3.5. Type of Decision Process

The following examples of paradigms of decision making illustrate the variety of decision-making processes. The first example is the rational (economic) paradigm, which posulates that decision processes attempt to maximize, *optimal solution,* the expected value of a decision by determining payoffs, costs, and risks for alternatives.[13] A second paradigm asserts that the decision-making process is one of finding the *first cost-effective alternative* by using simple heuristics rather than optimal search techniques.[14] A third paradigm describes decision making as a *process of successive limited comparisons* to reach a consensus on one alternative.[15]

3.6. Information Source

Some systems employ a generative process whereby current trends are captured by multiple variables and information about the future is extrapolated based on these trends. This approach requires the decision maker to identify both those variables which describe the current "state" of the environment and those extrapolation functions to describe "current trends." Based on these facilities the DSS "generates" the decision information. Other systems employ a search process whereby analytic-based search procedures are applied to large databases to filter out the desired information. The decision maker thus directs the search and construction of information based on stored data.

3.7. Number of Decision Processes

Many early DSSs offered a decision model employing a single decision process. As decision makers begin to depend on DSSs, they often find that they need the flexibility of employing multiple decision-making processes to solve problems.

With such a diversity of functional requirements, it would be difficult to identify the functional requirements of a "typical" DSS. However, the proliferation of microcomputer-based DSS packages has done a lot to popularize the concept of decision support systems. The term "spreadsheet package" is often used to reference these systems. Generally, these packages allow a decision maker to show the financial impact over several time periods of alternative decisions, assumptions, and uncertain events. For example, the impact on earnings per share (EPS) or return-on-investment (ROI) of alternative manufacturing policies would be important supporting evidence for a strategic choice.

The computational framework of spreadsheet packages is a table of rows and columns, sometimes extending to multiple dimensions. For example, a simple application may involve rows as the revenue and expense items of a profit-and-loss statement and columns as successive time periods. The third and fourth dimensions of an extended profit-and-loss statement could reflect several product lines and several geographic regions for a diversified enterprise. Access to and analysis of this so-called data cube underlies most spreadsheet applications. Model development involves specifying the relationships among these rows and columns. For example, the following formulas define some simplistic relationships:

$$\text{Sales} = \text{Estimated Growth} \times \text{Previous Sales}$$
$$\text{Cost of Sales} = 0.84 \times \text{Sales}$$

Perhaps the single most useful feature of these spreadsheet packages is the modification of data and relationships to show the implications of changes to one or more variables. Continuing the earlier illustration, a model developed with one of these packages could recompute the relevant measures (EPS or ROI) if sales grew periodically at say either 5% or 6% or 8% and/or if cost of goods were 0.82 or 0.80 instead of 0.84 times sales revenue. An obvious extension of this feature is to increment over a range of possible values for certain key assumptions. This procedure would automatically repeat the planning calculations, for example, for sales growth ranging from 0% through 20% in increments of two percentage points.

Most larger DSS packages include a feature whereby the sensitivity of a key "what-if" question on some critical measure is directly calculated. In this example, the real question might be, "At what rate must sales grow for EPS to double within five years?" or "What is the highest cost of goods as a fraction of sales for ROI to be at least 20%?" A quick answer to such questions can show management whether the desired objectives are within the realm of reasonable assumptions.

A logical extension of these "what-if" and iteration functions is the display of risk. For example, suppose the estimated sales growth rate mentioned above could be described by a probability distribution centered at 8%, with a minimum of 0% and maximum of 20%. Likewise, the cost-of-goods fraction might be described as centering near 0.82 and ranging from 0.75 to 0.90. Sample values from various probability distributions could be drawn, and the system could calculate perhaps several hundred resulting values for the specified measure of merit, like the EPS or ROI. The resulting distribution of such values displays for management the relative likelihood of achieving various levels.

4. DSS Architecture

As a class of systems, DSSs are as functionally diverse as the problem areas to which the technology has been applied. Nevertheless, there are a number of architectural features which they have in common. The DSS architecture described in Figure 1 is basically similar to the architectures

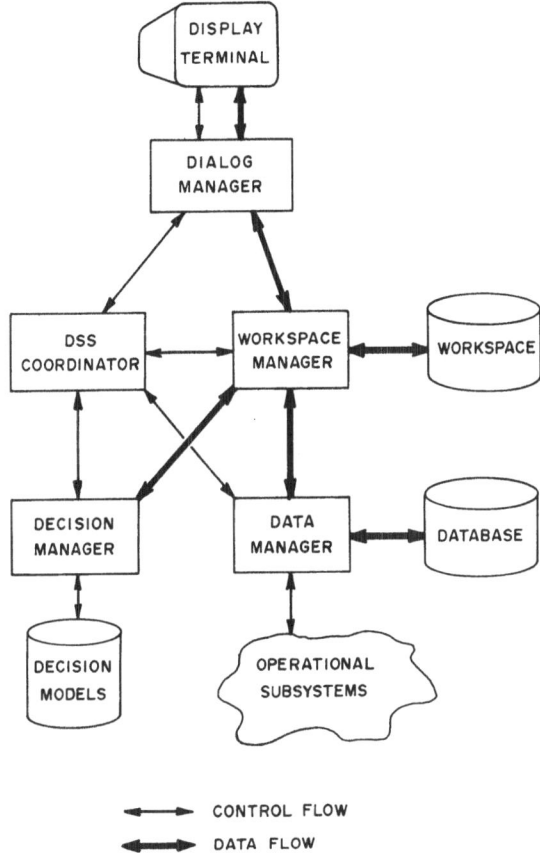

FIGURE 1. DSS architecture.

that have been described in numerous research articles and textbooks of decision support.

The behavior of a system based on this DSS architecture can be described as follows.

When the system is started by a user, the DSS Coordinator, via the Dialog Manager, asks the user which decision model is to be used. When the user identifies a specific model, the relevant model is called from the model base by the DSS Coordinator. Then, execution of that model begins. Upon completion of that model, the DSS Coordinator again asks the user which decision model to use next.

The source of data to each decision model can be found in the workspace. The model processes data according to its inherent algorithm, entrusting basic manipulation of data within the workspace to the Workspace Manager. The functionality of a decision model varies with the class of decisions for which it was designed. It could, for example, perform spreadsheet analysis, statistical analysis, linear programming, or math logic. Each decision model is capable of processing a particular class of data objects with which the Workspace Manager must be familiar. For example, spreadsheet analysis models assume an n-dimensional matrix (data cube) whereas math logic models assume a system of production rules.

The Workspace Manager permits the user to manipulate each type of data object according to a type-dependent set of operators. For example, one can redefine the attributes or dimension of a data cube or create sublists of production rules.

The Dialog Manager controls the conversation between the user and the other system components, isolating the format of the conversation from these components. The Dialog Manager may employ one of several conversational formats, e.g., menus, command language, query language, graphics, at different points in the conversation. The Dialog Manager also controls the graphic display of workspace data on the display terminal. This may require the conversion of data from, say, a data cube format to such display-oriented formats as graphs, lists, or reports.

The management of stored data, both internal and external, is handled by the Data Manager. Its primary responsibility involves support for ad hoc, trial and error, information searches. Conversion of data from its stored representation to whatever data object class is required by a particular decision model is another important feature of the Data Manager. In the case of systems limited to a single decision model, data independence is sacrificed for the simplicity of not having to perform such conversions. The Data Manager together with the underlying database are the keys to bridging the gap between the decision support subsystem and the operational subsystems.

5. Conclusion

Advances in computer technology have made possible the development of decision support systems designed to be used directly by non-computer-skilled executives, planners, and managers. From the diversity of existing systems a generalized DSS architecture has emerged which focuses future systems research in the areas of interactive computer interfaces, knowledge engineering, and decision model management.

Computer-based DSSs provide managers with new decision-making opportunities. Management acceptance is particularly keen among information service-oriented managers. However, it is expected that the demand for integrated manufacturing and office systems will necessitate the application of decision support technology in all areas of information system automation. Each new application will yield insight regarding the automation of decision-making paradigms and the processes by which the corporate data resource can be made to yield strategic information.

References

1. S. ALTER, *Decision Support Systems*, Addison-Wesley, Reading, Massachusetts, 1980.
2. R. H. SPRAGUE, JR. and E. D. CARLSON, *Building Effective Decision Support Systems*, Prentice-Hall, Englewood Cliffs, New Jersey, 1982.
3. E. D. CARLSON, An approach for designing decision support systems, *Data Base* **10**, 3 (1979).
4. Proceedings of the 13th Annual Hawaii International Conference on System Sciences, 1980.
5. User successes with decision support systems, International Data Corporation report ISPS No. M82-14, 1982.
6. C. WOLF and J. TRELEAVEN, Decision support systems: Management finally gets its hands on the data, *Manage. Technol.* 44–51 (1983).
7. P. G. W. KEEN and M. S. SCOTT MORTON, *Decision Support Systems: An Organizational Perspective*, Addison-Wesley, Reading, Massachusetts, 1978.
8. H. C. LUCAS, JR. and J. A. TURNER, A corporate strategy for the control of information processing, *Sloan Manage. Rev.* **23**, 25–36 (1982).
9. R. MINICUCCI, DSS: Computer wizardry for the executive, *Today's Office*, **18**, 36–46.
10. J. F. ROCKART and M. E. TREACY, The CEO goes on-line, *Harvard Bus. Rev.* **60**, 82–88 (1982).
11. R. N. ANTHONY, Planning and control systems: A framework for analysis, Graduate School of Business Administration, Harvard University, 1965.
12. H. A. SIMON, *The New Science of Management Decisions*, Harper and Row, New York, 1960.
13. R. M. CYERT and J. G. MARCH, *A Behavioral Theory of the Firm*, Prentice-Hall, Englewood Cliffs, New Jersey, 1963.
14. R. M. CYERT, H. A. SIMON, and THROW, Observations of a business decision, *J. Bus.* **29**, 237–248 (1956).
15. C. E. LINDBLOM, The science of muddling through, *Public Admin. Rev.* **19**, 79–88 (1959).

INDEX